D0906184

METHODS AND
REAGENTS FOR
GREEN CHEMISTRY

BICENTENNIAL
1807
WILEY
2007
BICENTENNIAL

THE WILEY BICENTENNIAL–KNOWLEDGE FOR GENERATIONS

*E*ach generation has its unique needs and aspirations. When Charles Wiley first opened his small printing shop in lower Manhattan in 1807, it was a generation of boundless potential searching for an identity. And we were there, helping to define a new American literary tradition. Over half a century later, in the midst of the Second Industrial Revolution, it was a generation focused on building the future. Once again, we were there, supplying the critical scientific, technical, and engineering knowledge that helped frame the world. Throughout the 20th Century, and into the new millennium, nations began to reach out beyond their own borders and a new international community was born. Wiley was there, expanding its operations around the world to enable a global exchange of ideas, opinions, and know-how.

For 200 years, Wiley has been an integral part of each generation's journey, enabling the flow of information and understanding necessary to meet their needs and fulfill their aspirations. Today, bold new technologies are changing the way we live and learn. Wiley will be there, providing you the must-have knowledge you need to imagine new worlds, new possibilities, and new opportunities.

Generations come and go, but you can always count on Wiley to provide you the knowledge you need, when and where you need it!

WILLIAM J. PESCE
PRESIDENT AND CHIEF EXECUTIVE OFFICER

PETER BOOTH WILEY
CHAIRMAN OF THE BOARD

METHODS AND REAGENTS FOR GREEN CHEMISTRY

An Introduction

Edited by

PIETRO TUNDO
ALVISE PEROSA
FULVIO ZECCHINI
The Ca' Foscari University of Venice and National Interuniversity Consortium, "Chemistry for the Environment" (INCA), Venice, Italy

WILEY-INTERSCIENCE
A John Wiley & Sons, **Inc.**, Publication

Published by John Wiley & Sons, Inc., Hoboken, New Jersey
Published simultaneously in Canada

For general information on our other products and services or for technical support, please contact our Customer Care Department within the United States at (877) 762-2974, outside the United States at (317) 572-3993 or fax (317) 572-4002.

Wiley also publishes its books in a variety of electronic formats. Some content that appears in print may not be available in electronic formats. For more information about Wiley products, visit our web site at www.wiley.com.

Library of Congress Cataloging-in-Publication Data:

Methods and reagents for green chemistry / edited by Pietro Tundo, Alvise Perosa, Fulvio Zecchini.
    p. cm.
  ISBN 978-0-471-75400-8
1. Environmental chemistry–Industrial applications 2. Environmental management
3. Chemical tests and reagents. I. Tundo, Pietro, 1945-II. Perosa, Alvise, 1965-III. Zecchini, Fulvio, 1968
  TP155.2.E58M48 2007
  660'.286- -dc22   2006052558

Printed in the United States of America

10 9 8 7 6 5 4 3 2 1

To my family, who support me every day and make my work easier.
P.T.

The three editors would like to thank Dr. Lara Clemenza, who acted as co-editor of the collection of lectures from the Summer School on Green Chemistry from which this volume is derived.
We are sure that the publication of this book would not have been possible without her precious and much appreciated contribution.

CONTENTS

FOREWORD

The Summer School on Green Chemistry was founded in 1998, in the wake of the growing interest in green chemistry among the chemical community. For the first time it was being recognized by chemists that there could be—and had to be—mutual understanding and collaboration between (A) the players involved in chemical production, and (B) representatives from all the social categories concerned with safeguarding the environment and human health. It appeared clear that the existing gap could be bridged best by young chemists able to redesign chemical production so it was safe, environmentally friendly, socially acceptable, and profitable. In short: green. The Summer School on Green Chemistry was devised by the Italian Interuniversity Consortium "Chemistry for the Environment" (INCA, www.unive.it/inca) as a high-level training school for young chemists to meet this goal.

The school became reality in 1998 with a grant from the European Commission's IV Framework Programme (FP) Training and Mobility of Researchers (TMR) program, and continued within the V FP as part of the improving program, as well as through funding from INCA, the Italian Ministry for Foreign Affairs, NATO, and INTAS. At the time the present volume goes to press (2007) the school has continued as a NATO Advanced Study Institute.

The innovative approach to the design of clean chemical reactions and processes has proved very successful, as shown by the increase in the number of applicants to the school year after year. From 1998 to 2005 nearly 500 chemistry researchers, between the ages of 25 and 35, from both academic and industrial backgrounds, have attended the school.

The success of the school can be judged by the large amount of positive feedback we as organizers have received over time. Many of the participants, after returning home, either continued research in green chemistry with a broader understanding of the issues, or started applying the green chemistry principles to their research.

Many students have benefited from meeting some of the teachers at the school, by visiting them in their laboratories, and by establishing collaborations among themselves and with the research groups represented at the school. Numerous friendships also have been established. All these links make up a wide web of people with a common interest in green chemistry, a network that has spread over through most of Europe, and beyond.

The school was established in Venice, Italy. For most of the young participants it was the first time they visited the city, which provided the perfect setting for informal and pleasant contact among all participants.

After the first three sessions of the Summer School on Green Chemistry it became apparent that a textbook was needed, since the lecture notes handed out to the participants represented the only comprehensive printed material existing on the subject. Thanks to the editorial effort by the teachers, with the support of INCA, the first edition of the volume, "*GREEN CHEMISTRY—A Collection of Lectures from the Summer Schools on Green Chemistry*," was produced in 2001. This book was based on the lecture notes, plus some explanatory text. The volume was made available on the Internet and handed out to the students attending the school that year. It was updated and enlarged twice, in 2002 and 2004, by incorporating new and revised chapters.

The Summer School on Green Chemistry has proved to be a stepping stone in the careers of many young researchers who wished to combine state-of-the-art research in chemistry with environmental awareness. It has also been central to a spontaneous network of scientists who practice green chemistry, and who found common ground for research, collaborations, and the teaching of green chemistry.

The present volume is the consolidation of eight years of work, during which new developments and a deeper understanding of green chemistry have developed. Hopefully, it will provide food-for-thought for the reader.

ALVISE PEROSA

The Ca' Foscari University of Venice
and National Interuniversity Consortium,
"Chemistry for the Environment"

PREFACE

In 2005, Yves Chauvin (Institut Français du Pétrole), Robert Grubbs (California Institute of Technology), and Richard Schrock (Massachusetts Institute of Technology) were the recipients of the Nobel Prize in Chemistry, "For the development of the metathesis method in organic synthesis." Motivation explicitly states: "This represents a great step forward for 'green chemistry,' reducing potentially hazardous waste through smarter production. Metathesis is an example of how important basic science has been applied for the benefit of man, society and the environment."

To my knowledge, this was the first time the Royal Swedish Academy of Science, with the preceding statement, extended the existence of a tight connection between Science and Ethics. For the sake of correctness, however, the 2001 Nobel laureates in chemistry (Knowles, Noyori, and Sharpless) came from the area of green chemistry. Their awards were for the new chiral syntheses in green manufacture and the discovery of improved "clean" ways to produce pharmaceuticals, an industry that is still one of the highest polluters.

Actually, chemists have always had the benefits of chemistry for society in mind. This was clearly illustrated by Giacomo Ciamician (Trieste, 1857; Bologna, 1922) in the following futuristic sentence published almost one hundred years ago (*Science*, *36*, 385 (1912)):

> On arid lands there will spring up industrial colonies without smoke and without smokestacks; forests of glass tubes will extend over the plains and glass buildings will rise everywhere; inside of these will take place the photochemical processes that hitherto have been the guarded secret of the plants, but that will have been mastered by human industry which will know how to make them bear even more abundant fruit than nature, for nature is not in a hurry but mankind is.

Because of this assertion, Ciamician can be considered the father of green chemistry, sharing with today's conception of this discipline the same disapproval of pollution, the same care for mankind, and the same intent to use natural resources. And the dream can today come true, thanks to modern technologies and to wider societal awareness and recognition.

Green chemistry is currently being acknowledged at scientist conventions, such as the recent European Science Open Forum (ESOF 2006), held in Munich, on July 15–18, 2006. ESOF 2006 was the second pan-European General Science Meeting. Its purpose was to promote interaction and dialogue between science and the general public. Green chemistry achieved the same recognition level as other more popular scientific disciplines, such as astronomy, natural disaster prevention, biodiversity, genomics, evolution, and medical science. It was acknowledged to be one of the main options for safeguarding the environmental, as evidenced by the basic enquiry: "Is green chemistry a real option?" This question clearly shows what the media and society want to know from chemists. And a positive answer to this question was given by the session entitled: Green Chemistry: A Tool for Socio-Economic Development and Environmental Protection.

At the same time, a few scientific networks have been established to foster the development of research through high-level capacity building in science, the improvement of regulatory frameworks and public policy design, the enhancement of public outreach and education, and other interventions. Two such organization have recently been created: the International Green Network (IGN) and the Mediterranean Green Network (MEGREC). A brief description of these organizations may clarify the purposes and benefits of this discipline.

The IGN mission includes research, coordination, and sponsorship of scientific collaborations, targeted training for a new generation of scientists, and the support of sustainable development. IGN consists of eight research centers, one in each of the G8 countries, and it will accelerate movement toward a sustainable-energy and materials economy, by bringing together scientists, engineers, research institutions, firms, analysts, and government regulators. IGN will provide know-how, coordination, and sponsorship for scientific collaborations, proper training for the new generation of chemists, and support for sustainable use of chemistry in developing nations. In addition, it will assist industrial production in G8 nations, foster the development of novel competitive technologies, and address such issues as climate change and energy, as well as other environmental concerns, from a chemical standpoint.

MEGREC constitutes a platform for the development of research and training in green chemistry in the countries of the Mediterranean basin, with focus on water management, the exploitation of local natural resources, the production and use of fertilizers, and monitoring and reducing the presence of toxic compounds in the food chain. With a clear focus on priorities for local areas, but with the extended know-how of all the partners.

Such recent developments show how science can also positively relate to ethics, thanks to green chemistry. Green chemistry represents a strategic challenge for the present and the future of the chemical industry, its development being mostly linked to the interrelated needs of and benefits for environment, economy, and society that must be initially approached through new ideas in fundamental research.

The scientific content of green chemistry can be easily taken for the aims of IGN, whose main research topics are: energy, green manufacture, life-cycle analysis, pollution prevention, food security, and chemical resources management.

In order to produce the expected and desired results, programs and strategies must be devised for the development and application of chemistry, and must involve explicit support from national governments to networked organizations that are involved in research, educational/academic, and industrial systems. This interaction is fundamental to the production of long-term and durable benefits.

By considering the opinions of the civil society and tackling the questions concerning chemical production it raises, the governments can achieve a relevant positive result: the merging of the consensus of the academic world and industry with that of youth and public opinion, which are increasingly focusing their attention on the environment and human health protection.

This book covers three leading topics of green chemistry: green reagents, alternative reaction conditions, and green catalysis. It is the culmination of more than 10 years of research in this field. I therefore thank the many authors who contributed to this volume, who year after year were constantly present as expert lecturers at the yearly Summer School on Green Chemistry (Venice), promoted and organized by the Consorzio Interuniversitario "Chimica per l'Ambiente" (Chemistry for the Environment), INCA, and enthusiastically exchanged their expertise with colleagues and students throughout the world.

Finally, if you wonder why the word "Introduction" is included in the title of this book, it is because research in this field is far from completed, and chemists have a long way to go before they meet and satisfy the needs of the environment, economy, and society.

PIETRO TUNDO

The Ca' Foscari University of Venice
and National Interuniversity Consortium,
"Chemistry for the Environment"

CONTRIBUTORS

Angelo Albini, Department of Organic Chemistry, Università degli Studi di Pavia, Via Taramelli 10, 27100 Pavia, Italy

Nicola Ballarini, Dipartimento di Chimica Industriale e Dei Materiali, Università di Bologna, Viale Risorgimento 4, 40136 Bologna, Italy

David StC. Black, School of Chemistry, The University of New South Wales, Sydney, NSW 2052, Australia

Fabrizio Cavani, Dipartimento di Chimica Industriale e Dei Materiali, Università di Bologna, Viale Risorgimento 4, 40136 Bologna, Italy

Hélène Degrand, Dipartimento di Chimica Industriale e Dei Materiali, Università di Bologna, Viale Risorgimento 4, 40136 Bologna, Italy

J. L. Dubois, ARKEMA, Centre de Recherche Rhône-Alpes, 69493 Pierre Bénite, France

Jan B. F. N. Engberts, Stratingh Institute, University of Groningen, Groningen, The Netherlands

Eric Etienne, Dipartimento di Chimica Industriale e Dei Materiali, Università di Bologna, Viale Risorgimento 4, 40136 Bologna, Italy

Ernst Anton Feicht, Institut für Ökologische Chemie, GSF-Forschungszentrum für Umwelt und Gesundheit, D-85764 Neuhergerg, München, Germany

Jean Luc Guillaume, Dow Europe GmbH, Bachtobelstrasse 3, 8810 Horgen, Switzerland

Michel Guisnet, University of Poitiers, Poitiers, UMR 6503, Faculté des Sciences, 40 avenue du recteur Pineau, 86022 Poitiers cedex, France

Martin Held, Institute of Biotechnology, Eidgenössische Technische Hochschule Zürich, CH 8093 Zürich, Switzerland

Dieter Lenoir, Institut für Ökologische Chemie, GSF-Forschungszentrum für Umwelt and Gesundheit, D-85764 Neuherberg, München, Germany

Zhi Li, Institute of Biotechnology, Eidenössische Technische Hochschule Zürich, CH 8093 Zürich, Switzerland

Frieder W. Lichtenthaler, Institute of Organic Chemistry, Technische Universität Darmstadt, D-64287 Darmstadt, Germany

José M. López Nieto, Intituto Tecnologìa Quimica, UPV-CSIC, Avda. Los Naranjos s/n, 46022 Valencia, Spain

Renata Mathys, Institute of Biotechnology, Eidgenössische Technische Hochschule Zürich, CH 8093 Zürich, Switzerland

Marchela Pandelova, Institut für Ökologische Chemie, GSF-Forschungszentrum für Umwelt and Gesundheit, D-85764 Neuherberg, München, Germany

Sven Panke, Institute of Biotechnology, Eidgenössische Technische Hochschule Zürich, CH 8093 Zürich, Switzerland

Alvise Perosa, The Ca' Foscari University of Venice and National Interuniversity Consortium, "Chemistry for the Environment," Venice, Italy

Anne Pigamo, Dipartimento di Chimica Industriale e Dei Materiali, Università di Bologna, Viale Risorgimento 4, 40136 Bologna, Italy

Natalia V. Plechkova, QUILL Centre, The Queen's University of Belfast, Belfast BT9 5AG, Northern Ireland, United Kingdom

Andrew Schmid, Institute of Biotechnology, Eidgenössische Technische Hochschule Zürich, CH 8093 Zürich, Switzerland

Karl-Werner Schramm, Institut für Okologische Chemie, GSF-Forschungszentrum für Umwelt and Gesundheit, D-85764 Neuherberg, München, Germany

Kenneth R. Seddon, QUILL Centre, The Queen's University of Belfast, Belfast BT9 5AG, Northern Ireland, United Kingdom

Maurizio Selva, The Ca' Foscari University of Venice and National Interuniversity Consortium, "Chemistry for the Environment," Venice, Italy

Roger A. Sheldon, Delft University of Technology, Delft, The Netherlands

Johan Thoen, Dow Benelux B.V., Terneuzen, The Netherlands

Ferruccio Trifirò, Dipartimento di Chimica Industriale e Dei Materiali, Università di Bologna, Viale Risorgimento 4, 40136 Bologna, Italy

Pietro Tundo, The Ca' Foscari University of Venice and National Interuniversity Consortium, "Chemistry for the Environment," Venice, Italy

Ivar Ugi, Institute of Organic Chemistry and Biochemistry, Technische Universität Munchen, Lichtenbergstrasse 4, D-85747 Garching, Germany

Birgit Werner, Institute of Organic Chemistry and Biochemistry, Technische Universitat München, Lichtenbergstrasse 4, D-85747 Garching, Germany

Bernard Witholt, Institute of Biotechnology, Eidenössische Technische Hochschule Zürich, CH 8093 Zürich, Switzerland

Sergei Zinovyev, The Ca' Foscari University of Venice and National Interuniversity Consortium, "Chemistry for the Environment," Venice, Italy

PART 1

GREEN REAGENTS

1

THE FOUR-COMPONENT REACTION AND OTHER MULTICOMPONENT REACTIONS OF THE ISOCYANIDES

IVAR UGI AND BIRGIT WERNER

Institute of Organic Chemistry and Biochemistry, Technische Universität München, Germany

INTRODUCTION

The usual syntheses of products from three or more educts require several preparative processes, and its intermediate or final product must be isolated and purified after each reaction. As the number of steps increase, the amounts of solvents and the preparative work grows, while the yields of products decrease and more and more solvents and by-products must be removed. In such reactions, scarcely all optimal aspects of green chemistry can be accomplished simultaneously.

Practically irreversible multicomponent reactions (MCRs), like the Ugi 4-component reaction (U-4CR), can usually fulfill all essential aspects of green chemistry. Their products can be formed directly, requiring minimal work by just mixing three to nine educts. Often minimal amounts of solvents are needed, and almost quantitative yields of pure products are frequently formed.

The chemistry of the isocyanide U-4CR was introduced in the late 1950s, but this reaction was relatively little used for more than three decades, only around 1995 almost suddenly it was recovered by the chemical industry.[1]

In the last few years the variability of educts and products of the U-4CR has essentially increased, so that by now the majority of new products have been prepared. The U-4CR allows the preparation of more different types of products than

Methods and Reagents for Green Chemistry: An Introduction, Edited by Pietro Tundo, Alvise Perosa, and Fulvio Zecchini
Copyright © 2007 John Wiley & Sons, Inc.

any other reaction. If such a product with desirable properties—a lead structure—is found, large amounts of related compounds can be prepared easily by the U-4CR and similar reactions.

It is barely possible to still find novel reactions of one or two components, whereas the chemistry of the MCRs is not yet exhausted. Still, many new combination of up to nine different types of MCR educts can form new types of products that can totally differ from the already known chemistry.[1]

1.1   THE CLASSICAL MCRs

Chemical reactions are in principle equilibria between one or two educts and products. In practice, the preferred preparative reactions proceed irreversibly. Syntheses of products from three or more educts are usually sequences of preparative steps, where after each reaction step its intermediate or final product must be isolated and purified while the yield decreases. Exceptions can be the reactions of three components on solid surfaces and also some MCRs with α-additions of intermediate cations and anions onto the isocyanides.[1,2]

Besides the usual chemistry, an increasing number of chemical compounds can be prepared by MCRs just by mixing more than two educts.[3–5] Such processes do not proceed simultaneously, but they correspond to collections of subreactions, whose final steps form the products. Any product that can be prepared by an MCR whose last step is practically irreversible requires considerably less work and is obtained in a much higher yield than by any conventional multistep synthesis.

Three basic types of MCRs are now known.[5] The MCRs of type I are collections of equilibrating subreactions. In type II the educts and intermediate products equilibrate, but their final products are irreversibly formed. The MCRs of type III correspond to sequences of practically irreversible reactions that proceed from the educts to the product.[6]

In 1960 Hellmann and Opitz[7] introduced their *α-Aminoalkylierung* book, wherein they mentioned that the majority of the "name reactions" by MCRs belong together since they have in common their essential features. This collection of 3CRs can be considered as Hellmann–Opitz 3-component reactions, (HO-3CRs). They are either α-aminoalkylations of nucleophiles of MCR type I, or they form intermediate products that react with further bifunctional educts into heterocycles by 4CRs of type II. Their last step is always a ring closure that proceeds irreversibly.

This MCR chemistry began in 1850 when the Strecker reaction S-3CR[8] of ammonia, aldehydes, and hydrogen cyanide was introduced. Since 1912 the Mannich reaction M-3CR[9] of secondary amines, formaldehyde, and β-protonated ketones is used.

The MCRs of type II forming heterocycles begin with α-aminoalkylations of nucleophilic compounds, and subsequently these products react further with bifunctional educts whose last step is always an irreversible ring formation. Such reactions were introduced in 1882 by Hantzsch[10] and by Radziszewski.[11] Shortly after this Biginelli[12] also entered a similar type of forming heterocycle by MCRs. In the 1920s Bucherer and Bergs[13] began to produce hydantoin derivatives by BB-4CRs. This reaction begins with an S-3CR whose product then reacts with CO_2 and forms irreversibly the hydantoin. The products of the S-3CR and of the BB-4CR can both be hydrolized into α-aminoacids, but the synthesis via the BB-4CR is used preferentially, since this leads to products of higher purity and with higher yields.

In the early Gatterman's preparative chemistry book,[14] the one-pot synthesis of dihydropyridine derivatives like those formed by the Hantzsch reaction was one of practical laboratory exercises.

Schildberg and Fleckenstein observed that calciumantagonists can advantageously influence the peripheric vessels and those of the heart.[15] With the 4-aryldihydropyridine-3,5-dicarboxylic esters 4 (Scheme 1.1) that have such effects, the first pharmaceutical products synthesized by Hantzsch reactions were independently introduced by the Bayer AG[16] and Smith Cline & French.[17]

As the last classical MCR in the 1950s, Asinger[18] introduced the 3CRs and 4CRs to form thiazole derivatives. It seems that these A-MCRs can belong to type I or to type II.

In preparative chemistry only a few MCRs of type III are known;[6] however, in living cells, the collections of the biochemical compounds are formed by MCRs of type III. In that case the formation of the individual products proceeds by subreactions that are accelerated by the enzymes present in the suitable areas within the living cells. The resulting collections of products can be considered to be their libraries.

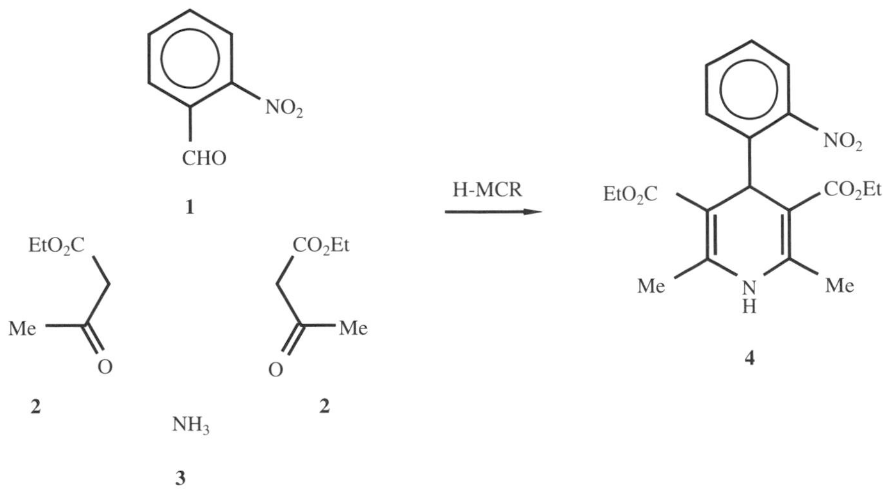

Scheme 1.1 Hantzsch synthesis of 4-aryldihydropyridine-3,5-dicarboxylic esters.

1.2   THE FIRST CENTURY OF THE ISOCYANIDES

The chemistry of the isocyanides[3] began when, in 1859 Lieke[19] formed allyl iso-cyanide from allyl iodide and silver cyanide, and when, in 1866 Meyer[20] produced in the same way 1-isocyano-1-desoxy-glucose. In 1867, Gautier[21a] used this procedure to prepare alkylisocyanides, and Hofmann[22] introduced the formation of isocyanides from primary amines, chloroform, and potassium hydroxyde. Gautier[3,21b] also tried to prepare an isocyanide by dehydrating an amine formiate via its formylamine using phosphorus pentoxide, but this process produced no isocyanide. Gautier had not yet realized that acidic media destroyed the isocyanides.

However, for a whole century the chemistry of the isocyanides remained as a rather empty part of organic chemistry, since they were not yet easily available, and furthermore they had a very unpleasant smell. At that time, only 12 isocyanides had been prepared and only a few of their reactions had been investigated.[3]

In the 1890s, Nef[23] mentioned that the functional group —NC of the isocyanides contains a divalent carbon atom C^{II}, and therefore there is a large difference between their chemistry and that of the other chemical compounds that contain only tetravalent carbon atoms C^{IV}. Any synthesis of isocyanides corresponds to a conversion of C^{IV} into C^{II}, and all chemical reactions of isocyanides correspond to transitions of the carbon atoms C^{II} into C^{IV}.

In this period, the most important reactions of the isocyanides were the formations of tetrazole derivatives from isocyanides and hydrazoic acid, a process introduced in 1910 by Oliveri-Mandala and Alagna,[24] and then in 1921 Passerini introduced the reaction (P-3CR),[25] which was the first 3-component reaction of the isocyanides. In the 1940s Baker,[26] and later Dewar,[27] proposed mechanisms of the P-3CR. The important role of the intermediate hydrogen bond between the carboxylic acid and the carbonyl compound in suitable solvents was mentioned.[4]

In 1948, Rothe[4,28] discovered the first naturally occurring isocyanide in the *Penicillum notatum Westling* and in the *Penicillum chrysogenum*. This compound was soon used as the antibiotic *xanthocillin* **5a**. Later Hagedorn and Tönjes[29] prepared its *O,O'*-dimethylether of *xanthocillin* **5b** by dehydrating its *N,N'*-diformylamine with phenylsulfonylchloride in pyridine (Scheme 1.2). Since

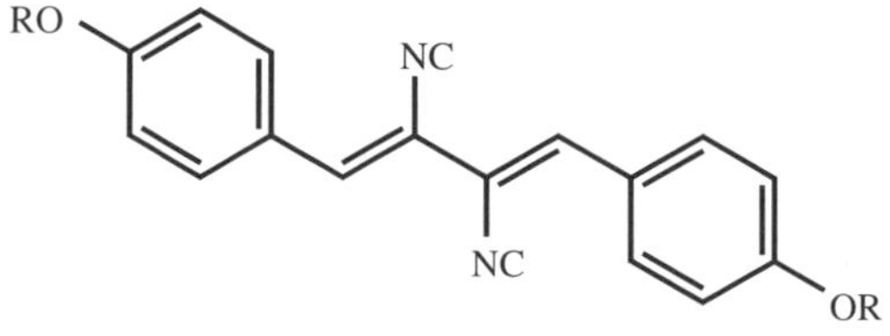

5a: R = H
5b: R = Me

Scheme 1.2   Xanthocillin.

1973 an increasing number of naturally occurring isocyanides has been found in plants and living cells.[30]

1.3 THE MODERN CHEMISTRY OF THE ISOCYANIDES

A new era of the isocyanide chemistry began in 1958 when the isocyanides became generally available by dehydrating the corresponding formylamines in the presence of suitable bases (Scheme 1.3).[4] A systematic search for the most suitable dehydrating reagent revealed early on that phosgene[31] is excellent for this purpose. Later, when phosgene transportation was not allowed anymore, it was locally produced from triphosgene.[32] Also diphosgene[33] and phosphorus oxychloride,[4] can be used, particularly in the presence of di-isopropylamine.[34] Baldwin et al.[35] prepared naturally occurring epoxy-isocyanides from the corresponding formylamines by dehydrating the latter with trifluoromethyl sulfonic acid anhydride in the presence of di-isopropylamine.

In the 1971 book *Isonitrile Chemistry*[3] 325 isocyanides were mentioned, and almost all of them had been prepared by dehydration of formylamines.

After some model reactions, Ugi et al.[3a−d] accomplished a new way of preparing Xylocaine® by one of the first U-4CRs. In 1944 Xylocaine **12**[36] (Scheme 1.4) was introduced by the A. B. Astra company in Sweden, and since then Xylocaine has been one of the most often used local anasthetics, particularly by dentists. In its early period, A. B. Astra patented 26 chemical methods of preparing **12**.

In January 1959, Ugi and co-workers decided to prepare **12** from diethylamine **9**, formaldehyde **10**, and 2,6-xylylisoxcyanide **11**. Initially they considered this as a variation of the Mannich reaction.[10] In their first experiment they noticed that this reaction is so exothermic that an immediate mixing of the educts can initiate an explosion,[3,37] and it was realized that this reaction was in reality a 4-component reaction in which water **7** also participates.

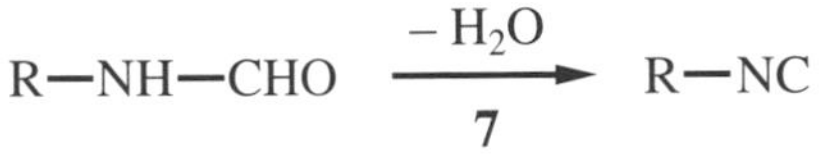

$$R-NH-CHO \xrightarrow[\text{7}]{-H_2O} R-NC$$

6   **8**

Scheme 1.3   General formation of isocyanides.

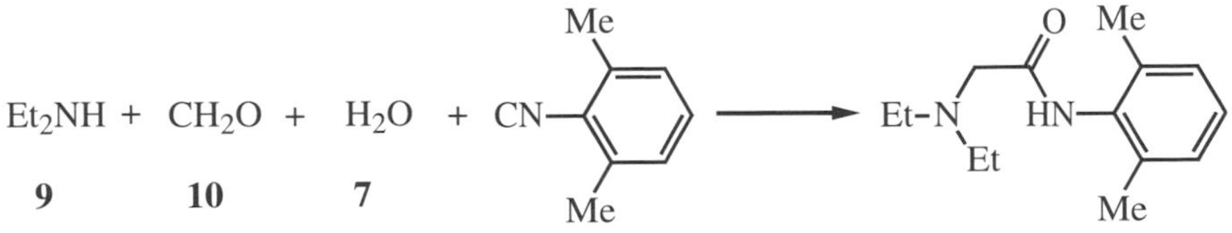

$$Et_2NH + CH_2O + H_2O + CN-\bigcirc \longrightarrow$$

9   **10**   **7**   **11**   **12**

Scheme 1.4   Four-component reaction of Xylocaine®.

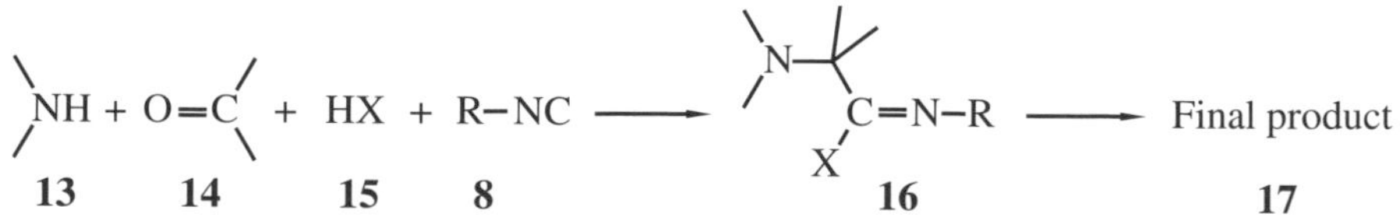

$$\underset{\mathbf{13}}{\overset{\diagdown}{\diagup}}NH + \underset{\mathbf{14}}{\overset{\diagup}{O=C}\diagdown} + \underset{\mathbf{15}}{HX} + \underset{\mathbf{8}}{R-NC} \longrightarrow \underset{\mathbf{16}}{\overset{\diagdown N-\diagup}{\underset{X}{\diagup}C=N-R}} \longrightarrow \underset{\mathbf{17}}{\text{Final product}}$$

Scheme 1.5 The Ugi reaction.

During the first month of this experiment, it was realized that this reaction is extremely variable. Thus, diverse amines (ammonia, primary and secondary amines, hydrazine derivatives, hydroxylamines) **13**, carbonyl compounds (aldehydes, ketones) **14**, acid components **15** or their anions (H_2O, $Na_2S_2O_3$, H_2Se, R_2NH, RHN-CN, HN_3, HNCO, HNCS, RCO_2H, RCOSH, $ROCO_2H$, etc.), and the isocyanides **8**[3,4,38] could form the α-adducts **16** that rearrange into their products **17** (Scheme 1.5).

Since 1962, this reaction has been called the Ugi reaction,[4a] or it is abbreviated as the U-CC,[38a] or as the U-4CR.[38b] The U-4CR can formally be considered to be a union,[39] 4CR = HO-3CR ∪ P-3CR **18** (Scheme 1.6), of the HO-3CR and the P-3CR that have in common the carbonyl compounds and acids, while the HO-3CR also needs an amine and the P-3CR an isocyanide.

In each type of chemical reaction, the skeleton of the product is characteristic, and only its substituents can be different, whereas in the U-4CR and related reactions of the isocyanides the skeleton of the products can also include different types of amines and acid components. This is illustrated by the eight skeletally different products in Scheme 1.7. Besides these compounds, many other types of compounds also can be prepared by the U-4CR.

Ordinary chemical reactions have their "scopes and limitations" for various reasons. Many sterically crowded products cannot be formed by conventional syntheses, but they can still be prepared by the U-4CR. Thus, the product **22**[40] can be formed only by the U-4CR, (Scheme 1.8).

The U-4CR forms its products by less work and in higher yields than other syntheses. The U-4CR is nowadays one of the most often used chemical reaction

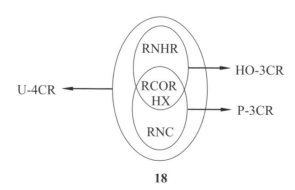

18

Scheme 1.6 The U-4CR as a union of the HO-3CR and the P-3CR.

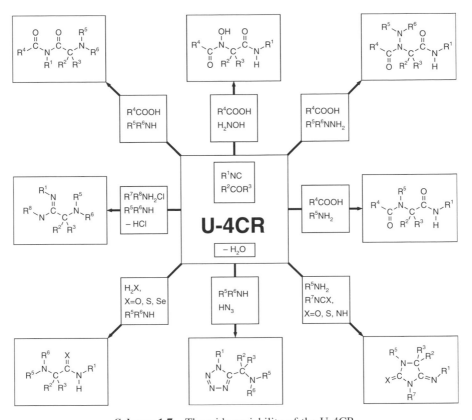

Scheme 1.7 The wide variability of the U-4CR.

for the formation of chemical libraries. These libraries had already been proposed in 1961 (Ref. 3, p.149; Ref. 41), but only in the 1980s the chemical industry has recognized the advantages of the libraries.[5,42,43]

In ordinary reactions where two educts participate, 10 different components of each educt type can form 100 constitutionally different products. The U-4CR can form 10,000 different products when 10 different starting materials of each type of educt[44] are involved. In this way, libraries of an extremely high number

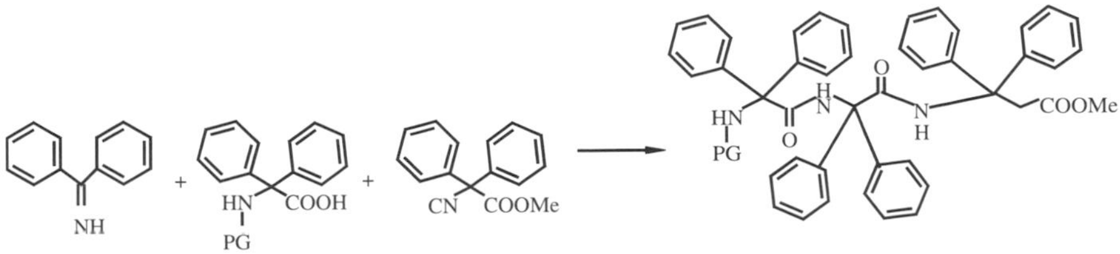

19   **20**   **21**   **22**

Scheme 1.8 Synthesis of a sterically extremely hindered product by the U-4CR.

of products can be formed via the U-4CR. Combined with other combinatorial methods, the search for new desirable products can thus be accomplished particularly well.

A product of the U-4CR is only formed in a good yield and purity if the optimal reaction conditions are used. The U-4CR proceeds faster and in a higher yield, when the amine component and the carbonyl compound are precondensed, and the acid component and the isocyanide are added later.[44] Very often methanol or trifluoroethanol are suitable solvents, but sometimes a variety of other solvents can be used. Furthermore, the sequence of the educts and their concentrations must be optimal and a suitable temperature of the reaction must be used. In many cases, the U-4CR can be improved by a catalyst.[1,45]

In a special case, the reaction mechanism of the U-4CR was investigated.[3,44,46] The aldehyde and chiral amine were precondensed into the Schiff-base isobutyraldehyde-(S)-α-phenylethylamine that was reacted with benzoic acid and *tert*-butylisocyanide in methanol at O°C. In one series of experiments, the dependence of the electrical conductivity of this Schiff-base and the carboxylic acid was determined, and in a second series of experiment, the relation between the educt concentrations and the ratio of diastereoisomeric products caused by competing different stereoselective U-4CRs was investigated. The ratio of the diastereomeric products was determined by their optical rotations.[44] The large collection of numerical values of these experimental data were evaluated by a mathematically based computer program. It was found that four pairs of stereoselective processes compete and, depending on the concentrations of the educts, one or the other diastereomeric product is preferentially formed. This knowledge made it generally possible to find the optimal conditions of the U-4CR by fewer experiments than usual.

Rather early it was recognized how much easier natural products and related compounds can be prepared by the U-4CR,[1,4] but the advantages of searching for new desirable pharmaceutical and plant-protecting compounds became evident only during the last few years, when industry began to produce the U-4CR products.[1,42]

For a whole decade a research group at Hofmann-LaRoche AG tried, without success, to find suitable thrombine inhibitors by the coventional methods. But only in 1995 Weber et al.[47] discovered two such desired products, **23a** and **23b** (Scheme 1.9), when they used libraries of 4-CR products for their systematically planned search, which also included mathematically oriented methods.

Recently, the Merck Research Laboratory demonstrated an important example.[48] Initially the HIV protease inhibitor Crixivan™ (MK 639) **29** (Scheme 1.10) could not be prepared very well by a complicated conventional multistep synthesis, but **29** became available when it was prepared by an easier synthesis, whose essential step was accomplished by a U-4CR.

Park et al.[49] used U-4CR libraries to prepare Ras-Raf protein-binding compounds like **30** that are active against HIV. The patented product **31** has been formed by Lockhoff[50] at the Bayer AG using a U-4CR of four different protected glucose derivatives that were later deprotected. The product **32** of

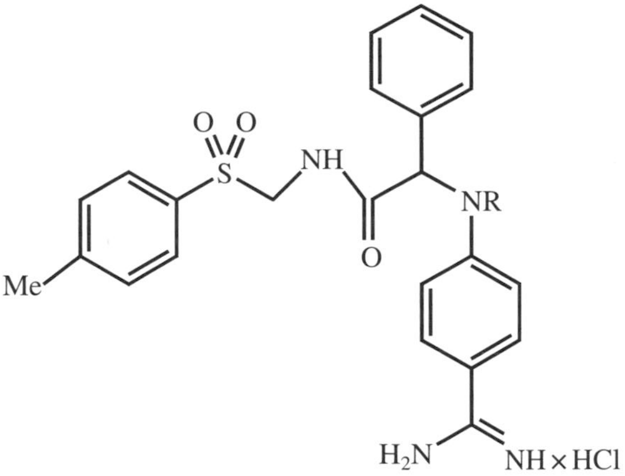

23   (a: R = H)

(b: R = *p*-hydroxy-benzoyl)

Scheme 1.9   Thrombine inhibitors by Weber et al.

Dömling et al.[51] can be prepared very easily by the U-4CR. This compound is related to the PNA compounds of Nielsen.[52]

Many cyclic products have been formed by U-4CRs from multifunctional educts. This is illustrated here by a few examples (Scheme 1.12).

The synthesis of the penicillin-related compound **39**, introduced in 1962, begins with an A-4CR of **37a**, which is hydrolized into **37b**. This undergoes a U-4CR with isopropyl-isocyanide **38** and forms **39**.[53] During the following decades, a large variety of antibiotically active β-lactam derivatives was produced.[54] Recently **42**, compound **43**, and one of its stereoisomers were stereospecifically prepared by U-4CRs.[55]

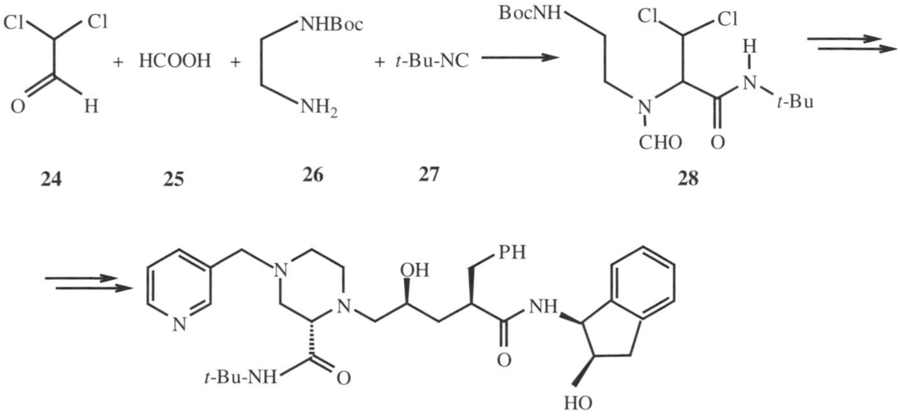

Scheme 1.10   Synthesis of Crixivan™ (MK 639).

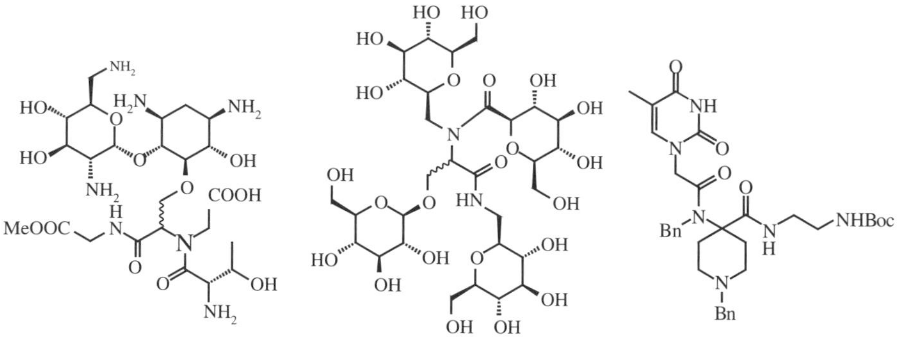

Scheme 1.11 Biologically active compounds synthesized via U-4CR.

A variety of cyclic products have been prepared from educts containing carbonyl as well as carboxylic groups. Thus, Hanusch-Kompa and Ugi[56,57] prepared a large number of five-membered cyclic gamma-lactam compounds like **44** from levulinic acid. Other carbonylic acids can lead to compounds like **45**, which is made from phthalaldehyde acid, valine methylester, and *tert*-butylisocyanide. The products like **46** and **47** can result from the U-4CR and further cyclization.

In addition, the six- to eight-membered lactams like **48**, **49**,[58] and **50**[59] have been formed from amines, carbonyl-carboxylic acids, and isocyanides (Scheme 1.14).

Product **56** (Scheme 1.15) with a particularly complicated structure was prepared by the U-4CR of **51**–**54**, followed by a few further steps.[60]

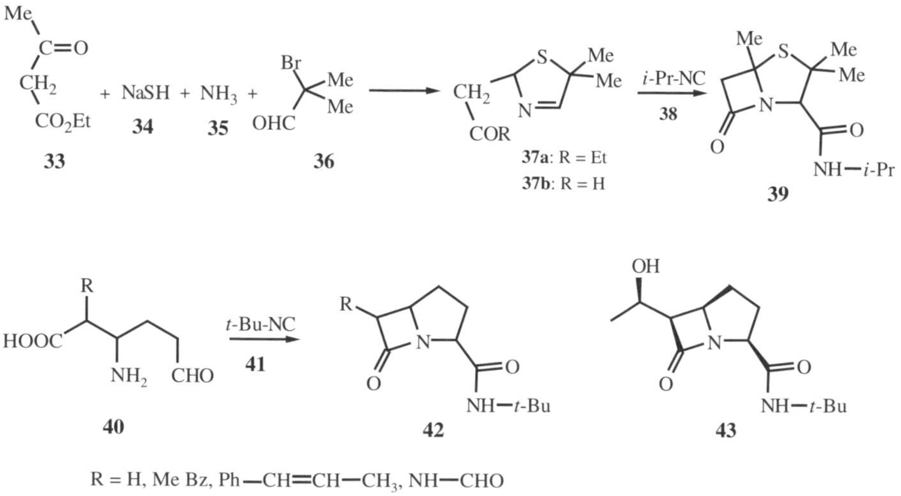

R = H, Me Bz, Ph—CH=CH—CH₃, NH—CHO

Scheme 1.12 U-4CR products of multifunctional educts.

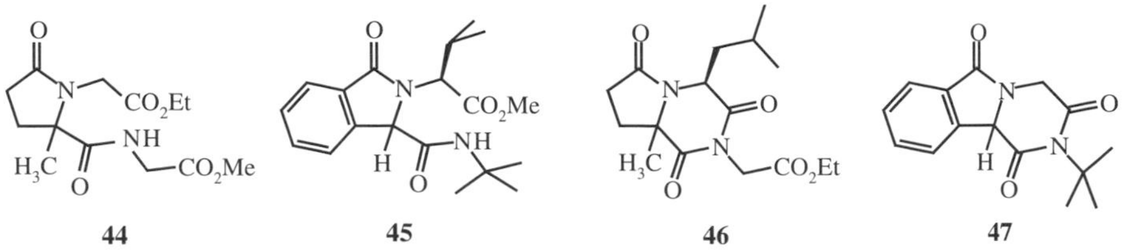

Scheme 1.13   U-4CR products from carbonyl groups containing carboxylic acids.

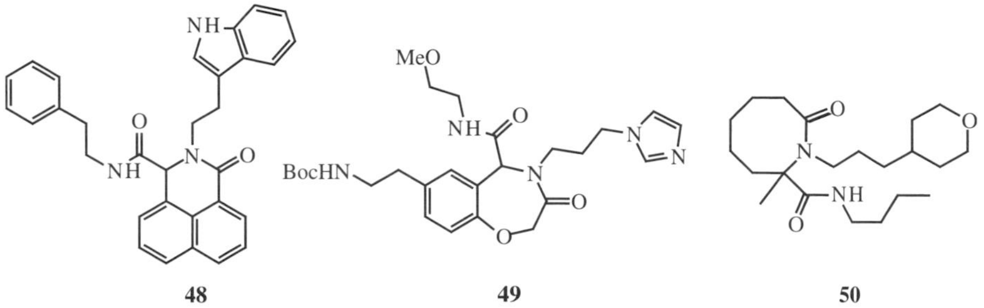

Scheme 1.14   Six- to eight-membered lactams formed via U-4CR.

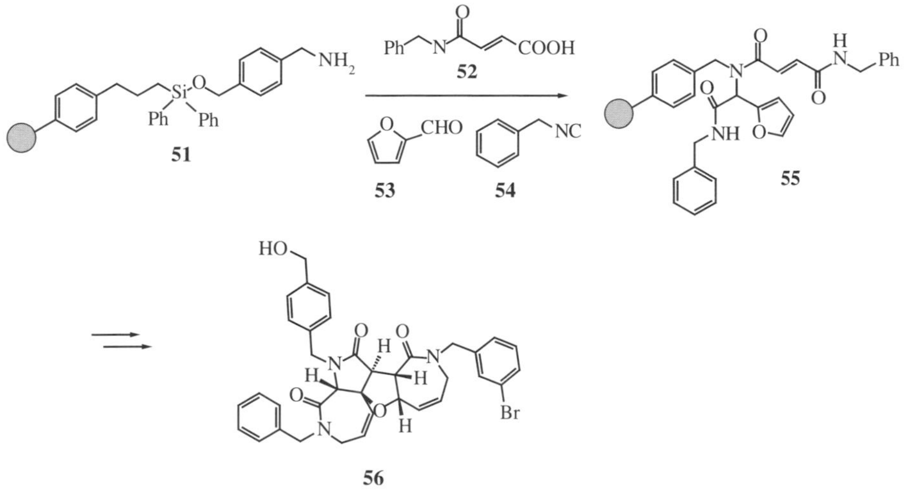

Scheme 1.15   Solid-phase synthesis of a polycyclic ring system employing a U-4CR.

1.4   STEREOSELECTIVE U-4CR

After the U-4CR had been introduced, it was soon recognized that this reaction can form diastereomeric products from chiral amine components,[61,62] for example, chiral α-ferrocenyl-alkylamines.

As the latter were not easily accessible by chemical synthesis at that time,[44] new methods of preparing these ferrocene derivatives were developed and introduced in 1969.[63] It was then proved that the U-4CRs of chiral α-ferrocenyl-alkylamines can form diastereomeric α-aminoacid derivatives stereoselectively, and it was further shown that after the reaction the α-ferrocenyl groups of the products can be replaced by protons, thus resynthesizing the chiral α-ferrocenyl-alkylamines simultaneously.[44] Later, the development of this ferrocene chemistry was given up since such syntheses cannot form the products in sufficient quantity and stereoselective purity.[64]

In 1988 Kunz and Pfrengle[65] introduced the preparation of chiral amino acid derivatives by the U-4CR in the presence of 2,3,4,6-tetra-O-pivaloyl-β-D-galactopyranosylamine, **57**, in the presence of $ZnCl_2$-etherate as catalyst. They obtained excellent stereoselectivity and high yields of their products. One of the disadvantages of such U-4CRs is that only formic acid can be used as the acid component, and the auxiliary group of the products can only be removed by half-concentrated hot methanolic HCl.

A few years later Goebel and Ugi[66] formed α-aminoacid derivatives by the U-4CR with tetra-O-alkyl-1-glucopyranosylamines, **58**, where any carboxylic acid component can participate. Lehnhoff and Ugi[67] used the U-4CR with 1-amino-2-deoxy-2-N-acetylamino-3,4,6-tri-O-acetyl-β-D-glucopyranose, **59**, whose large variety of products could be formed stereoselectively in excellent yields. The desired selective cleavage of the auxiliary groups of these products was equally unefficient.

Zychlinski[68] prepared 1-amino-5-deoxy-5-acetamido-2,3,4-tri-O-acetyl-β-D-glucopyranose **60** by a synthesis of 11 steps. This amine component undergoes the U-4CRs very well and the products are cleavable by water, but unfortunately they are not very stable.

Ross and Ugi[43] prepared 1-amino-5-deoxy-5-thio-2,3,4-tri-O-isobutanoyl-β-D-xylopyranose **61a** from xylose via the 5-desoxy-5-thio-D-xylopyranose. The U-4CRs of this amine form α-aminoacid derivatives stereoselectively and in excellent yields. These products have the advantage that their products are stable and their auxiliary group 5-desoxy-5-thio-D-xylopyranose can be cleaved off selectively by mercury(II) acetate and trifluoroacetic acid. The expected steric structure of the corresponding U-4CR product was confirmed by X-ray measurement.[69]

Since Ugi is now an emeritus and he and his co-workers cannot continue their experimental studies, we propose that the analog 1-amino-5-desoxy-5-thio-D-xylopyranose **61b** should be prepared and be used as a reagent of U-4CRs. It has a good chance to form stereoselectively high yields of products whose auxiliary group can be selectively removed.

In 1985 Kochetkov et al.[70] introduced the preparation of 1-amino-carbohydrates from xylose, glucose and 2-acetylamino-glucose just by adding ammonia, and later they improved the preparation of pyranosylamines by using additional ammonium carbonate.[71]

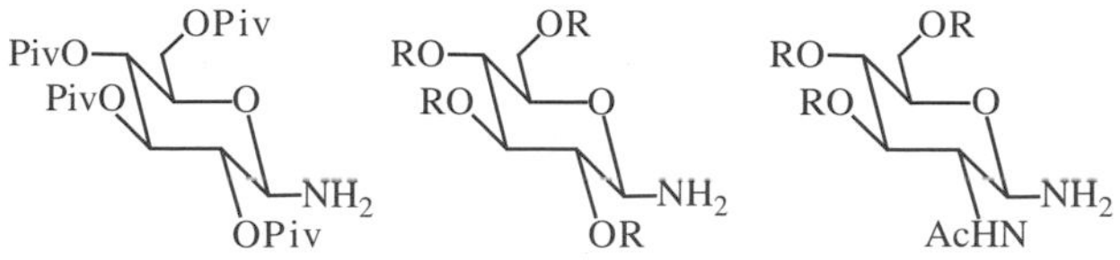

57 **58** (R = Me, Et, Am, Bz) **59** (R = H, Ac)

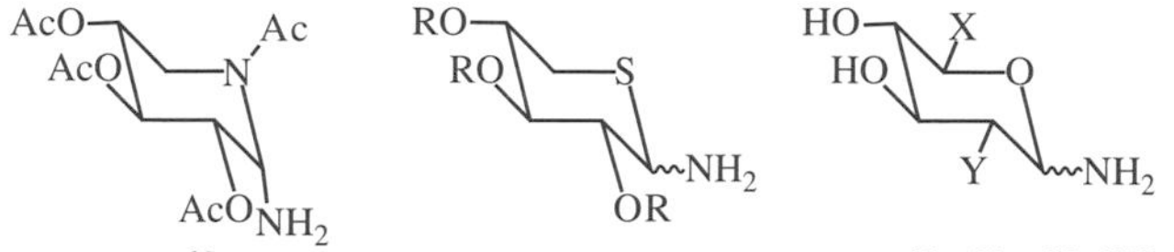

60 **61a**: R = Piv **61b**: R = H **62**: X = H, CH_2OH; Y = NH–CO–R with R = Me, O*t*-Bu, Ph, 4-MeO-Ph

Scheme 1.16 Chiral amino carbohydrates employed in the U-4CR.

Drabik et al.[45] prepared these and additional 1-amino-carbohydrates, **62**, which were used as the amine components of the U-4CRs. Thus, α-aminoacid derivatives could be prepared stereoselectively in good yields. The stereoselectivity and yields resulted especially well if 0.1 gram equivalent (eq.) $ZnCl_2$–OEt_2 or $CeCl_3$–$7H_2O$ or $ZrCl_4$ were used as catalysts. Among these, $CeCl_3$–$7H_2O$ had a particularly good influence, so that 99% yields and stereoselectivities of 99% d.e. can result.

The auxiliary carbohydrate parts of the products could be removed only moderately. The most efficient cleavages were achieved when the U-4CR products of the amine component **62** were treated with 1 M HCl in methanol at 40°C for 19 h. In that case, yields up to 30% could be achieved; for Y = NH-CO-4-MeOPh in compound **62**, the cleavage rate could be increased up to a yield of 46%.

1.5 THE UNIONS OF THE U-4CR AND FURTHER REACTIONS

Already in the early days of the U-4CR, several types of 5CRs were found.[3,72] It was also observed long ago that an autoxydizing 6-component reaction of two isocyanides took place besides the main U-4CR,[73] and the structure of one of these by-products was determined by an X-ray measurement.[74] The reaction mechanism of such autoxydation was determined[75] by the assistance of the computer program RAIN.[76] At that time it was not yet known that the MCRs of isocyanides with more than four educts proceed by different reaction mechanisms.

The new era of isocyanide chemistry was determined by two aspects. First, it was the formation of products by MCRs with high numbers of educts and second, the recently initiated search for new desirable products in libraries of MCR

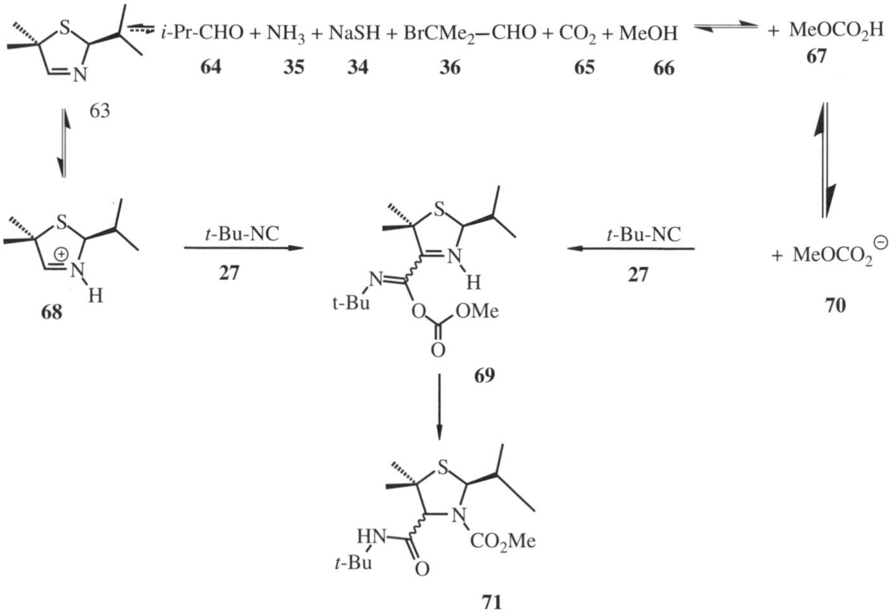
i-Pr-CHO + NH3 + NaSH + BrCMe2—CHO + CO2 + MeOH
64
35
34
36
65
66
+ MeOCO2H
67
63
t-Bu-NC
27
t-Bu-NC
27
+ MeOCO2
70
68
H
t-Bu
OMe
H
69
HN
t-Bu
CO2Me
71

Scheme 1.17 The first 7-CR.

products.[1] In 1995, the chemical industry began to search for new compounds in the libraries of products formed by the U-4CR and related reactions.

Using this new technology, a single chemist can now form more than 20,000 new compounds a day, whereas before a good chemist could accomplish up to 10,000 syntheses in the 40 years of his or her professional life. MCRs are especially suitable for the formation of libraries, since they have the big advantage that their products can be prepared with a minimum of work, chemical compounds and energy, and in essentially higher yields than by conventional methods.

In 1993 the first MCR composed of seven educts was introduced,[77] and it was soon recognized that such higher MCRs are usually unions[39] of the U-4CR and additional reactions.[38] In the first 7-CR, the intermediate **63** was formed by an A-4CR and underwent with the equilibrating product **67** the α-addition of the cations and ions onto the isocyanide **27**. Finally, this α-adduct, **69**, rearranges into the final product, **71** (Scheme 1.17).

The variety of educts and products of the higher MCRs is illustrated here. Product **72** (Scheme 1.18) is formed from the five functional groups of lysine, benzaldehyde, and tert-butylisocyanide.[78] The synthesis of **73** is achieved with hydrazine, furanaldehyde, malonic acid, and the isocyano methylester of acetic acid,[57,79] compound **74** results from the reaction of benzylamine, 5-methyl-2-furanaldehyde, maleic acid mono-ethylester, and benzylisocyanide.[80] Zhu et al.[81] prepared a variety of related products, such as, **75**, from *O*-amino-methyl cinnamate, heptanal, and α-isocyano α-benzyl acetamides.

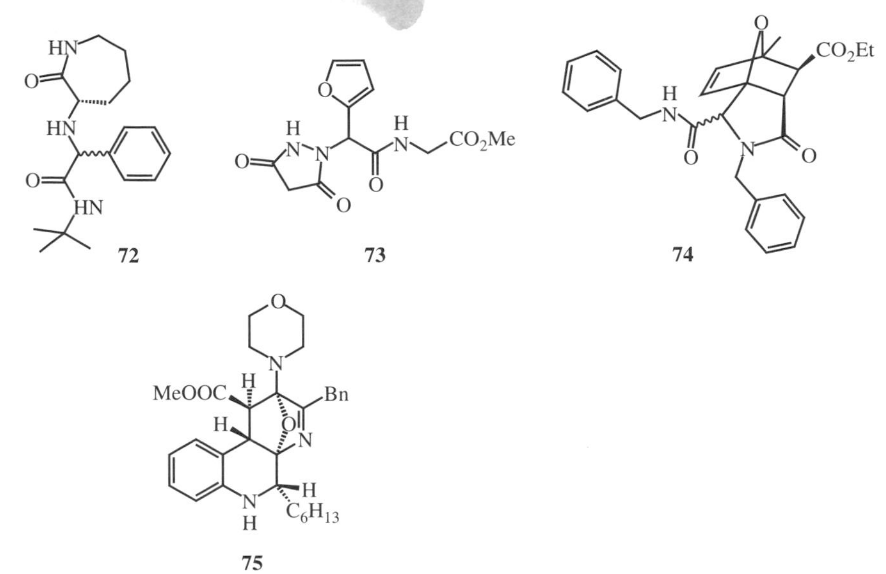

Scheme 1.18 Products of higher MCRs.

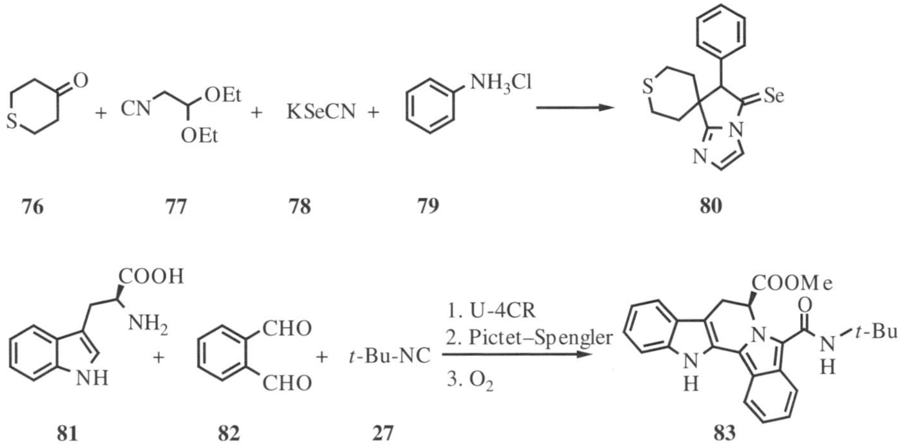

Scheme 1.19 Polycyclic products of higher MCRs.

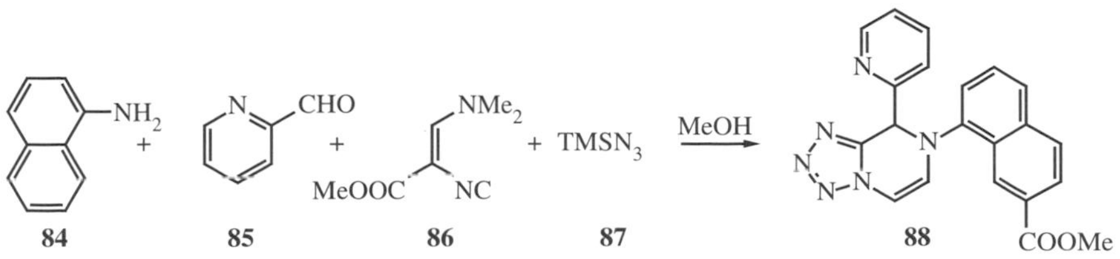

Scheme 1.20 Heterocycle formation by MCRs.

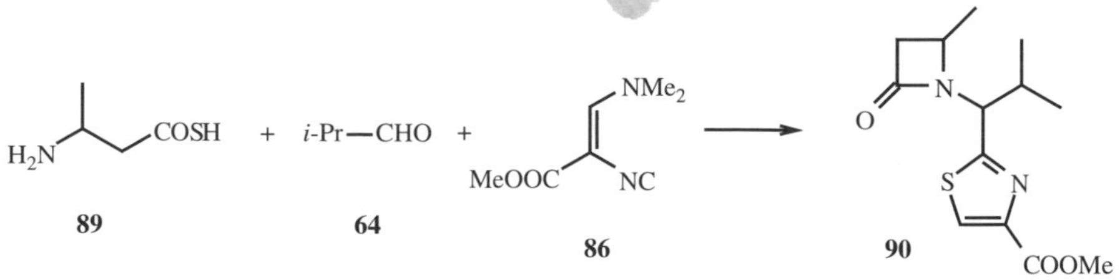

Scheme 1.21   MCR employing a thioacid.

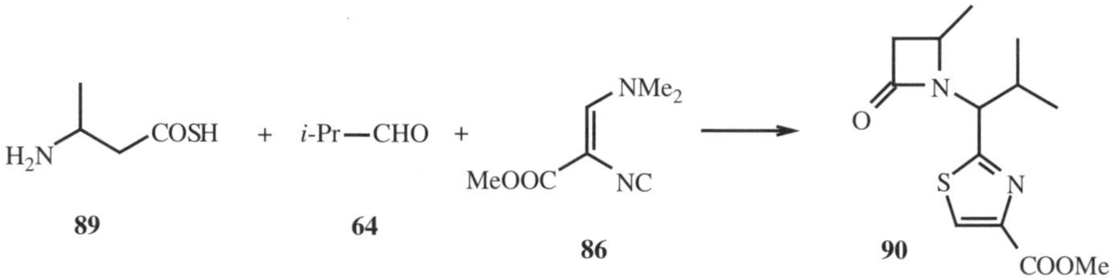

Scheme 1.22   Examples of MCRs with up to nine reacting functional groups.

In the last decade, Bossio et al.[82] have formed cyclic products of many different types by using a variety of new MCRs. Thus, **80** was made from **76–79** (Scheme 1.19). Recently, Dömling and Chi[83] prepared **83** from **81**, **82**, and **27**, and synthesized similar polycyclic products from other α-aminoacids with **82** and **27**.

In 1979, Schöllkopf et al.[84] formed α-isocyano-β-dimethylamino-acryl methyl esters **86**, and Bienaymé prepared many similar isocyanides,[85] which can undergo a variety of heterocycles, forming reactions like the synthesis of **88**[86] from **84–87** (Scheme 1.20).

Dömling et al.[87] made react β-amino butyric thioacid, **89**, the isobutyraldehyde **64**, and **86** into the product **90**, which simultaneously contains a ß-lactam group and a thiazole system.

A variety of MCRs with seven to nine functional groups of several pairs of educts can be carried out, as is illustrated by the four subsequent reactions.[88–90] (Scheme 1.22).

REFERENCES

1. A. Dömling, I. Ugi, *Angew. Chem.*, **112**, 3300 (2000); *Angew. Chem. Int. Ed. Engl.*, **39**, 3168 (2000).

2. D. Janezic, M. Hodoscek, I. Ugi, *Internet Electron. J. Mol. Des.*, **1**, 293 (2002).

3a. I. Ugi, *Isonitrile Chemistry*, p. xi Academic Press, New York, 1971.

3b. D. Marquarding, G. Gokel, P. Hoffman, I. Ugi, in *Isonitrile Chemistry*, I. Ugi, Ed. p. 133, Academic Press, New York, 1971.

3c. G. Gokel, G. Lüdke, I. Ugi., in *Isonitrile Chemistry*, I. Ugi, Ed. p. 145, Academic Press, New York, 1971.

3d. G. Gokel, P. Hoffmann, H. Kleimann, K. Klusacek, G. Lüdke, D. Marquarding, I. Ugi, in *Isonitrile Chemistry*, I. Ugi, Ed. p. 201, Academic Press, New York, 1971.

4. I. Ugi, S. Lohberger, R. Karl, in *Comprehensive Organic Synthesis: Selectivity for Synthetic Efficiency*, B. M. Trost, C. H. Heathcock, Eds., Vol. 2, Chap. 4.6, p. 1083, Pergamon, Oxford, 1991, (a) p. 1090.

5. I. Ugi, *J. Prakt. Chem.*, **339**, 499 (1997).

6. J. Chattopadhyaya, A. Dömling, K. Lorenz, et al., *Nucleosides & Nucleotides,* **16**, 843 (1997).

7. H. Hellmann, G. Opitz, α-*Aminoalkylierung*, Verlag Chemie, Weinheim, 1960.

8. A. Strecker, *Ann. Chem.*, **75**, 27 (1850).

9. C. Mannich, I. Krötsche, *Arch. Pharm.*, **250**, 647 (1912); R. Adams, *Organic Reaction*, F. F. Blick, Ed., Vol. 1, p. 303, John Wiley & Sons, New York, 1942.

10. A. Hantzsch, *Liebigs Ann. Chem.*, **219**, 1 (1892); *Ber. Dtsch. Chem. Ges.*, **23**, 1474 (1890); see also, C. Böttinger, *Liebigs Ann. Chem.*, **308**, 122 (1981); U. Eisener, J. Kuthan, *Chem. Rev.*, **72**, 1 (1972).

11. B. Radziszewski, *Ber. Dtsch. Chem. Ges.*, **15**, 1499, 2706 (1882).

12. P. Biginelli, *Ber. Dtsch. Chem. Ges.* **24**, 1317, 2962 (1891); **26**, 447 (1893).

13. Bergs, H. *German Patent No.* 566094 (1929); *Chem. Abstr.* 27 1001 (1933); H. T. Bucherer, W. Steiner, *J. Prakt. Chem.*, **140**, 291 (1934); H. T. Bucherer, H. Barsch, *J. Prakt. Chem.*, **140**, 151 (1934).

14. L. Gattermann, *Die Praxis des Organischen Chemikers*, Walter Du Gruiter & Co, Berlin, 1894.

15. F. W. Schildberg, A. Fleckenstein, *Pflügers Arch. Gesamte Physiol. Menschen Tiere*, **283**, 137 (1965); A. Fleckenstein, *Med. Klin.*, **70**, 1665, (1975).

16. F. Bossert, W. Vater, Südafr. Patent No. 6801, 482 (1968); *Bayer AG, Chem. Abstr.*, **70**, 96641 d (1969); *Naturwissenschaften,* **58**, 578 (1971); W. Vater, G. Kroneberg, F. Hoffmeister, H. Kaller, K. Meng, A. Oberdorf, W. Puls, K. Schloßmann, K. Stoepel, *Arzneim. Forsch.*, **22**, 124 (1972).

17. B. Loev, M. M. Goodman, K. M. Snader, R. Tedeschi, E. Macho, *J. Med. Chem.*, **17**, 956 (1974).

18. (a) F. Asinger, *Angew. Chem.*, **68**, 413 (1956); F. Asinger, M. Thiel, E. Pallas, *Liebigs Ann. Chem.*, **602**, 37 (1957); (b) F. Asinger, M. Thiel, *Angew. Chem.*, **70**, 667 (1958); (c) F. Asinger, W. Leuchtenberg, H. Offermanns, *Chem. Ztg.*, **94**, 6105 (1974); F. Asinger, K. H. Gluzek, *Monats. Chem.*, **114**, 47 (1983).

19. W. Lieke, *Liebigs Ann. Chem.*, **112**, 316 (1859).

20. E. Meyer, *J. Prakt. Chem.*, **67**, 147 (1866).

21. A. Gautier, *Liebigs Ann. Chem.*, **142**, 289 (1867); *Ann. Chim. (Paris)*, **17**, 193, 203 (1869).

22. A. W. Hofmann, *Ber. Dtsch. Chem. Ges.*, **3**, 63 (1870); see also, W. P. Weber, G. W. Gokel, I. Ugi, *Angew. Chem.*, **84**, 587 (1972); *Angew. Chem. Int. Ed. Engl.*, **11**, 530 (1972).

23. U. Nef, *Liebigs Ann. Chem.*, **270**, 267 (1892); *Liebigs Ann. Chem.*, **309**, 126 (1899).

24. E. Oliveri-Mandala, B. Alagna, *Gazz. Chim. Ital.*, **40 II**, 441 (1910); B. N. Zimmerman, R. A. Olafson, *Tetrahedron Lett.*, 5081 (1969).

25. M. Passerini, *Gazz. Chim. Ital.*, **51 II**, 126, 181 (1921); M. Passerini, G. Ragni, *Gazz. Chim. Ital.*, **61**, 964 (1931).

26. R. H. Baker, L. E. Linn, *J. Am. Chem. Soc.*, **70**, 3721 (1948); R. H. Baker, A. H. Schlesinger, *J. Am. Chem. Soc.*, **73**, 699 (1951).

27. M. J. S. Dewar, *Chem. Soc.*, **67**, 1499 (1945); *Theory of Organic Chemistry*, Oxford University Press (Clarendon), London and New York, 1949.

28. W. Rothe, *Pharmazie*, **5**, 190 (1950).

29. I. Hagedorn, H. Tönjes, *Pharmazie*, **11**, 409 (1956); **12**, 567 (1957).

30. P. J. Scheuer, *Acc. Chem. Res.*, **25**, 433 (1992); C. W. J. Chang, P. J. Scheuer, *Top. Curr. Chem.*, **167**, 33 (1993).

31. I. Ugi, R. Meyr, *Angew. Chem.*, **70**, 702 (1958); I. Ugi, R. Meyr, *Chem. Ber.*, **93**, 239 (1960); I. Ugi, U. Fetzer, U. Eholzer, H. Knupfer, K. Offermann, *Angew. Chem.*, **77**, 492 (1965); *Angew. Chem. Int. Ed. Engl.*, **4**, 452 (1965).

32. H. Eckert, B. Forster, *Angew. Chem.*, **99**, 922 (1987); *Angew. Chem. Int. Ed. Engl.*, **26**, 1221 (1987).

33. G. Skorna, I. Ugi, *Angew. Chem.*, **89**, 267 (1977); *Angew. Chem. Int. Ed. Engl.*, **16**, 259 (1977).

34. R. Obrecht, R. Herrmann, I. Ugi, *Synthesis*, 400 (1985).

35. J. E. Baldwin, A. E. Derome, L. D. Field, P. T. Gallagher, A. A. Taha, V. Thaller, D. Brewer, A. Taylor, *J. Chem. Soc., Chem. Commun.*, **1981**, 1227; J. E. Baldwin, D. R. Kelly, C. B. Ziegler, *J. Chem. Soc., Chem. Commun.*, **1984**, 133; J. E. Baldwin, M. A. Adlington, J. Chandrogianni, M. S. Edenborough, J. W. Keeping, C. B. Ziegler, *Chem. Soc. Chem. Commun.*, **1985**, 816; J. E. Baldwin, I. A. O'Neil, *Tetrahedron Lett.*, **31**, 2047 (1990); I. A. O'Neil, J. E. Baldwin, *Synlett*, **1990**, 603.

36. K. Lindquest, S. Sundling, *Xylocaine*, A. B. Astra, Södertälje, Sweden, 1993; R. Dahlbom, *Det händen på lullen* (A. R. Astra 1940–1960), A. B. Astra, Södertälje, Sweden, 1993.

37. I. Ugi, R. Meyr, U. Fetzer, C. Steinbrückner, *Angew. Chem.*, **71**, 386 (1959).

38. (a) I. Ugi, A. Dömling, W. Hörl, *Endeavour*, **18**, 115 (1994); (b) I. Ugi, A. Dömling, W. Hörl, *GIT Fachz. Lab.*, **38**, 430 (1994).

39. S. Mac Lane, G. Birkhoff, *Algebra*, Macmillan Company, New York, 1967, p. 3: Given sets of R and S have the intersection $R \cap S$ with the common elements R and S. This means $R \cap S = \{x \mid x \subset R$ and $x \subset S\}$, whereas a union $R \cup S$ is $R \cup S = \{x \mid x \subset R$ or $x \subset S\}$.

40. T. Yamada, Y. Omote, Y. Yamanaka, T. Miyazawa, S. Kuwata, *Synthesis*, 991 (1998).

41. (a) G. Gokel, G. Lüdke, I. Ugi, in *Isonitrile Chemistry*, I. Ugi, Ed., Academic Press, New York, 1971; (b) I. Ugi, C. Steinbrückner, *Chem. Ber.*, **94**, 734 (1961).

42. F. Balkenhohl, C. V. Buschen-Hünnefeld, A. Lanshy, C. Zechel, *Angew. Chem.*, **108**, 3436 (1996); *Angew. Chem. Int. Ed. Engl.*, **35**, 2288 (1996).

43. G. Ross, I. Ugi, *Can. J. Chem.*, **79**, 1934 (2001).

44. I. Ugi, D. Marquarding, R. Urban, *Chemistry and Biochemistry of Amino Acids, Peptides and Proteins*, Vol. 6, p. 245, B. Weinstein, Ed., Marcel Dekker, New York, 1982.

45. J. Drabik, J. Achatz, I. Ugi, *Proc. Estonian Acad. Sci. Chem.*, **51**, 156 (2002).

46. I. Ugi, G. Kaufhold, *Liebigs Ann. Chem.*, **709**, 11 (1967).

47. L. Weber, S. Waltbaum, C. Broger, K. Gubernator, *Angew. Chem.*, **107**, 2452 (1995); *Angew. Chem. Int. Ed. Engl.*, **34**, 2280 (1995).

48. K. Rossen, P. J. Pye, L. M. DiMichele, R. P. Volante, P. J. Reider, *Tetrahedron Lett.*, **39**, 6823 (1998).

49. W. K. C. Park, M. Auer, H. Jaschke, C. H. Wong, *J. Am. Chem. Soc.*, **118**, 10150 (1996).

50. O. Lockhoff, (Bayer AG) DE-A 196 36538, 1996; O. Lockhoff, *Angew. Chem.*, **110**, 3634 (1998); *Angew. Chem. Int. Ed. Engl.*, **37**, 3436 (1998).

51. A. Dömling, K.-Z. Chi, M. Barrere, *Bioorg. Med. Chem. Lett.*, **9**, 2871 (1999).

52. P. E. Nielsen, M. Egholm, R. H. Berg, O. Buchardt, *Science*, **254**, 1497 (1991).

53. I. Ugi, E. Wischhöfer, *Chem. Ber.*, **95**, 136 (1962).

54. I. Ugi, H. Eckert, *Natural Product Chemistry*, Vol. 12, pp. 113–143, Science Publ. 1000 AE, Amsterdam, The Netherlands, 1992.

55. C. Burdack, Doctoral Thesis, Technical University of Munich, 2001.

56. C. Hanusch-Kompa, I. Ugi, *Tetrahedron Lett.*, **39**, 2725 (1998).

57. C. Hanusch-Kompa, Doctoral Thesis, Technical University of Munich, 1997.

58. J. Zhang, A. Jaconbson, J. R. Rusche, W. Herlihy, *J. Org. Chem.*, **64**, 1074 (1999).

59. K. M. Short, B. W. Ching, A. M. M. Mjalli, *Tetrahedron*, **53**, 3183 (1997); K. M. Short, A. M. M. Mjalli, *Tetrahedron Lett.*, **38**, 359 (1997); G. C. B. Harriman, *Tetrahedron Lett.*, **38**, 5591 (1997).

60. D. Lee, J. K. Sello, S.-L. Schreiber, *Org. Lett.*, **2**, 709 (2000).

61. I. Ugi, *Angew. Chem.*, **74**, 9 (1962); *Angew. Chem. Int. Ed. Engl.*, **1**, 8 (1962).

62. I. Ugi, K. Offermann, *Angew. Chem.*, **75**, 917 (1963); *Angew. Chem. Int. Ed. Engl.*, **2**, 624 (1963).

63. G. Wagner, R. Herrmann, in *Ferrocenes*, A. Togni, T. Hayashi, Eds., VCH, Weinheim, 1995.

64. I. Ugi, *Rec. Chem. Progr.*, **30**, 289 (1969); A. Demharter, I. Ugi, *J. Prakt. Chem.*, **335**, 244 (1993).

65. H. Kunz, W. Pfrengle, *J. Am. Chem. Soc.*, **110**, 651 (1988); *Tetrahedron*, **44**, 5487 (1988).

66. M. Goebel, I. Ugi, *Tetrahedron Lett.*, **36**, 6043 (1995); M. Goebel, H.-G. Nothofer, G. Ross, I. Ugi, *Tetrahedron*, **53**, 3123 (1997).

67. S. Lehnhoff, M. Goebel, R. M. Karl, R. Klösel, I. Ugi, *Angew. Chem.*, **107**, 1208 (1995); *Angew. Chem. Int. Ed. Engl.*, **34**, 1104, (1995).

68. A. V. Zychlinski, in *MultiComponent Reactions & Combinatorial Chemistry*, Z. Hippe, I. Ugi, Eds., pp. 28–30, German-Polish Workshop, Rzeszów, Sept. 1997; University of Technology, Rzeszów/Technical University, Munich, 1998, p. 31.

69. G. Ross, E. Herdtweck, I. Ugi, *Tetrahedron*, **58**, 6127 (2002).

70. L. K. Likhosherstov, O. S. Novikova, V. A. Derivitkava, N. K. Kochetkov, *Carbohydr. Res.*, **146**, C1 (1986).

71. L. K. Likhosherstov, O. S. Novikova, V. N. Shbaev, N. K. Kochetkov, *Russ. Chem. Bull.*, **45**, 1760 (1996).

72. I. Ugi, *Proc. Estonian Acad. Sci. Chem.*, **44**, 237 (1995).

73. G. George, Doctoral Thesis, Technical University of Munich, 1974.

74. A. Gieren, B. Dederer, G. George, et al., *Tetrahedron Lett.*, **18**, 1503 (1977).

75. S. Lohberger, E. Fontain, I. Ugi, et al., *New J. Chem.*, **15**, 915 (1991).

76. I. Ugi, M. Wochner, E. Fontain, et al., *Concepts and Applications of Molecular Similarity*, M. A. Johnson, G. M. Maggiora, Eds., pp. 239–288, John Wiley & Sons, New york, 1990; I. Ugi, J. Bauer, E. Fontain, *Personal Computers for Chemists*, J. Zupa, Ed., pp. 135–154, Elsevier, Amsterdam, The Netherlands, 1990.

77. A. Dömling, I. Ugi, *Angew. Chem.*, **105**, 634 (1993); *Angew. Chem. Int. Ed. Engl.*, **34**, 1104 (1995).

78. I. Ugi, A. Demharter, A. Hörl, T. Schmid, *Tetrahedron* 52, 11657 (1996).

79. C. Hanusch, I. Ugi, *ARKIVOC*, in press.

80. K. Paulvannan, *Tetrahedron Lett.*, **40**, 1851 (1999).

81. E. Gonzales-Zamora, A. Fayol, M. Bois-Choussy, et al., *Chem. Commun.*, 1684 (2001).

82. R. Bossio, R. Marcaccini, R. Pepino, *Liebigs Ann. Chem.*, 1229 (1993).

83. A. Dömling, K. Chi, unpublished.

84. U. Schöllkopf, P.-H. Porsch, H.-H. Lau, *Liebigs Ann. Chem.*, 1444 (1979).

85. H. Bienaymé, *Tetrahedron Lett.*, **39**, 4255 (1998).

86. H. Bienaymé, K. Bouzid, *Tetrahedron Lett.*, **39**, 2735 (1998).

87. J. Kolb, B. Beck, A. Dömling, *Tetrahedron Lett.*, **44**, 3679 (2003).

88. B. E. Ebert, Doctoral Thesis, Technical University of Munich, 1998.

89. B. E. Ebert, I. Ugi, M. Grosche et al., *Tetrahedron*, **54**, 11887 (1998).

90. I. K. Ugi, B. Ebert, W. Hörl, *Chemosphere*, **43**, 75 (2001).

2

CARBOHYDRATES AS RENEWABLE RAW MATERIALS: A MAJOR CHALLENGE OF GREEN CHEMISTRY

FRIEDER W. LICHTENTHALER

Institute of Organic Chemistry, Technische Universität Darmstadt, Darmstadt, Germany

A raw material as feedstock should be renewable rather than depleting wherever technically and economically practicable.[1]

INTRODUCTION

Coal, oil, and natural gas, the fossil resources built up over eons, are not only our main energy suppliers but they are also raw materials for a large variety of man-made products ranging from gasoline and diesel oil to bulk, intermediate, and fine chemicals. However, as our fossil raw materials are irrevocably decreasing and as the pressure on our environment is building up, the progressive changeover of chemical industry to renewable feedstocks for their raw materials emerges as an inevitable necessity,[2–4] that is, it will have to increasingly proceed to the raw materials basis that prevailed before natural gas and oil outpaced all other sources.

The present overreliance of the chemical industry on fossil raw materials has its foreseeable limits, as these materials are depleting and are irreplaceable. The basic question today is not "When will affordable fossil fuels be exhausted?"—fossil

Methods and Reagents for Green Chemistry: An Introduction, Edited by Pietro Tundo, Alvise Perosa, and Fulvio Zecchini
Copyright © 2007 John Wiley & Sons, Inc.

oil will be around for a long time, even if it has to be isolated eventually from oliferous rocks or shale—but "When will the end of cheap oil be?" or, stated more appropriately, "When will fossil raw materials have become so expensive that biofeedstocks are an economically competitive alternative?" Experts realistically prognosticate this for 2040.[5]

The transition to a more biobased production system is hampered by a variety of obstacles: Fossil raw materials are not only more economic at present, but the process technology for their conversion into organic chemicals is exceedingly well developed and basically different from that required for transforming biobased raw materials into products with industrial application profiles. This situation originates from the inherently different chemical structures of the two types of raw materials. Compared to coal, oil, and natural gas, terrestrial biomass is considerably more complex, constituting a multifaceted array of low and high-molecular-weight products: sugars, hydroxy and amino acids, lipids, and biopolymers such as cellulose, hemicelluloses, chitin, starch, lignin, and proteins. By far the most important class of organic compounds in terms of volume produced are carbohydrates, as they represent roughly 75% of the annually renewable biomass of about 180 billion tons (Figure 2.1). Of these, only a minor fraction (ca. 4%) is used by man, the rest decays and recycles along natural pathways.

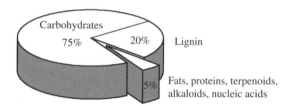

Figure 2.1   Renewable biomass (180 billion tons/year).

Thus, carbohydrates, a single class of natural products—aside from their traditional uses for food, lumber, paper, and heat—are the major biofeedstocks from which to develop the industrially and economically viable organic chemicals that are to replace those derived from petrochemical sources.

The bulk of the annually renewable carbohydrate biomass is polysaccharides, yet their nonfood utilization is confined to the textile, paper, and coating industries, either as such or in the form of simple esters and ethers. Organic commodity chemicals, however, are usually of low molecular weight, so they are more expediently obtained from low-molecular-weight carbohydrates than from polysaccharides. Accordingly, the constituent repeating units of these polysaccharides—*glucose* (cellulose, starch), *fructose* (inulin), *xylose* (hemicelluloses), or disaccharide versions thereof, most notably *sucrose*—are the actual carbohydrate raw materials for organic chemicals with tailor-made industrial applications: they are inexpensive, ton-scale accessible, and provide a resulting chemistry that is better worked out and more variable than that of their polymers.

2.1 AVAILABILITY OF MONO- AND DISACCHARIDES

Table 2.1 lists the availability and bulk-quantity prices of the eight least expensive sugars—all well below about €10/kg—as compared to some sugar-derived, naturally occurring compounds and basic chemicals from petrochemical sources. The result is stunning, since the five cheapest sugars, some sugar alcohols, and

TABLE 2.1 Annual Production Volume and Prices of Simple Sugars and Sugar-Derived Alcohols and Acids as Compared to Some Petrochemically Derived Basic Chemicals and Solvents

	World Production[a] (metric t/year)	Price[b] (€/kg)
Sugars		
Sucrose	149,900,000	0.25
D-Glucose	30,000,000	0.30
Lactose	295,000	0.60
D-Fructose	60,000	1.00
Isomaltulose	70,000	2.00
Maltose	3,000	3.00
D-Xylose	10,000	4.50
L-Sorbose	60,000	7.50
Sugar Alcohols		
D-Sorbitol	650,000	1.80
Erythritol	30,000	2.25
Xylitol	30,000	5.00
D-Mannitol	30,000	8.00
Sugar-Derived Acids		
Citric acid	1,500,000	1.00
D-Gluconic acid	100,000	1.40
L-Lactic acid	150,000	1.75
L-Tartaric acid	35,000	6.00
L-Ascorbic acid	80,000	8.00
Amino Acids		
L-Glutamic acid	1,500,000	1.20
L-Lysine	740,000	2.00
Petrochemicals		
Ethylene	90,000,000	0.40
Propylene	45,000,000	0.35
Benzene	23,000,000	0.40
Terephthalic acid	12,000,000	0.70

(Continued)

TABLE 2.1   *Continued*

	World Production[a] (metric t/year)	Price[b] (€/kg)
Aniline	1,300,000	0.95
Acetaldehyde	900,000	1.10
Adipic acid	1,500,000	1.70
	Solvents	
Methanol	25,000,000	0.15
Toluene	6,500,000	0.25
Acetone	3,200,000	0.55

[a]Reliable data are only available for the world production of sucrose, the figure given referring to the crop cycle 2005/2006.[6] All other data are average values based on estimates from producers and/or suppliers, as the production volume of many products is not publicly available.
[b]Prices given are those attainable in early 2006 for bulk delivery of crystalline material (where applicable) based on pricing information from the sugar industry (sugars) and the *Chemical Market Reporter* (2006) (acids, basic chemicals, and solvents). The listings are intended as a benchmark rather than as a basis for negotiations between producers and customers. Quotations for less pure products are, in part, sizably lower, for example, the commercial sweetener "high fructose syrup," which contains up to 95% fructose, may readily be used for large-scale preparative purposes.

sugar-derived acids are not only cheaper than any other natural product, but they compare favorably with basic organic bulk chemicals such as acetaldehyde or aniline. Actually, the first three of these sugars—sucrose, glucose, and lactose—are in the price range of some of the standard organic solvents.

Despite their large-scale accessibility, the chemical industry, at present, utilizes these mono- and disaccharides as feedstock for organic chemicals only to a minor extent, which is amply documented by the fact that of the 100 major organic

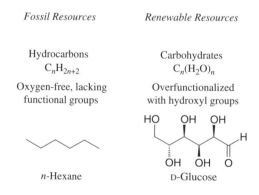

Figure 2.2   Hydrocarbons vs. carbohydrates: more than a play on words, as their names, taken literally, reveal the basic differences in their utilization as organic raw materials.

chemicals manufactured in the United States in 1995,[7] seven were derived from biofeedstocks, and of these only five—ethanol, sorbitol, citric acid, lysine, and glutamic acid—used sugars as the raw-materials source. The reasons, already alluded to, lie in the inherently different structure of carbohydrates and fossil raw materials, of which the essence is manifested in their structure-based names (Figure 2.2): Our fossil resources are *hydrocarbons*, distinctly hydrophobic, oxygen-free, and lacking functional groups; annual renewables are *carbohydrates*, overfunctionalized with hydroxyl groups and pronouncedly hydrophilic. Needless to say, that methods required for converting carbohydrates into viable industrial chemicals—reduction of oxygen content with introduction of C=C and C=O unsaturation—are diametrically opposed to those prevalent in the petrochemical industry.

Intense efforts within the last decade[8-11] to boost the acquisition of organic chemicals from the sugars listed in Table 2.1 have, so far, failed to bridge the conceptional, technological, and economic gap between hydrocarbons and carbohydrates as organic raw materials.

2.2 CURRENT NONFOOD INDUSTRIAL USES OF SUGARS

The current utilization of carbohydrates as a feedstock for the chemical industry—be it for bulk, commodity, intermediate, fine, or high-value-added speciality chemicals—is modest when considering their ready availability at low cost and the huge as yet unexploited potential. The examples currently realized on an industrial scale are outlined briefly.

2.2.1 Ethanol

With production of about 24 million tons (300 million hectoliters (hL)[12]), fermentation ethanol ("bioethanol") is the largest volume biobased chemical today. The principal organism for fermentation is *Saccharomyces cerevisiae*, an ascomycetous yeast that can grow on a wide variety carbohydrate feedstocks: sugar crops, and sugar-containing by-products, such as sugar cane, sugar beet, sorghum, molasses, and—after hydrolysis to glucose—starchy crops, such as corn, potatoes and grains, or cellulosic materials, for example, wood pulping sludges from pulp and paper mills.[13a]

The manufacturing costs are said to be roughly the same as those for its production from ethylene at a comparable plant size.[13b] The large growth in production of industrial-grade fermentation ethanol within recent years is less due to its use as a solvent and starting material for follow-up chemicals such as acetaldehyde, ethyl esters (e.g., EtOAc), and ethers (Et_2O)—these mostly result from ethylene-based processing lines—but from its high potential as a fuel additive. It is either directly admixed to standard gasoline to the extent of 5%, or indirectly in the form of ETBE (ethyl *t*-butyl ether) in proportions of up to 15%; however, a hefty government subsidy is required (repeal of the gasoline tax) to remain competitive. The growth opportunities for fuel-grade bioethanol are enormous and are predicted to increase substantially within the next five years.

2.2.2 Furfural

With an annual production of about 250 000 tons, furfural (2-furfuraldehyde) appears to be the only unsaturated large-volume organic chemical prepared from carbohydrate sources. The technical process involves exposure of agricultural or forestry wastes, that is, the hemicelluloses contained therein consisting up to 25% of D-xylose polysaccharides (xylosans), to aqueous acid and fairly high temperatures, the xylosans first being hydrolyzed, then undergoing acid-induced cyclodehydration.[14]

The chemistry of furfural is well developed, providing a host of versatile industrial chemicals by simple straightforward operations (Scheme 2.1): furfuryl alcohol (**2**) and its tetrahydro derivative **1** (hydrogenation), furfurylamine **3** (reductive amination), furoic acid **4** (oxidation), and furanacrylic acid **5** (Perkin reaction), or furylidene ketones **6** (aldol condensations). Furfural is also the key chemical for the commercial production of furan (through catalytic decarbonylation) and tetrahydrofuran (**8**) (hydrogenation), thereby providing a biomass-based alternative to its petrochemical production via dehydration of 1,4-butanediol.[14] Additional importance of these furanic chemicals stems from their ring-cleavage chemistry,[15] which has led to a variety of other established chemicals such as fumaric, maleic, and levulinic acid, the latter being a by-product of its production.[16]

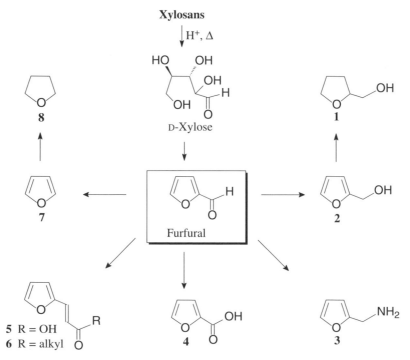

Scheme 2.1 Furanic commodity chemicals derived from pentosans in agricultural wastes (corn cobs, oat hulls, bagasse, wood chips).

Currently, the bulk of the furfural produced is used as a selective solvent in the refining of lubricating oil, and, together with furfurylalcohol in condensations with formaldehyde, phenol, acetone or urea to yield resins of complex, ill-defined structures, yet excellent thermosetting properties, most notably high corrosion resistance, low fire hazard, and extreme physical strength;[14] they are extensively used in the foundry industry as cores for high-quality castings.

2.2.3   D-Sorbitol (≡D-Glucitol)

Readily produced by hydrogenation of D-glucose,[17] the main consumer of the sizable annual production of D-sorbitol (Table 2.1) is the food industry, primarily as a noncaloric sweetening agent; as a key intermediate for the production of ascorbic acid (vitamin C),[18] it has important nonfood applications due to its moisture conditioning, softening, and plastifying properties. These entail its use in adhesives, paper, printing, textiles, cellulose-based foils, and pharmaceutical formulations.

Other nonfood applications of D-sorbitol result from etherification and poly-condensation reactions providing biodegradable polyetherpolyols used for soft polyurethane foams and melamine/formaldehyde or phenol resins. Sizable amounts of D-sorbitol also enter into the production of the sorbitan ester surfactants (cf. later in this chapter).

2.2.4   Lactic Acid → Polylactic Acid (PLA)

Large amounts of D-glucose—in crude form as obtainable from corn, potatoes, or molasses by acid hydrolysis—enter industrial fermentation processes in the

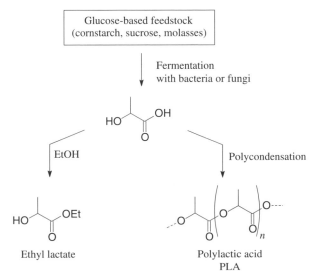

Scheme 2.2   Production of lactic acid and its uses.

production of lactic acid (cf. Scheme 2.2), citric acid, and various amino acids, such as L-lysine or L-glutamic acid. While the major use of these products is in food and related industries, recent nonfood uses of lactic acid have made it a large-scale, organic commodity chemical. Most of it is subsequently polymerized via its cyclic dimer (lactide) to polylactic acid,[19,20] a high molecular weight polyester.

Due to its high strength, polylactic acid (PLA) can be fabricated into fibers, films, and rods that are fully biodegradable ($\rightarrow$ lactic acid, CO_2) and compostable, since they degrade within 45–60 days. Accordingly, PLA and copolymers of lactic and glycolic acid are of particular significance for food packaging and in agricultural or gardening applications; they are also highly suitable materials for surgical implants and sutures, as they are bioresorbable.

Since 1989, Cargill, has invested some \$750 million to develop and commercialize polylactic acid (tradename: NatureWorks). Its Nebraska plant, with an annual capacity of 140,000 metric tons, opened in 2002.[19–21] Thus, polylactides, combining favorable economics with green sustainability, are poised to compete in large-volume markets that are now the domain of thermoplastic polymers derived from petrochemical sources.

Another green development based on lactic acid is its ethyl ester (Vertec[TM]) that has been marketed for applications in specialty coatings, inks, and straight-use cleaning because of their high performance and versatility.[22] As a very benign solvent— green, readily biodegradable, and an excellent toxicology record it has the potential to displace various petrochemically based solvents such as acetone, dimethyl formamide (DMF), toluene, or N-methylpyrrolidone in industrial processes.

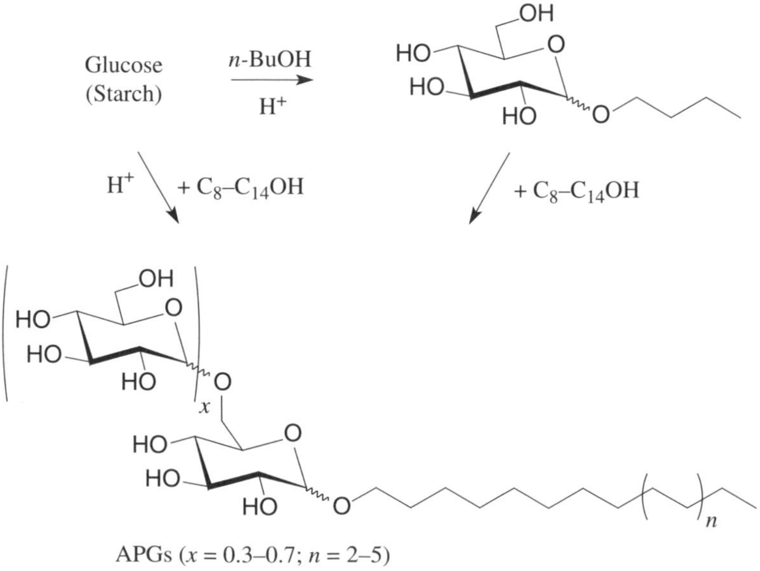

APGs (x = 0.3–0.7; n = 2–5)

Scheme 2.3   Synthesis of alkyl polyglucosides (APGs).

2.2.5 Sugar-Based Surfactants

The use of cheap, bulk-scale accessible sugars as the hydrophilic component, and fatty acids or fatty alcohol as the lipophilic part, yields nonionic surfactants, which are nontoxic, nonskin-irritating, and fully biodegradable. The industrially relevant surfactants of this type[23] are fatty acid esters of sorbitol (sorbitan esters[17]) and of sucrose,[24] fatty acid amides of 1-methylamino-1-deoxy-D-glucitol (NMGAs), and, most apparent in terms of volume produced, fatty alcohol glucosides, the so-called *alkyl polyglucosides* (*APGs*).[25] The latter are produced

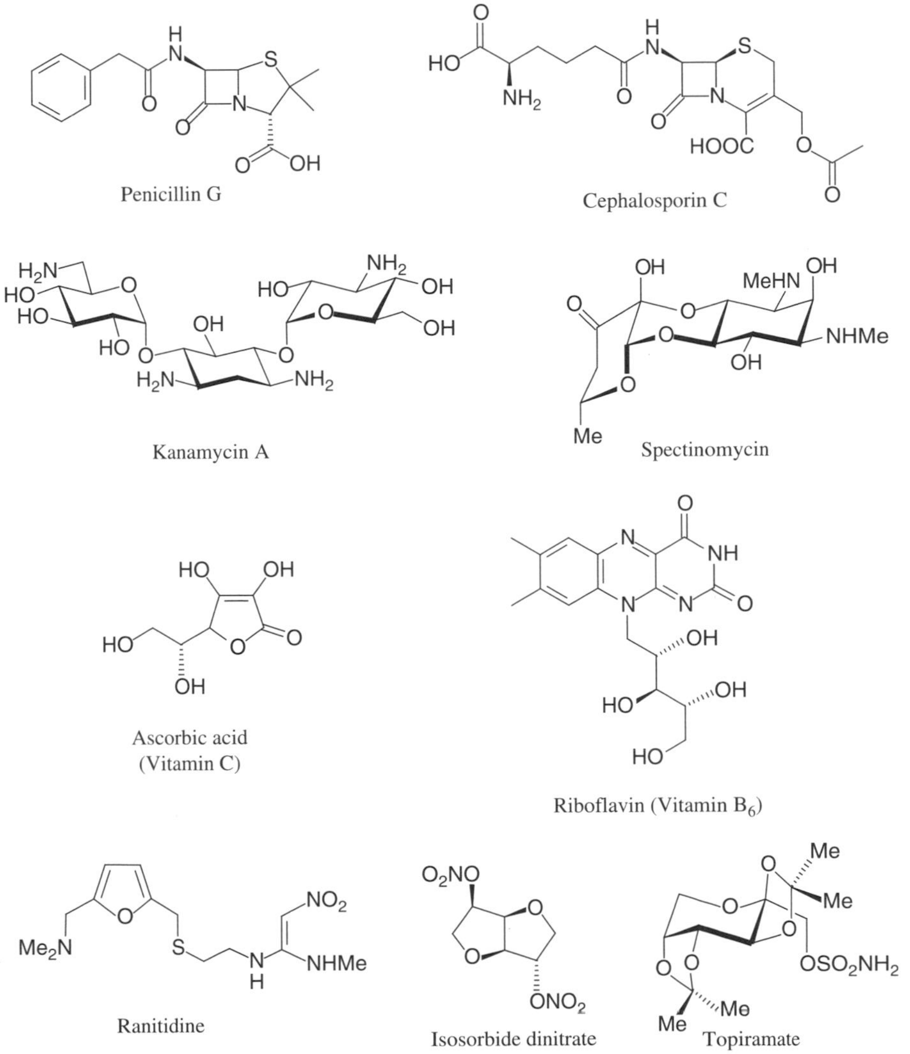

Penicillin G

Cephalosporin C

Kanamycin A

Spectinomycin

Ascorbic acid
(Vitamin C)

Riboflavin (Vitamin B₆)

Ranitidine

Isosorbide dinitrate

Topiramate

Figure 2.3 Sugar-derived high-value-added products: antibiotics, vitamins, and pharmaceuticals.

by several companies—most notably by Cognis, with a capacity in the 50,000-t/y range—and are by far the most important nonionic surfactants. They represent fatty alcohol glucosides with an alcohol chain length normally between C_8 and C_{14}. Their industrial synthesis either comprises a direct acid-catalyzed Fischer glycosidation of glucose (in the form of a syrupy starch hydrolysate) or starch itself. The alternate process consists of two stages, the first being Fischer glycosidation with *n*-butanol to butyl glycosides, which are subsequently subjected to acid-promoted transacetalization.[25]

The resulting product mixtures (Scheme 2.3) contain mostly α-D-glucosides and are marketed as such. APGs are not skin-irritating, have good foaming properties, and are completely biodegradable; hence, they are widely used in hand dishwashing detergents and in formulations of shampoos, hair conditioners, and other personal-care products.[24,25]

2.2.6 Pharmaceuticals and Vitamins

Aside from the enormous amount of sugars, mostly glucose and sucrose, that flows into the fermentative production of amino and hydroxy acids (cf. Table 2.1)—a substantial part of which is for food use—a significant volume of these sugars enters into the fermentation processes of high–value-added products: antibiotics and vitamins, much too complex in their structures to be generated by chemical synthesis. Figure 2.3 lists a number of representative examples: penicillins and cephalosporins, with an estimated world production in the 75,000-t/y range, the aminoglycoside antibiotics of the kamamycin and spectinomycin type, or the recently optimized bioprocesses for the bulk-scale production of vitamins C and B_6.

Some sugar-derived drugs obtained by chemical means have also achieved some importance, for example, ranitidine (Zantac®), an inhibitor of gastric acid secretion—one of the top 30 drugs based on sales[26]—isosorbide dinitrate, a coronary vasodilatator,[27] or topiramate, a fructose-derived anticolvulsant drug with high antiepileptic efficacy.[28]

2.3 TOWARD FURTHER SUGAR-BASED CHEMICALS: POTENTIAL DEVELOPMENT LINES

Considering the large-scale, low-cost availability of the basic biomass-sugars listed in Table 2.1, their present nonfood use by the chemical industry is modest, that is, the huge feedstock potential of carbohydrates in general, and of low-molecular-weight sugars in particular, is largely untapped. In view of the necessity of the chemical industry to somehow effect the changeover from fossil raw materials to biofeedstocks—that is, primarily, to carbohydrates as these are more accessible from agricultural crops and waste materials than any other natural products—their further development as industrial products is one of the major challenges of green chemistry. Thus, the fundamental basic and applied research objectives of the near future—hopefully, incorporated into the European Commission's 7th Framework Programme—must be to systematically improve existing methods for either chemical or enzymatic conversions of carbohydrates into industrially viable chemicals and materials, and to develop new ones.

The major directions that broad-scale exploratory research toward carbohydrate-based nonfood products will have to take, are—as far as conceivable today—outlined in the following for the four key sugars of biomass: the "royal carbohydrate"[29] sucrose, D-glucose, D-fructose, and D-xylose.

2.3.1 Nonfood Valorization of Glucose: Potential Development Lines

2.3.1.1 Chemical Conversions. Although D-glucose is the component sugar of cellulose and starch, only the latter is the raw material for its commercial

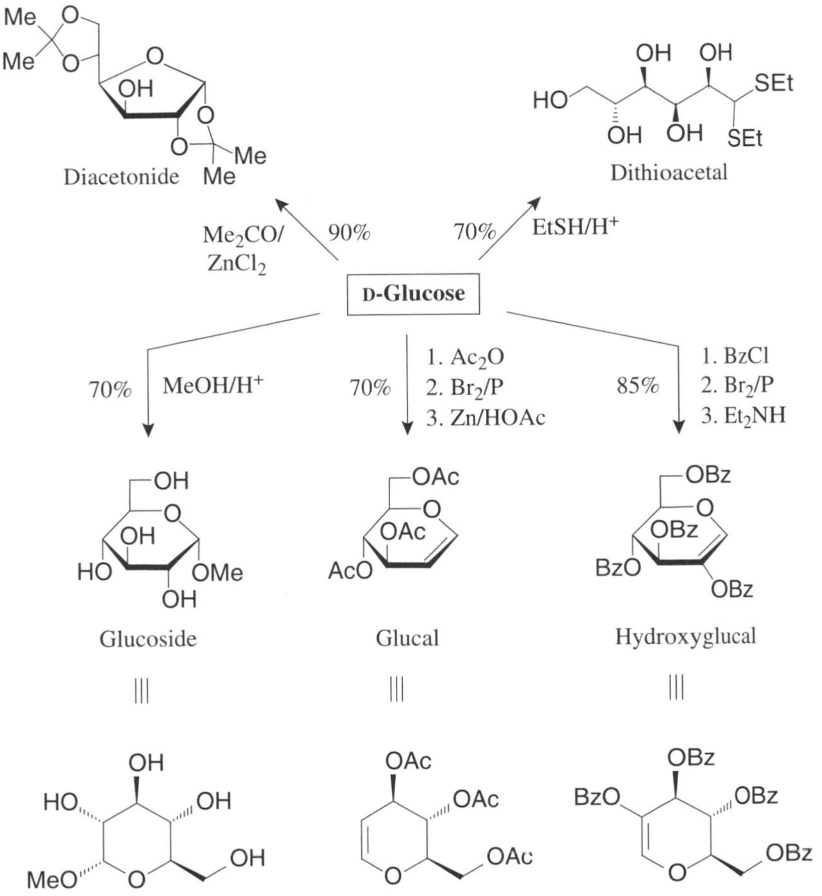

Scheme 2.4 Accessible, tautomerically fixed D-glucose derivatives with which to embark toward versatile building blocks.[32]

production.[30] The chemistry of D-glucose is exceedingly well developed, its basic ensuing reactions going back to Emil Fischer, who in the 1890s, on the basis of oxidative and reductive conversions and through the synthesis of glucose by cyanohydrin extension of arabinose, succeeded in figuring out its (relative) configuration.[31] As a reducing sugar, D-glucose can form pyranoid, furanoid, and acyclic tautomers, so for ensuing straightforward reactions, the tautomeric form has to be fixed first (Scheme 2.4): isopropylidenation leads to the furanoid diacetonide, mercaptalization to an *acyclic* dithioacetal, pyranoid structures may be effectively generated in the form of glucosides, esters of glucal, and hydroxyglucal.[32]

Another simple, one-step entry from D-glucose to highly substituted furans involves their $ZnCl_2$-mediated reaction with acetylacetone.[33] Since only the first two sugar carbons of D-glucose contribute to the formation of the furan, a distinctly hydrophilic tetrahydroxybutyl side chain is produced; this chain can be shortened oxidatively to a dicarboxylic acid or a variety of other furanic building blocks (Scheme 2.5). By contrast, under mildly basic conditions D-glucose reacts with pentane-2,4-dione in an entirely different way, producing 2-C-glucosyl-propanone via C-addition and subsequent retroaldol-type elimination of acetate.[34] Because this conversion can be performed with unprotected sugar and with simple reagents in aqueous solution, it fully complies with green and sustainable principles.[1] The procedure is equally operable with other monosaccharides, and, thus, one of the cleanest and most efficient preparative entries into the area of C-glycosides, which as stable "mimics" of the usual O-glycosides, command major interest as glycosidase inhibitors.[35]

Despite the easy accessibility of these "entry products," and their fairly well-developed ensuing chemistry, their development toward industrial intermediates

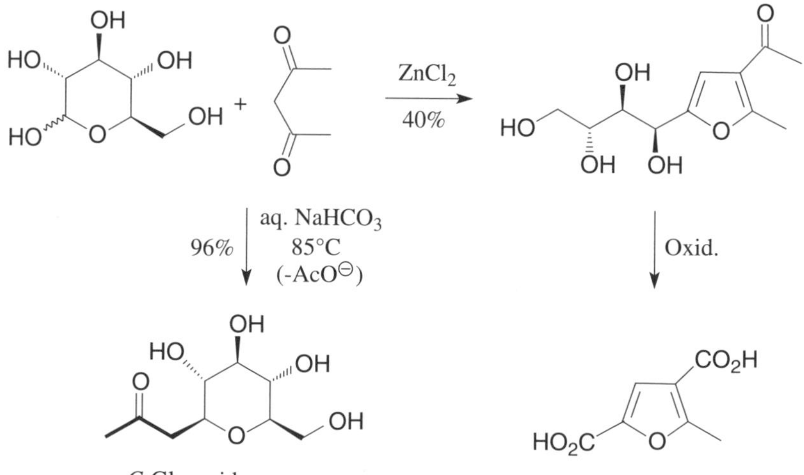

Scheme 2.5 One-pot conversions of D-glucose into hydrophilic furans[33] or, alternatively, into C-glucosides by reaction with acetylacetone.[34]

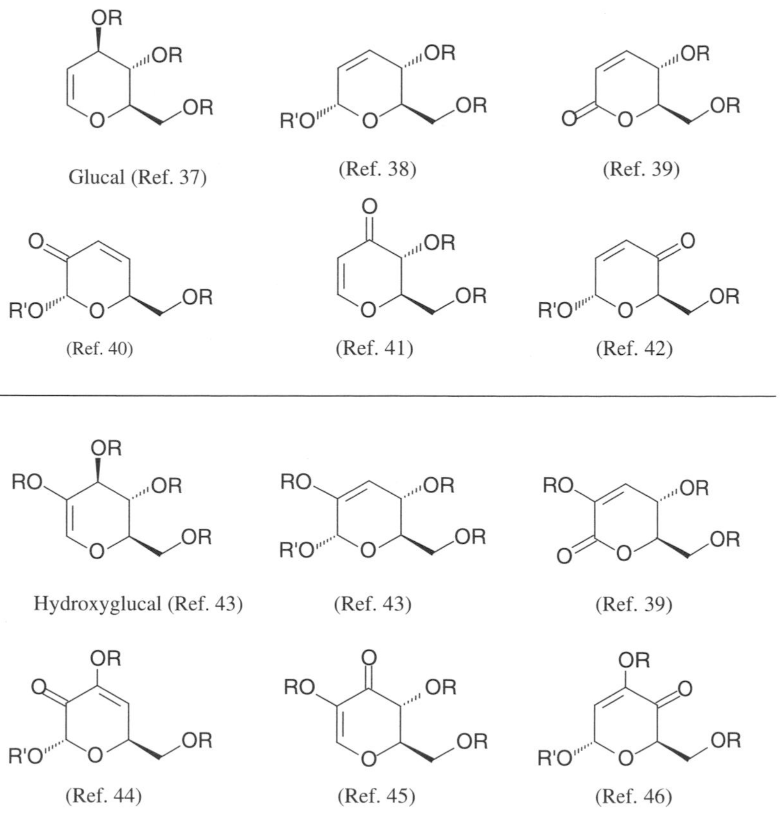

Glucal (Ref. 37)   (Ref. 38)   (Ref. 39)

(Ref. 40)   (Ref. 41)   (Ref. 42)

Hydroxyglucal (Ref. 43)   (Ref. 43)   (Ref. 39)

(Ref. 44)   (Ref. 45)   (Ref. 46)

R = acyl; R' = alkyl

Figure 2.4   Enantiomerically pure six-carbon building blocks accessible from D-glucose via glucal (upper half) or hydroxyglucal esters (lower entries) as the key intermediates. All products require no more than 3 to 5 straightforward steps from D-glucose.[37–46]

is exceedingly modest. Nevertheless, to emphasize their potential toward industrial intermediates, be it as enantiopure building blocks for the synthesis of noncarbohydrate natural products[36] or for agrochemicals and/or high-value added pharmaceuticals, a highly versatile array of *dihydropyrans* and *dihydropyranones* is given in Figure 2.4, all of which are derivable from D-glucose (via the glucal and hydroxyglucal esters) in no more than three to five straightforward steps. As in each of these products, at least two of the asymmetric centers of the D-glucose are retained, they are enantiomerically pure, and are thus ideal six-carbon building blocks for the synthesis of pharmaceuticals in enantiopure form.

Levoglucosenone, a bicyclic dihydropyranone, is accessible even more directly by vacuum pyrolysis of waste paper.[47] Although the yield attainable is relatively low—levoglucosan is also formed, the amount depending on the exact conditions

(Scheme 2.6)—relatively large amounts can be amassed quickly; levoglucosenone has been used for the synthesis of a diverse variety of natural products in enantio-pure form.[48]

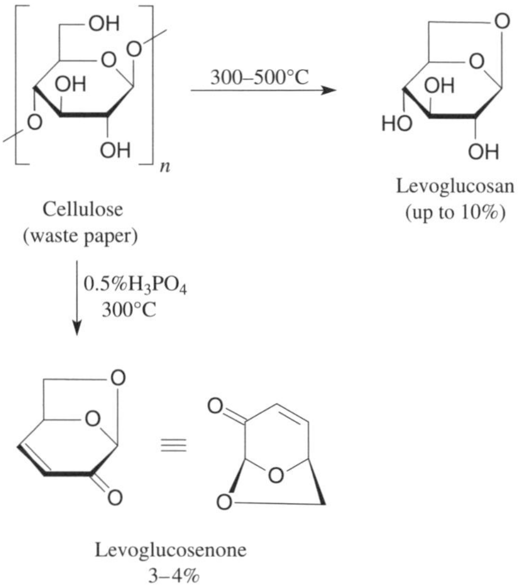

Scheme 2.6   High vacuum pyrolysis of cellulose.[48]

Kojic acid, a γ-pyrone, is readily obtained from D-glucose, either enzymati-cally by *Aspergillus oxyzae* growing on steamed rice,[49] chemically via pyranoid 3,2-enolones.[36,50] A structurally similar α-pyrone can be effectively generated by oxidation of glucose to D-gluconic acid and acetylation.[51] At present, both, are of little significance as six-carbon building blocks, despite a surprisingly effective route to cyclopentanoid products,[52] which is surmised to have industrial potential.

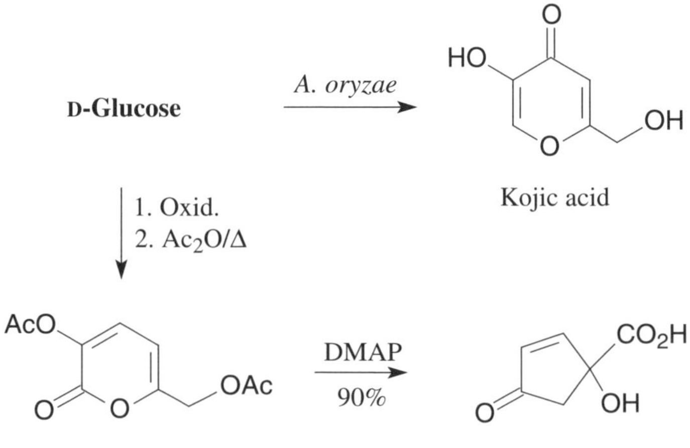

Glucose-Derived Carboxylic Acids. There are several carboxylic acids derivable from D-glucose by chemical means that have broad potential as versatile intermediate chemicals for biorefinery platforms (Scheme 2.7): D-*gluconic acid*, the large quantity produced by oxidation of glucose[53] (cf. Table 2.1) being used in the food, beverage, and pharmaceutical industries, yet also for removing calcareous and rust deposits from metals surfaces (due to its complexing properties) D-*glucaric acid* and *levulinic acid*.

D-*Glucaric acid*, directly produced by nitric oxidation of glucose or starch,[54] is usually isolated as its 1,4-lactone. The technical barrier to its large-scale production mainly includes development of an efficient and selective oxidation technology to eliminate the need for nitric acid as the oxidant. Because it represents a tetrahydroxy-adipic acid, D-glucaric acid is of similar utility as adipic acid for the generation of polyesters and polyamides (see later in this chapter).

Levulinic acid and formic acid are end products of the acidic and thermal decomposition of lignocellulosic material, their multistep formation from the hexoses contained therein proceeding through hydroxymethylfurfural (HMF) as the key intermediate, while the hemicellulosic part, mostly xylans, produces furfural.[55] A commercially viable fractionation technology for the specific

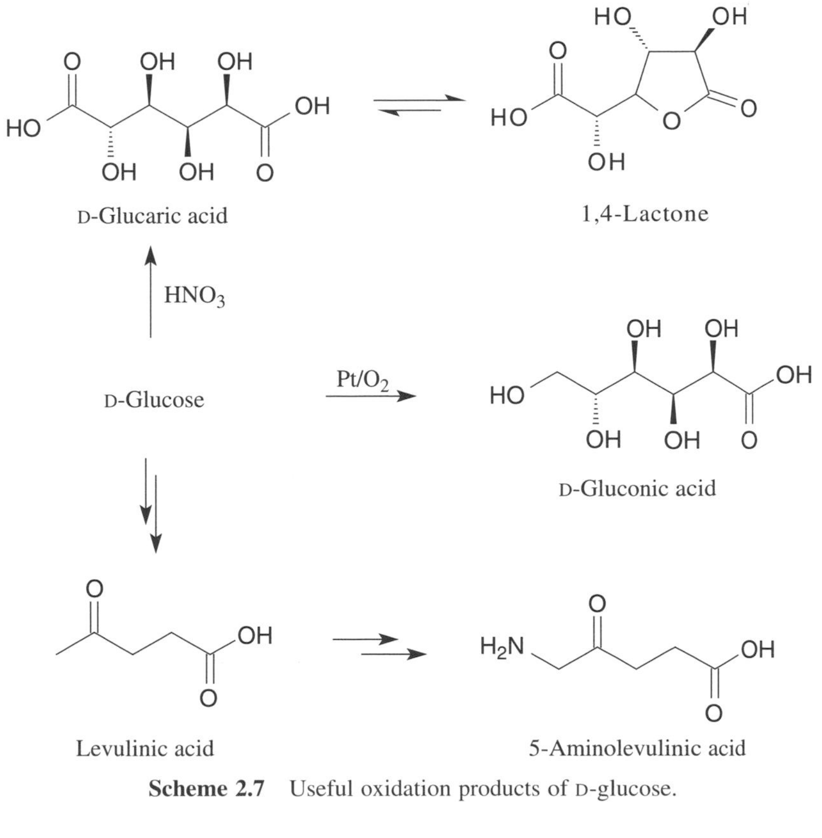

Scheme 2.7 Useful oxidation products of D-glucose.

acquisition of levulinic acid has been developed,[56] rendering it an attractive option for a biorefinery platform chemical.[57]

Levulinic acid is a starting material for a large number of higher-value products, because it can be converted through established procedures into acrylic and succinic acids, pyrrolidines, diphenolic acid (which has the potential of replacing bisphenol A in the manufacture of polycarbonate resins), or 5-aminolevulinic acid used in agriculture as a herbicide and a growth-promoting factor for plants.

Hydrocarbons from D-Glucose?   Being a six-carbon commodity graciously provided by Nature, albeit "overhydroxylated," a full deoxygenation of glucose (or other hexoses) formally leads to *n*-hexane, which is usable as a liquid fuel. If such a process could be made practically feasible—this author is well aware that some will say "Never"—it would certainly exceed glucose-derived ethanol as a biofuel (additive), inasmuch as fermentation cuts the six-carbon chain into ethanol and CO_2, while deoxygenation implies full atom economy by retaining it.

Recent investigations aimed at establishing such a deoxygenation process have met with some success, yet are admittedly far from industrial implementation. Sorbitol, for example, readily accessible from glucose through catalytic hydrogenation,[17] can be tailored to produce a mixture of alkanes consisting primarily of butane, pentane, and hexane, by exposing an aqueous solution to a metal (Pt or Pd) and solid acid catalysts (SiO_2-Al_2O_3) and hydrogen at 225°C.[58] The process is complex, as it not only entails a series of dehydrations and hydrogenations to eventually provide *n*-hexane, but also dehydrogenations and C—C fissions to produce pentane and butane.

C_4–C_6 Alkanes are highly volatile and, hence, of low value as a transportation fuel or a fuel additive. Since high-quality fuels require the generation of liquid hydrocarbons, the fructose-derived HMF and acetone have been converted into their mono- (C_9) and bis-aldols (C_{15}), which on SiO_2-Al_2O_3/Pt-catalyzed dehydration/hydrogenation produce C_9–C_{15} alkanes (Scheme 2.8).[59] A major drawback of this approach, however, is the fact that HMF, de facto, is a fructose-derived product,[60] and is not producible in an industrially viable price frame at present (*vide infra*, Section 2.3.2).

2.3.1.2 Valorization of D-Glucose Through Microbial Conversions. Some experts predict that biotechnology will produce up to 20% of the industrial chemicals by 2010—from currently 5%.[61,62] Undoubtedly, such an increase will receive its major thrust from the various genetically engineered bioprocesses currently in industrial pipelines, most notably those that involve the bioconversion of D-glucose—Nature's principal sugar for essentially any biotransformation—into industrially important C_3–C_5-carboxylic acids apart from those already exploited (cf. Table 2.1) or into alcohols other than ethanol.[63]

Intense research and development efforts currently appear to go into the following chemicals:

3-hydroxypropionic acid
1,4-diacids (malic, fumaric, succinic)

itaconic acid

1,3- and 1,2-propanediol

of which the carboxylic acids—currently petroleum-based bulk commodities—are on the list of the 12 future sugar-derived platform chemicals[57] of the U.S. Department of Energy. If low-cost D-glucose-based fermentation routes can be developed and implemented on an industrial scale, there is a good chance of their production process being replaced along petrochemical channels.

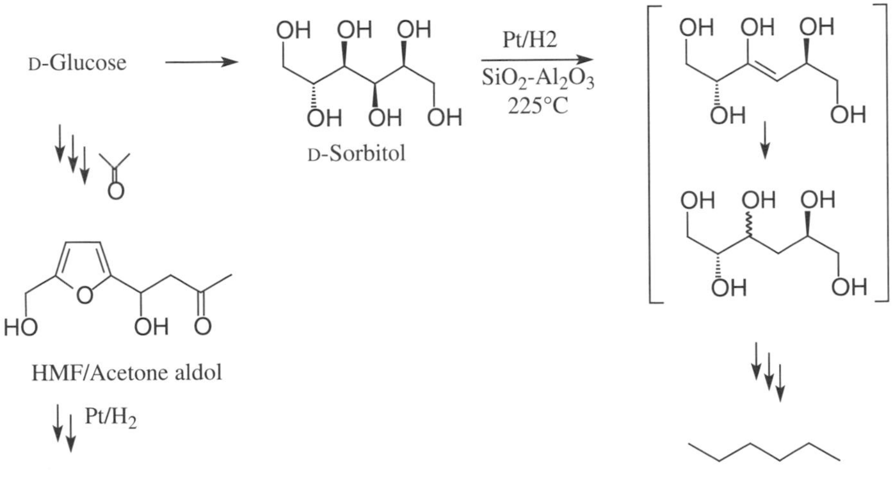

Scheme 2.8 Effectuation of the deoxygenation of D-glucose (or other sugars) to hydrocarbons.[58,59]

3-Hydroxypropionic Acid (3-HPA). Like the structurally isomeric lactic acid, 3-HPA constitutes a three-carbon building block with the potential of becoming a key intermediate for a variety of high-volume chemicals: malonic and acrylic acids, methacrylate, acrylonitrile, 1,3-propanediol, and so forth.[57b] Thus, Cargill is developing a low-cost fermentation route by metabolic engineering of the microbial biocatalyst that produces 3-HPA under anaerobic conditions,[64a] yet it will take another one or two years for the process to reach commercial viability.[64b]

Lactic acid (LA)
(2-Hydroxypropionic acid)

3-Hydroxypropionic acid
(3-HPA)

Unlike a product such as lactic acid, another of 3-HPA's appeals is that, at present, it is not manufactured commercially, either by chemical or biological means.

1,4-Diacids. The microbial generation of malic, fumaric, and succinic acid essentially implies Krebs cycle pathway engineering of biocatalytic organisms to overproduce oxaloacetate as the primary four-carbon diacid that subsequently undergoes reduction and dehydration processes (Scheme 2.9). The use of these four-carbon diacids as intermediate chemicals and the state of their desirable microbial production is briefly outlined.

The major portion of *malic acid* currently produced at an approximate 10,000 t/y is racemic, because it originates from petrochemically produced fumaric acid. The L-form can also be generated from fumaric acid by its hydration with immobilized cells of *Brevibacterium* or *Corynebacterium*.

Fumaric acid, a metabolite of many fungi, lichens moss and some plants, and mainly used as the diacid component in alkyd resins,[65] is produced commercially to some extent by fermentation of glucose in *Rhizopus arrhizus*,[66] yet productivity improvements appear essential for the product to be an option for replacing its petrochemical production by catalytic isomerization of maleic acid.

Succinic acid is used in producing food and pharmaceutical products, surfactants and detergents, biodegradable solvents and plastics, and ingredients to stimulate animal and plant growth. Although it is a common metabolite formed by plants, animals, and microorganisms, its current commercial production of 15,000 t/y is from petroleum, that is, by hydrogenation of fumaric or maleic acid. The major technical hurdles for succinic acid as a green, renewable, bulk-scale commodity chemical—1,4-butanediol, THF, γ-butyrolactone, or pyrrolidones are industrially relevant follow-up products—include the development of a very low-cost fermentation route from sugar feedstocks. Currently available anaerobic fermentations of glucose (Scheme 2.9) include a genetically cloned form of *Aspergillus succinoproducens*, an engineered *E. coli* strain developed by DOE laboratories,[57c] and a number of others[67]—processes that are currently under active development. Production costs are to be brought to or below $ 0.25/pound in order to match those via petrochemical channels.[57c]

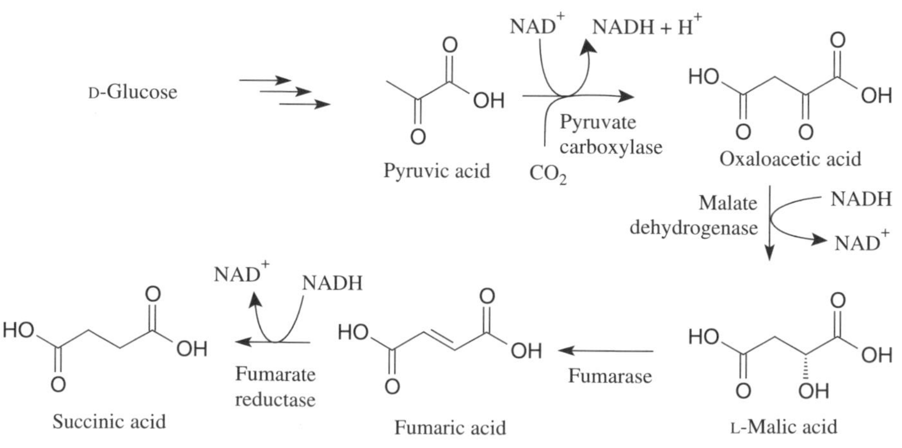

Scheme 2.9   Glycolytic pathway leading to the L-malic, fumaric, and succinic acids.

Itaconic Acid. Structurally an α-substituted methacrylic acid, itaconic acid constitutes a C_5 building block with significant market opportunities. It is currently produced via fungal fermentation at about 10,000 t/a[68] and mainly used as a specialty
comonomer in acrylic or methacrylic resins, as incorporation of small amounts of
itaconic acid into polyacrylonitrile significantly improve their dyeability.

Itaconic acid

To become a commodity chemical, though, productivity improvements with the
currently used fungi *Aspergillus terrous* and *Aspergillus itaconicus* are required,
and promising ameliorations appear to be in the making.[69] To be competitive to
analogous commodities, the crucial production price of about 0.25 $/lb has to be
reached[57d]—a significant technical challenge still to be solved.

1,3-Propanediol. Both the diol and the dicarboxylic acid components of poly-
trimethylene-terephthalate, a high performance polyester fiber with extensive
applications in textile apparel and carpeting, are currently manufactured from petrochemical raw materials.

Sorona® (Dupont[70])
Corterra® (Shell[71])

Poly-trimethylene-terephtalate

For the polyester's 1,3-propanediol portion, however, biobased alternatives
have been developed, relying on microbial conversions of glycerol,[72] a by-product
of biodiesel production, or of corn-derived glucose.[73] For the latter conversion,
DuPont has developed a biocatalyst, engineered by incorporating genes from
baker's yeast and *Klebsiella pneumoniae* into *E. coli*, which efficiently converts
corn-derived glucose in 1,3-propanediol.[70,73] The bioprocess, implemented on an
industrial scale in a Tennessee manufacturing plant by a DuPont/Tate & Lyle
joint venture, provided the first bulk quantities in November 2006.[74]

1,2-Propanediol. In its racemic form, 1,2-propanediol is a petroleum-based high-
volume chemical with an annual production of over 500,000 t, mostly used to
manufacture the unsaturated polyester resins, yet also featuring excellent
antifreeze properties. Enantiomerically pure (*R*)-1,2-propanediol accumulates
along two different pathways via DAHP (3-deoxy-D-*arabino*-heptulosonic acid
7-phosphate) and methylglyoxal, which then is reduced with either

hydroxyacetone or lactaldehyde as the intermediates. Both routes have been examined for their microbial production from glucose by means of genetically engineered biocatalysts, obtained by expressing glycerol dehydrogenase genes or by overexpressing the methylglyoxal synthase gene in *E. coli*.[75] Another approach implies inoculating silos with chopped whole-crop maize with *Lactobacillus buchneri*; after storing for four months, yields of 50 g/kg were reported.[76] Thus, prospects for elaborating an economically sound bioprocess look promising.

2.3.2 D-Fructose: Potentials for Nonfood Uses

The substantial amounts of this ketohexose are mainly prepared by base-catalyzed isomerization of starch-derived glucose,[77] yet may also are generated by hydrolysis of inulin, a fructooligosaccharide.[78] An aqueous solution of fructose—consisting of a mixture of all four cyclic tautomers (Figure 2.5), of which only the β-D-pyranose (β-*p*) form present to about 73% at room temperature is sweet[79]— about 1.5 times sweeter than an equimolar solution of sucrose; hence, it is widely used as a sweetener for beverages ("high fructose syrup").

The nonfood utilization of fructose is modest—not surprising, since its basic chemistry is more capricious and considerably less developed than that of glucose.[79a]

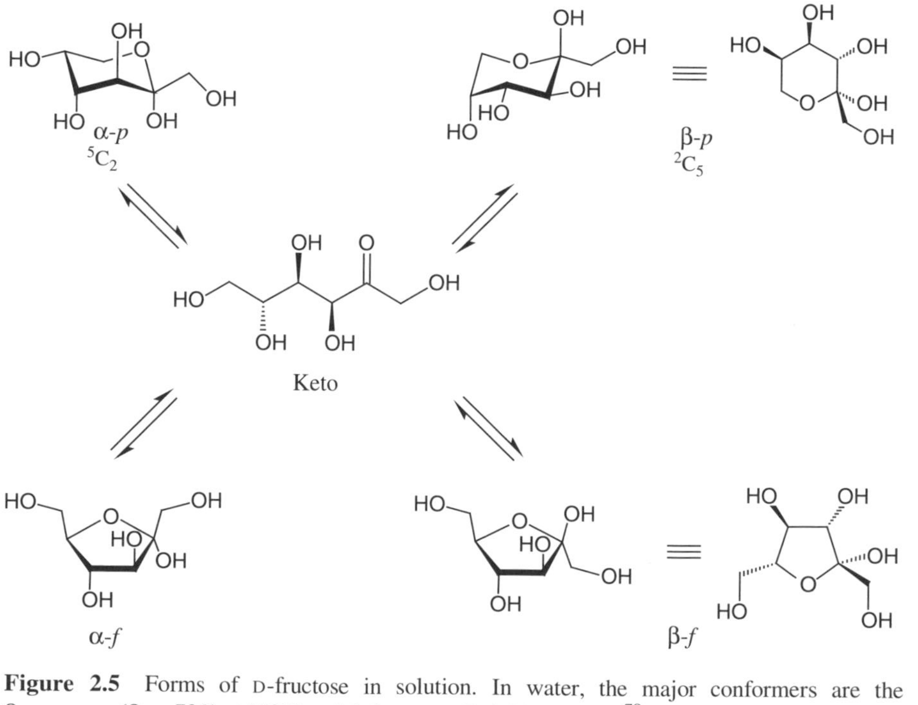

Figure 2.5 Forms of D-fructose in solution. In water, the major conformers are the β-pyranose (β-*p*; 73% at 25°C) and β-furanose (β-*f*, 20%) forms.[79] On crystallization in water, D-fructose exclusively adopts the $^2C_5$ chair conformation, as shown by X-ray analysis.[80]

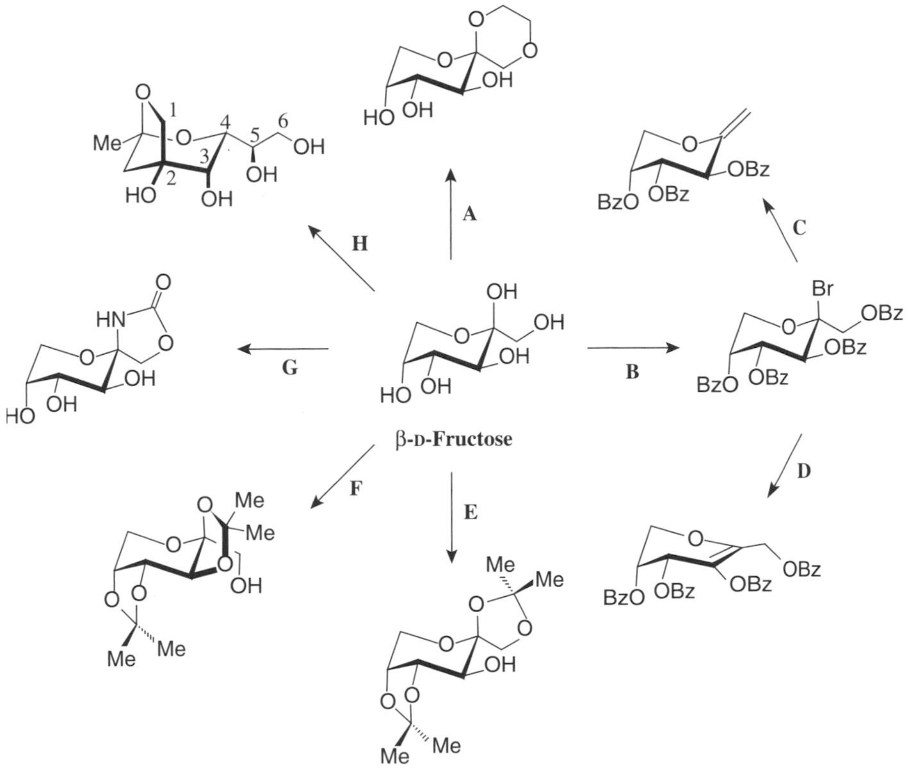

Scheme 2.10 Readily accessible fructose derivatives fixed in pyranoid form. Key for conditions: A: Glycol/H$^+$, 74%[81] B: BzCl/Pyr., $-10°C \to$ HBr, 63%[82] C: Zn/MIM, 90%[83] D: NaI/acetone, then 140°C in xylene, 53%[84] E: Me$_2$CO/cat. H$_2$SO$_4$, 58%[85] F: Me$_2$CO/ $> 5\%$ H$_2$SO$_4$[86] G: acetylacetone, aq. NaHCO$_3$, 85°C, 35%.[87]

Nevertheless, there are various "entry reactions" into simple pyranoid derivatives (Scheme 2.10) with which to exploit their industrial application potential.

Equally simple entries—in fact, one-pot reactions each—lead from D-fructose to N-heterocycles of the imidazol, pyrrole, and pyridine type (Scheme 2.11), all of which, due to their hydrophilic substitution patterns, are considered useful building blocks to pharmaceuticals.

By far the highest industrial potential for a fructose-based compound is to be attributed to HMF, which has been termed "a key substance between carbohydrate chemistry and mineral-oil-based industrial organic chemistry."[92] Like the bulk-scale commodities hexamethylenediamine and adipic acid, HMF represents a six-carbon compound with broad industrial application profiles. It is readily accessible from fructose or inulin hydrolysates by acid-induced elimination of three moles of water.[60] Even a pilot-plant-size process has been elaborated.[92]

HMF as such has been used for the manufacture of special phenolic resins, as acid catalysis induces its aldehyde and hydroxymethyl group to react with phenol.[98]

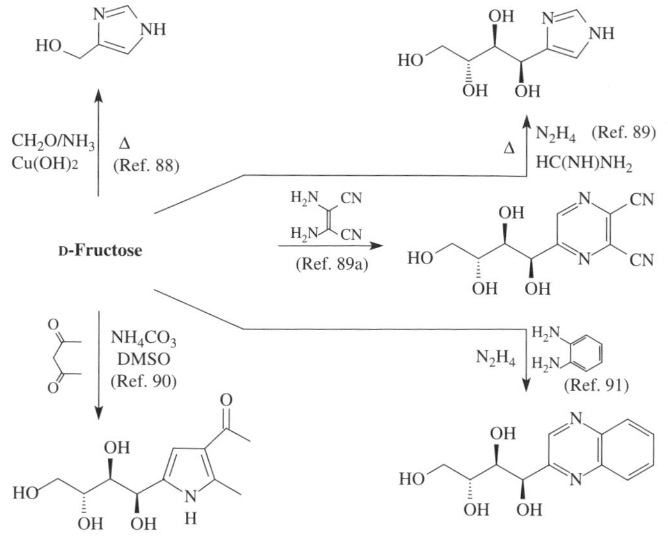

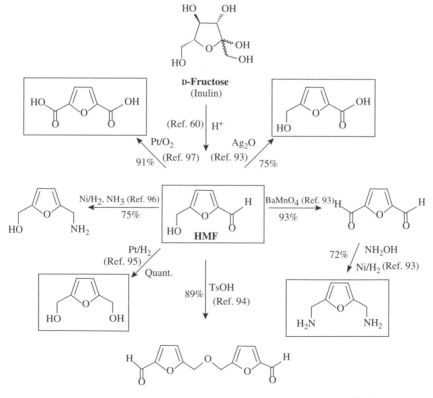

Scheme 2.11   Versatile *N*-heterocyclic building blocks derivable from fructose in one-pot reactions;[88−91] the hydrophilic side chain can be oxidatively shortened to elaborate aldehyde or carboxylic acid functions.

Scheme 2.12   Versatile intermediate chemicals from HMF.[93−97]

Of equally high industrial potential as intermediate chemicals are the various HMF-derived products for which well-worked-out, large-scale adaptable production protocols are available. Of these, the 5-hydroxymethyl-furoic acid, the 2,5-dicarboxylic acid, the 1,6-diamine, and the respective 1,6-diol (framed in Scheme 2.12) are the most versatile intermediate chemicals of high industrial potential, as they represent six-carbon monomers that could replace adipic acid, alkyldiols, or hexamethylenediamine in the production of polyamides and polyesters.

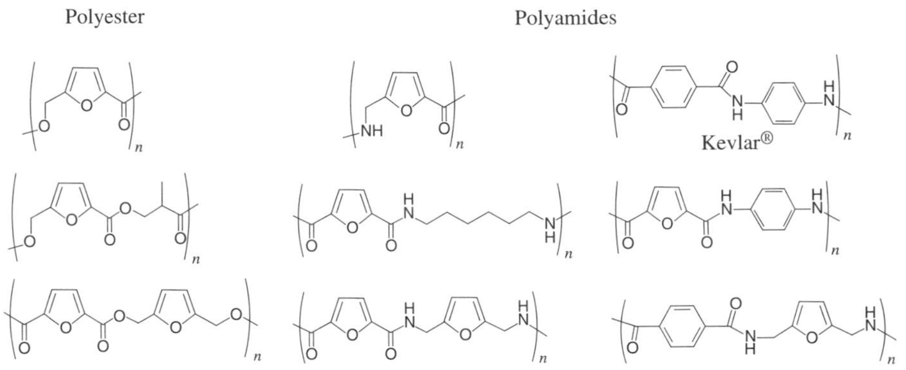

Polyester   Polyamides

Figure 2.6   Furanoic polyesters and polyamides of potential industrial significance.

Indeed, an impressive series of furanic polyesters and polyamides has been prepared[15] in which the furan-dicarboxylic acid replaces terephthalic and isophthalic acid in the present industrial products (cf. Figure 2.6), yet none has proved economically competitive to existing products. Thus, as of now, HMF, is not produced on an industrial scale. A tentative assessment of its economics as compared to petrochemical raw materials clearly unfolds the reasons underlying: ton prices of naphtha and ethylene are in the 150–400 € range, that of anilin (500 €/t), and of fructose in particular (~1000 €/t) are substantially higher, entailing an HMF-marketing price of at least 2500 €/t—too expensive at present for a bulk-scale industrial product. Accordingly, as long as the economic situation favors fossil raw materials, applications of HMF lie in high-value-added products, such as pharmaceuticals or special niche materials.

2.3.3   Potential Industrial Chemicals from D-Xylose

D-Xylose, also referred to as "wood sugar" because it is, in the form of its polysaccharide xylan, widely distributed in woods, straw, and other fibrous tissues, usually in close association with cellulose. Acid hydrolysis of such xylans in agricultural wastes (corn cobs,[99] cottonseed bulks, or other woody materials) split their β(1→4)-glycosidic linkages to D-xylose (Scheme 2.13). When more forcing acidic conditions are applied, furfural is the product due to dehydratizations (cf. see earlier in this chapter). Although D-xylose can thus be made cheaply (cf. Table 2.1 for its estimated present production), as insufficient uses have unfolded yet to make the manufacture of the sugar of commercial interest.

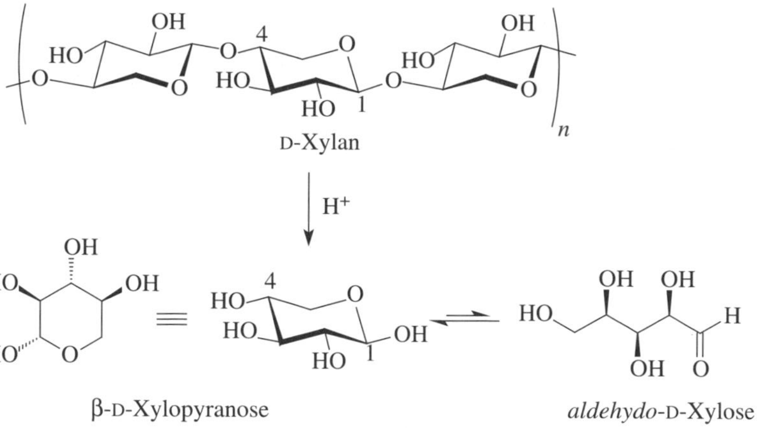

Scheme 2.13 Acquisition of D-xylose from xylans in woody materials.

Like D-glucose and D-fructose, however, D-xylose can be utilized—chemically or microbially—to generate a variety of interesting five-carbon chemicals other than furfural (*vide supra*) or xylitol, a noncaloric sweetener,[100] both being directly produced from xylan hydrolysates, that is, without the actual isolation of the sugar. Other readily accessible intermediate products of high preparative utility (Scheme 2.14) are the open-chain fixed dithioacetal,[101] the D-xylal,[102] and D-hydroxy-xylal esters,[103] or pyrazol or imidazol *N*-heterocycles with a hydrophilic trihydroxypropyl side chain.[89]

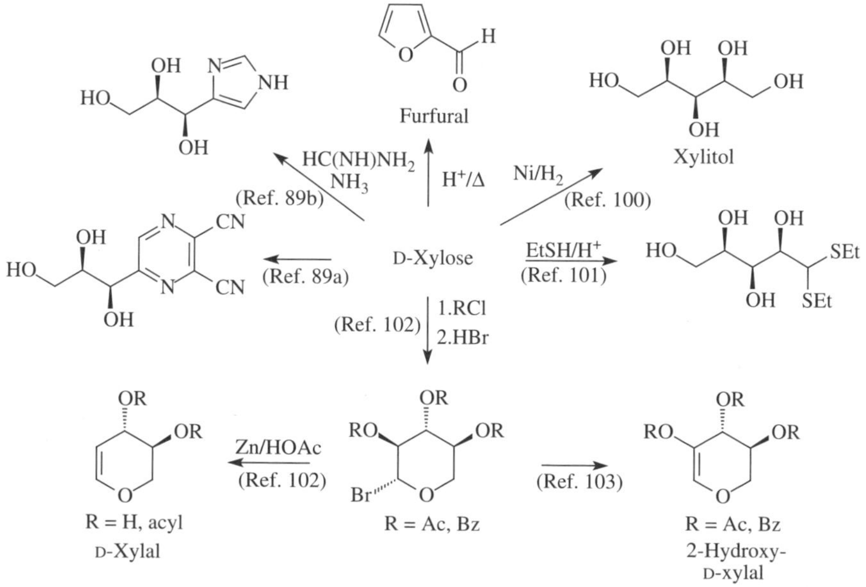

Scheme 2.14 Readily accessible five-carbon building blocks from D-xylose.

Another entry into useful five-carbon building blocks from D-xylose encompasses the expeditious four-step protocol for the 1-phenylpyrazol-3-carboxaldehyde with a 5-hydroxymethyl substituent (Scheme 2.15) and the various follow-up reactions feasible.[104,105]

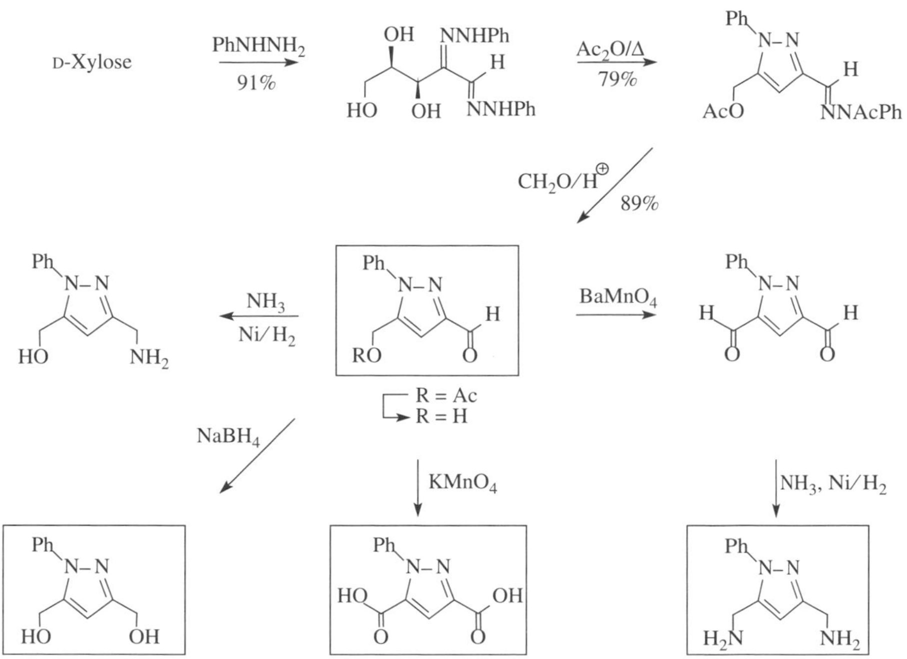

Scheme 2.15 Useful five-carbon building blocks from D-xylose.[104]

Aside from the multifaceted chemical conversions, there are sources to develop into industrially viable microbial conversions. 1,2,4-Butanetriol, for example, used as an intermediate chemical for alkyd resins and rocket fuels, is currently prepared commercially from malic acid by high-pressure hydrogenation or hydride reduction of its methyl ester. In a novel environmentally benign approach to this chemical, wood-derived D-xylose is microbially oxidized to D-xylonic acid, followed by a multistep conversion to the product effected by a biocatalyst specially engineered by inserting *Pseudomonas putida* plasmids into *E. coli*:[106]

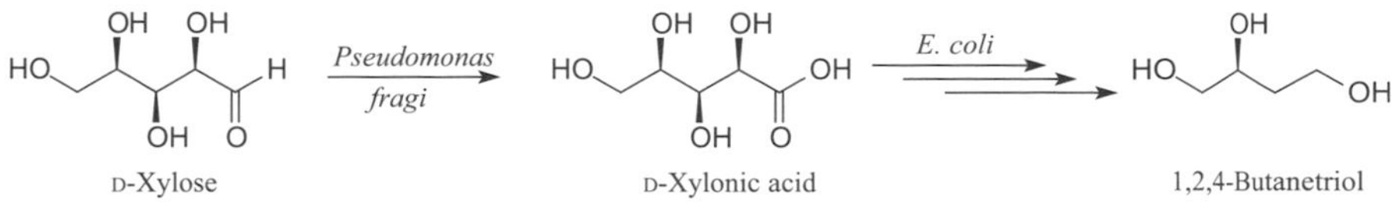

Although further metabolic engineering is required to increase product concentration and yields, the microbial generation of 1,2,4-butanetriol is a clear alternative to its acquisition by chemical procedures.

2.3.4   Nonfood Valorization of Sucrose

Sucrose, affectionately called "the royal carbohydrate,"[29] is a nonreducing dis-accharide, because its component sugars, D-glucose and D-fructose, are glycosidi-cally linked through their anomeric carbon atoms. Hence, it constitutes a β-D-fructofuranosyl α-D-glucopyranoside (Figure 2.7). It is widely distributed throughout the plant kingdom, is the main carbohydrate reserve and energy source, and an indispensable dietary material for humans. For centuries, sucrose has been the world's most plentifully produced organic compound of low molecu-lar mass (cf. Table 2.1). Due to the usual overproduction, and the potential to be

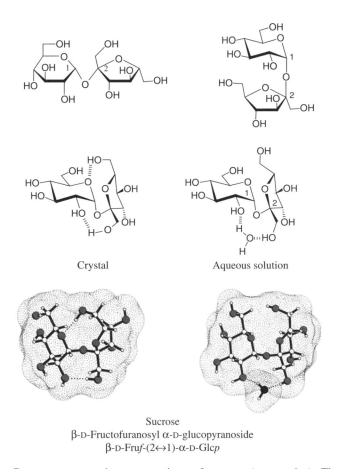

Crystal   Aqueous solution

Sucrose
β-D-Fructofuranosyl α-D-glucopyranoside
β-D-Fru*f*-(2↔1)-α-D-Glc*p*

Figure 2.7   Common structural representations of sucrose (top entries). The molecular geometry realized in the crystal is characterized by two intramolecular hydrogen bonds between the glucose and fructose portion[107] (center left). In an aqueous solution, the two sugar units are similarly disposed toward each other, which is caused by insertion of a water molecule between the glucosyl-2-OH and fructosyl-1-OH,[108,109]—a water bridge, so to say—fixed by hydrogen bonding (center right). The bottom entries show the solvent-accessible surfaces (dotted areas) of the crystal form (left) and the form adopted in water (right).[108,110]

producable on an even higher scale if required, it is, together with cellulose- or starch-derived glucose, the major carbohydrate feedstock of low molecular weight, from which to elaborate organic chemicals.

The resulting chemistry of sucrose is capricious.[111] The pronounced acid sensivity of the intersaccharidic linkage excludes any reaction that requires acidic conditions, and, featuring eight hydroxyl groups with only subtle reactivity differences, reactions with high regioselectivities for one or two of the OH-groups are few, in fact, mostly enzymatic.

2.3.4.1 Oxidation Products of Sucrose.

The essentially regiospecific oxidation by *Agrobacterium tumefaciens*, whose dehydrogenase exclusively generates 3^g-ketosucrose,[112] is the prototype of an entry reaction into modified sucroses. This ready access opened the way to manifold modifications at the 3^g-carbonyl function (Scheme 2.16).[113] Chemical oxidation proceeds less uniformly, for

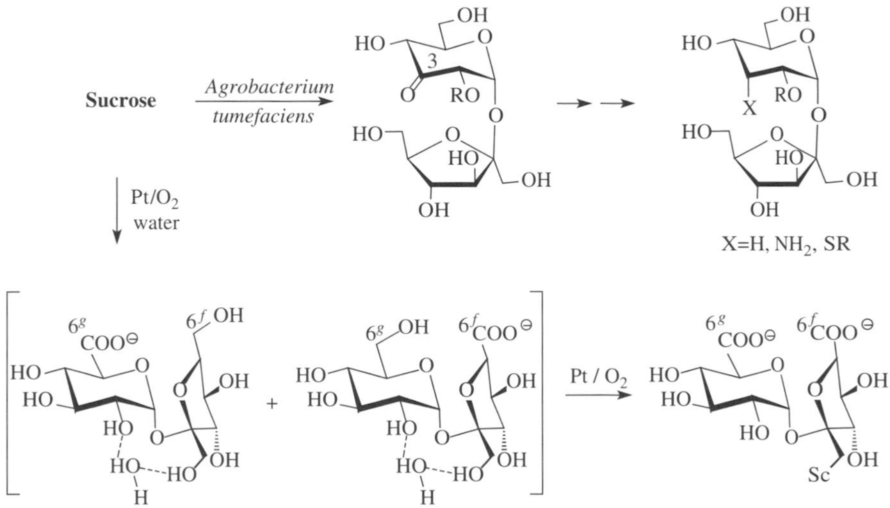

Scheme 2.16 Useful oxidation products of sucrose.

example agitation of an aqueous solution of pH 6.5–7.0 at 35°C with air in the presence of 0.5% Pt/C gave a 9:9:1 ratio of the 6^g-, 6^f-, and 1^f-saccharonic acids.[114] On further oxidation, particularly when using large amounts of the Pt catalyst and higher temperature (80–100°C), the preferred formation of the $6^g,6^f$-dicarboxylic acid has been observed,[115] which may be isolated in up to a 70% yield by continuous electrodialytic removal.[116]

Extended catalytic oxidation finally yields the 1^f-, 6^f-, 6^g-tricarboxylic acid, that is, all primary hydroxyl groups have yielded to oxidation.[117] An alternate useful oxidant to the tricarboxylate is the NaOCl/TEMPO system, which, on applying high-frequency ultrasound, produces the tricarboxylate in up to a 70% yield.[118]

These sucrose-derived carboxylic acids have potential as the acid components of polyesters and polyamides. On amidation of the methyl ester of sucrose-6^f,6^g-dicarboxylic acid with fat-amines, for example, surface-active diamides (left formula) with remarkable tensidometric properties are obtained, whereas reaction with hexamethylenediamine produces an interesting, highly hydrophilic polyamide (right):[119]

$(n = 1–4)$

2.3.4.2 Sucrose Esters. Sucrose esters have industrial interest in the area of surfactants,[24] bleaching boosters, cosmetics,[120] and fat substitutes.[121] Synthetically prepared[122] octa-fatty acid esters of sucrose have similar properties as the normal triglycerides, yet are not degraded by lipases, which entailed their marketing as noncaloric fat substitutes—after being approved by the U.S. Food and Drug Administration[123] under the name Olestra® or Olean®.[121]

Less highly esterified sucroses, usually mixtures with a high proportion of either mono-, di-, or tri-esters of variable regioisomeric distribution over the 2^g-, 6^g-, 6^f-, as well as other hydroxyls (see arrows in Figure 2.8), are cosmetic emulsifiers and have favorable surfactant properties, combining low toxicity, skin compatibility, and biodegradability. Currently, they are produced at an estimated 5000 t/a level, mainly in Japan,[24b] yet have the potential of becoming viable alternatives to the APG biodetergents if they become more selectively producible.

2.3.4.3 Sucrose Ethers. Being next to the anomeric center and intramolecularly hydrogen-bonded, the 2^g-OH of sucrose is the most acidic, which means it is deprotonated first under alkaline conditions, and thus preferentially yields to etherification. Benzylation with NaH/benzylbromide in DMF, for example, results in an 11:2:1 mixture of 2^g-O-benzyl-sucrose (Figure 2.8) and its 1-O- and 3^f-O-isomers.[124] Because the former is readily accessible, it proved to be a versatile intermediate for the generation of 2^g-modified sucroses, for example, the 2^g-keto and 2^g-deoxy derivatives as well as sucrosamine (2^g-amino-2^g-deoxy-sucrose),[124] whose application profiles remain to be investigated.

Of higher interest industrially is the etherification of sucrose with long-chain epoxides such as 1,2-epoxydodecane[125] or 1,2-epoxydodecan-3-ol,[126] which are performable as one-pot reactions in dimethyl sulfoxide (DMSO) and the presence

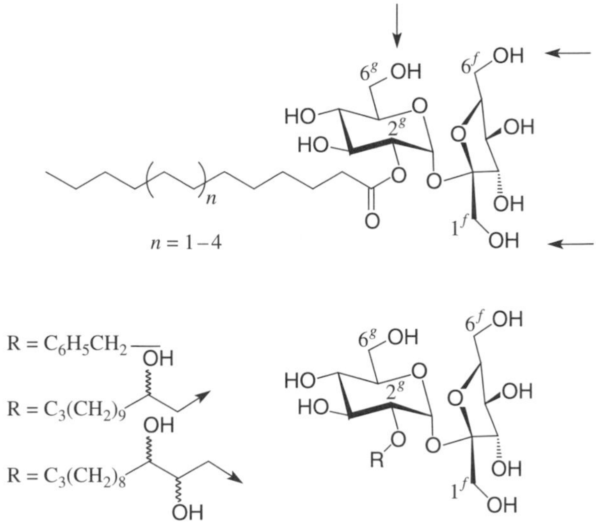

Figure 2.8 Sucrose monoesters and monoethers with useful surfactant properties.

of a base to provide sucrose monoethers with preferred regioselectivities of the 2^g-O- and 1^f-O-positions. Unlike sucrose esters, the long-chain ethers are resistant to alkaline conditions, which considerably extends their potential applications as nonionic surfactants. They also have promising liquid crystalline properties, their mesophases depending on the point where the fatty chain attaches to the sucrose.

The only large-scale application of sucrose ethers appears to be to use poly-O-(hydroxylpropyl) ethers, generated by alkoxylation with propylene oxide, as the polyol component for rigid polyurethanes[127]—sucrose itself gives only brittle ones—which are used primarily in cushioning applications. The structures of these products, that is, the positions at which sucrose is alkoxylated and then carbamoylated with diisocyanates, and the type(s) of cross-linking involved, are not well defined though.

2.3.4.4 Sucrose Conversion to Isomaltulose. As a 6-O-(α-D-glucosyl)-D-fructose, isomaltulose is isomeric with sucrose, from which it is produced at an industrial level (cf. Table 2.1)—for food reasons, as it is hydrogenated to an equimolar mixture of glucosyl-α(1→6)-glucitol and mannitol,[128] the low caloric sweetener isomalt.[129] As illustrated in Scheme 2.17, the industrial process involves a glucosyl shift from the 2^f-O of sucrose to the 6^f-OH, effected by action of an immobilized *Protaminobacter rubrum*-derived α(1→6)-glucosyltransferase. Having become most readily accessible in this way, isomaltulose developed into a lucrative target for generating disaccharide intermediates of industrial potential. Particularly relevant in this context are oxidative conversions, hydrogen peroxide as the oxidant leading to shortening of the fructose chain by four carbons to provide the glucoside of glycolic acid (GGA) in 40% yield.[130] Air oxidation in strongly alkaline solution (KOH), is less rigorous,

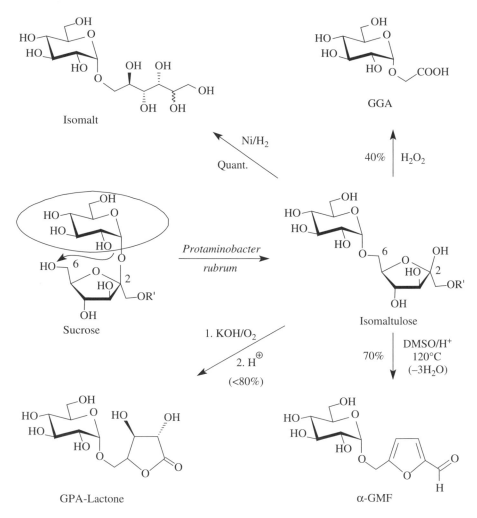

Scheme 2.17 Generation of isomaltulose from sucrose, and the follow-up products with industrial potential.

however, affording the potassium salt of the next lower aldonic acid, that is, glucosyl-α-(1→5)-D-arabinonic acid ('GPA') (Scheme 2.17), isolable as such, or upon neutralization, as the GPA-lactone in high yields each.[131]

Another, industrially relevant follow-up reaction of isomaltulose comprises its ready conversion into 5-(α-D-glucosyloxymethyl)-furfural ("α-GMF") by acidic dehydration of its fructose portion under conditions (acidic resin in DMSO, 120°C[132]) that retain the intersaccharidic linkage (Scheme 2.17). As this process can also be performed in a continuous-flow reactor,[132a] a most versatile building block is available from sucrose in two steps, of which the first is already industrially realized, and the second simple enough to be performed on a large scale.

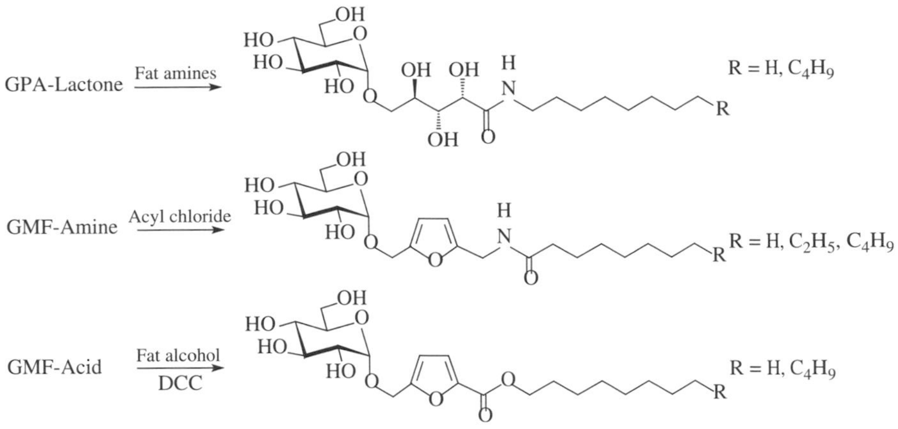

Scheme 2.18 Isomaltulose-derived products with surfactant and liquid crystalline properties.

Various products with industrial application profiles have been prepared from GPA and α-GMF (Scheme 2.18): Amidation of GPA-lactone with the C_8- and C_{12}- "fat amines," provided the GPA-amides,[131] which not only exhibit promising detergent profiles but also surprising liquid crystalline properties, such as S_{Ad}— phases over a broad temperature range.[133] As a glucosylated HMF, α-GMF provides a particularly rich chemistry:[132,134] aldol-type condensations provide derivatives with polymerizable double bonds that are expected to yield novel, hydrophilic polymers; oxidation and reductive amination generate the α-GMF-carboxylate and α-GMF-amine, respectively, which on esterification with long-chain alcohols or N-acylation with fatty acids afford a novel type of liquid crystals,[133] as the hydrophilic glucose part and the hydrophobic fat-alkyl residue are separated by an quasi-aromatic spacer; and they combine high surface activity with biocompatibility, making them promising candidates for biomedical applications.

2.3.4.5 Linear C—C-Polymers with Pendant Sucrose Residues.
The synthesis of sugars carrying O-linked residues with polymerizable double bonds ("vinyl-saccharides") and their radical or cationic copolymerization has been extensively pursued over the last 70 years,[135] with major emphasis on suitable derivatives of glucose and sucrose—the first example, the polymerization of 1,2:5,6-di-O-isopropylidene-3-O-vinyl-D-glucofuranose dating back to Reppe and Hecht in the 1930s.[136] Thus, a large series of mono-O- and di-O-substituted derivatives of sucrose—with polymerizable C=C double bonds in ester or ether moieties attached—have been prepared,[137,138] usually as mixtures with average degrees of substitution: esters of acrylic or methacrylic acid, or vinylbenzyl ethers mostly. Their polymerization as such, or copolymerization with the standard petroleum-based monomers (methyl methacrylate, methyl acrylate, acrylonitril, styrene, etc.), have led to a variety of interesting linear and cross-linked polymers with "sucrose

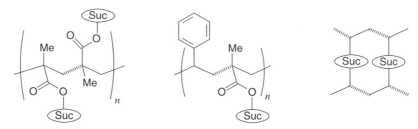

Figure 2.9   Idealized representation of a linear polymer resulting from radical polimerization of a mono-*O*-methacroyl-sucrose (left) and a 1:1 copolymerization product with styrene. Di-*O*-substituted vinyl-sucroses are deemed to lead to cross-linked polymers (right).

anchors" attached to the polymeric carbon chain(s),[137,138] as schematically represented in Figure 2.9. Various surface modifications of polymers by graft polymerizations with vinyl-sucroses have also been reported, for example, grafting of sucrose acrylate on polyvinyl chloride (PVC) films.[139]

Despite the highly versatile application profiles of polymers with adjunct sucrose (or other sugar) residues—their major asset is enhanced hydrophilicity as compared to their hydrophobic petroleum-derived counterparts—interest appears to be restricted to biomedical uses. Currently none is produced commercially, as the generation of vinyl-sucroses and their often capricious polymerization have made their use as commodity plastics uneconomical. Another reason is their limited biodegradability: only the sugar portion is biodegradable, with a polymeric carbon chain left over. Because biodegradability is a major issue today,[140] these polyvinylsaccharides are unlikely to become petrochemical substitution options in the near future.

2.4   CONCLUSION

The utilization in nonfoods of inexpensive, bulk-scale-accessible, low-molecular-weight carbohydrates—sucrose, glucose, xylose, and fructose being the most readily accessible—is at a rather modest level in terms of large-scale manufactured commodities currently on the market. The unusually diverse stock of readily accessible products described in this chapter, which covers a wide range of industrial application profiles, is mostly unexploited in its potentialities. The reasons are mostly economic as equivalent products based on petrochemical raw materials are simply cheaper. Nevertheless, a basic change in this scenario is clearly foreseeable. As depletion of our fossil raw materials is progressing, petrochemicals will inevitably increase in price, such that biobased products will eventually become competitive. Realistic prognoses[5] expect this to occur by the middle of this century at the latest.

In the meantime, it is imperative that carbohydrates be used systematically to achieve efficient, environmentally benign, and economical processes for their

Figure 2.10 Impatience with renewable resources.[142]

large-scale conversion into industrially viable products, be it bulk or intermediate chemicals, pharmaceuticals, or polymeric organic materials. In this endeavor, national and supranational funding institutions—in Europe, the corresponding EU bodies (in the European Commission's seventh framework program, hopefully) and/or the European Renewable Raw Materials Association[141]—will have to play a much more dynamic role than heretofore. One decisive action, of course, is the generous funding not only of applied but of basic research activities in this area, and this over a considerably longer time frame—5–10 years for promising projects, rather than expecting that marketable products be delivered within 3–5 years. Impatience with the development of renewable resources for high-value-added products (Figure 2.10) is futile if they are harvested too early.

Another key issue is the development of a concise, long term strategy that takes hold in academia *and* chemical industry. This strategy should not to be directed toward generating the very same basic chemicals from carbohydrates that are easily accessible from petrochemical sources, but toward the development of products with analogous industrial application profiles, with as little alteration of the carbohydrate structural framework as possible. Only then will economically sound biobased alternatives to petrochemicals—various potential examples are contained in this chapter—become available.

REFERENCES

1. Anastas, P. T.; Warner, J. C. *Green Chemistry: Theory and Practice*, Oxford Univ. Press, Oxford, 1998, pp. 30–48.

2. Okkerse, C.; van Bekkum, H. From fossil to green, *Green Chem.*, 1999, 107–114.

3. National Research Council, USA, *Priorities for Research and Commercialization of Biobased Industrial Products*, Natl. Acad. Sci. Press, Washington, DC, 2000.

4. U.S. Department of Energy (DOE): (a) *Vision for Bioenergy & Biobased Products in The United States*, October 2002; bioproducts-bioenergy.gov/pdfs/BioVision_03_ Web.pdf. (b) *Roadmap for Biomass Technologies in the United States*, December 2002; bioproducts-bioenergy.gov/pdfs/FinalBiomassRoadmap.pdf

5. (a) Campbell, C. J.; Laherrère, J. H. The end of cheap oil, *Sci. Am.*, March 1998, 60–65. (b) Attarian, J. The coming end of cheap oil: Hubbert's peak and beyond, *Soc. Contracts*, 2002, **12**, 276–286. (c) Klass, D. H. Fossil Fuel Reserves and Depletion, in *Biomass for Renewable Energy, Fuels and Chemicals*, Acad. Press, San Diego, 1998, pp. 410–419.

6. UN Food & Agriculture Organization, *World Sugar Production, 2005/06*; fao.org/ docrep/009/j7927e/j7927e07.htm.

7. Klass, D. H. Organic commodity chemicals from biomass, in *Biomass for Renewable Energy, Fuels, and Chemicals*, Acad. Press, San Diego, 1998, pp. 495–546.

8. *Carbohydrates as Organic Raw Materials*, VCH, Weinheim/New York: (a) Vol. I, Lichtenthaler, F. W. (Ed.), 1991, 365 pp; (b) Vol. II, Descotes, G. (Ed.), 1993, 278 pp.; (c) Vol. III, van Bekkum, H.; Röper, H.; Voragen, A. G. J. (Eds.), 1996, 358 pp.; (d) Vol. IV, Praznik, W. (Ed.), Wiener Univ. Verlag, Vienna, 1998, 292 pp.

9. Lichtenthaler, F. W.; Mondel, S. Perspectives in the use of low molecular weight carbohydrates as organic raw materials, *Pure Appl. Chem.*, 1997, **69**, 1853–1866.

10. Bozell, J. J. (Ed.), *Chemicals and Materials from Renewable Resources*, ACS Symposium Series No. 784, American Chemical Society, Washington, D.C., 2001, 226 pp.

11. (a) Lichtenthaler, F. W. Unsaturated *O*- and *N*-heterocycles from carbohydrate feedstocks, *Acc. Chem. Res.*, 2002, **35**, 728–737; (b) Lichtenthaler, F. W. Carbohydrates as Organic Raw Materials, *Ullmann's Encyclopedia Industrial Chem.*, 6th Ed., Vol. 6, 2002, pp. 262–273; Electronic Release, 7th Ed., chapt. 9, Wiley-VCH, Weinheim, 2007; (c) Lichtenthaler, F. W.; Peters, S. *Comptes Rend. Chim.*, 2004, **7**, 65–90.

12. Wirtschaftliche Vereinigung Zucker, Biokraftstoffe aus Zuckerrüben und Getreide; zuckerwirtschaft.de/pdf/Broschuere_BioE.pdf

13. (a) For a pertinent overview, see: Himmel, M. E.; Adney, W. A.; Baker, J. O.; et al. Advanced bioethanol production technologies, in *Fuels and Chemicals from Biomass*, Saha, B. C.; Woodward, J., Eds., ACS Symposium Series No. 666, American chemical Society, Washington D.C., 1997, pp. 2–45. (b) Goebel, O. Comparison of process

economics for synthetic and fermentation ethanol, *Ullmanńs Encyclopedia Industrial Chem.*, Electronic Release, 7th Ed., chap. 7, Wiley-VCH, Weinheim, 2007.

14. (a) McKillip, W. J.; Collin, G.; Höke, H.; Zeitsch, K. J. Furan and derivatives, *Ullmanńs Encyclopedia Industrial Chem.*, Electronic Release, 7th Ed., Wiley-VCH, Weinheim, 2007; (b) Zeitsch, K. J. *The Chemistry and Technology of Furfural and Its Many Byproducts*, Elsevier, Amsterdam, 2000, 374pp.

15. (a) Gandini, A.; Belgacem, M. N. Furans in polymer chemistry, *Prog. Polym. Sci.*, 1997, **22**, 1203–1379; (b) Moreau, C.; Belgacem, M. N.; Gandini, A., Substituted furans from carbohydrates and ensuing polymers, *Topics Catal.*, 2004, **27**, 11–30.

16. Klingler, F. D. Levulinic acid, *Ullmann's Encyclopedia Industrial Chem.*, Electronic Release, 7th Ed., Wiley-VCH, Weinheim, 2007.

17. Vogel, R. Sorbitol, *Ullmann's Encyclopedia Industrial Chem.*, 5th Ed., Vol. A25, 1994, pp. 418–423.

18. Oster, B.; Fechtel, W. Vitamin C, *Ullmanńs Encyclopedia Industrial Chem.*, 5th Ed., Vol. A27, 1996, pp. 547–559.

19. McCoy, M. Seeking biomaterials, *Chem. Eng. News*, 2003, **81**(8), 18; 2003, **81**(45), 17–18.

20. Ritter, S. K. Green chemistry progress report, *Chem. Eng. News*, 2002, **80**(47), 20.

21. natureworksllc.com/Product-And-Applications.aspx

22. vertecbiosolvents.com/

23. For a pertinent review, see: Hill, K.; Rhode, O. Sugar-based surfactants for consumer products and technical applications, *Lipid/Fett*, 1999, **101**, 23–33.

24. (a) Desai, N. B. Esters of sucrose and glucose as cosmetic materials, *Cosmetics & Toiletries*, 1990, **105**, 99–107; (b) Mitsubishi-Kagaku Foods Corp., Sugar esters; mfc.co.jp/english/whatsse.htm

25. von Rybinski, W.; Hill, K. Alkyl polyglycosides—properties and applications of a new class of surfactants, *Angew. Chem. Int. Ed.*, 1998, **37**, 1328–1345.

26. McGuire, J. L.; Hasskare, H.; et al. Pharmaceuticals: general survey and development, *Ullmann's Encyclopedia Industrial Chem.*, Electronic Release, 7th Ed., Wiley-VCH, Weinheim, 2007.

27. Silviri, L. A.; DeAngelis, N. J. Isosorbide dinitrate, *Anal. Profiles Drug Subst.*, 1975, **4**, 225–244.

28. Maryanoff, B. E.; Nortey, S. O.; Gardocki, S. O.; et al. Anticonvulsant sulfamates. 2,3:4,5-Di-*O*-isopropylidene-β-D-fructopyranose sulfamate, *J. Med. Chem.*, 1987, **30**, 880–887.

29. Hugill, A., *Introductory Dedicational Metaphor to Sugar and All That. A History of Tate & Lyle*, Gentry Books, London, 1978.

30. Schenck, F. W. Glucose and glucose-containing syrups, *Ullmann's Encyclopedia of Industrial Chem.*, 4th Ed., Vol. A12, 1989, pp. 457–476.

31. Lichtenthaler, F. W. Emil Fischer's establishment of the configuration of sugars—a centennial tribute, *Angew. Chem. Int. Ed.*, 1992, **31**, 1541–1556.

32. For useful preparative procedures, see: *Methods Carbohydr. Chem.*, 1963, **2**, 318–325; 326–328; 405–408; 427–430.

33. Garcia-Gonzáles, F. *Adv. Carbohydr. Chem.*, 1956, **11**, 97–143.

34. (a) Rodrigues, F.; Canac, Y.; Lubineau, A. *Chem. Commun.*, 2000, 2049-2059; (b) Riemann, I.; Papadopoulos, M. A.; Knorst, M.; Fessner, W.-D. *Aust. J. Chem.*, 2002, **55**, 147–154.

35. Chapleur, Y. (Ed.), *Carbohydrate Mimics*, Wiley-VCH, Weinheim, 1998, 604 pp., and refs. cited therein.

36. Lichtenthaler, F. W. Building blocks from sugars and their use in natural product synthesis, in R. Scheffold, Ed., *Modern Synthetic Methods*, Vol. 6, VCH, Weinheim, 1992, pp. 273–376.

37. Roth, W.; Pigman, W. *Methods Carbohydr. Chem.*, 1963, **2**, 405–408.

38. Ferrier, R. J.; Prasad, N. *J. Chem. Soc. (C)*, 1969, 570–575.

39. Lichtenthaler, F. W.; Rönninger, S.; Jarglis, P. *Liebigs Ann. Chem.*, 1989, 1153–1161.

40. Hanessian, S.; Faucher, A. M.; Leger, S. *Tetrahedron*, 1990, **46**, 231–243.

41. Czerneckí, S.; Víjayakuraman, K.; Ville, G. *J. Org. Chem.*, 1986, **51**, 5472–5475.

42. Fraser-Reid, B.; McLean, A.; Usherwood, E. W.; Yunker, M. *Can. J. Chem.*, 1970, **48**, 2877–2884.

43. Ferrier, R. J. *Methods Carbohydr. Chem.*, 1972, **6**, 307–311.

44. Lichtenthaler, F. W.; Kraska, U. *Carbohydr. Res.*, 1977, **58**, 363–377.

45. Lichtenthaler, F. W.; Nishiyama, S.; Weimer, T. *Liebigs Ann.*, 1989, 1163–1170.

46. Lichtenthaler, F. W.; Ogawa, S.; Heidel, P. *Chem. Ber.*, 1977, **110**, 3324–3332.

47. Shafizadeh, F.; Furneaux, R.; Stevenson, T. *Carbohydr. Res.*, 1979, **71**, 169–191.

48. Witczak, Z. J. *Pure Appl. Chem.*, 1994, **66**, 2189–2192.

49. Beélik, A. *Adv. Carbohydr. Chem.*, 1956, **11**, 145–183.

50. Lichtenthaler, F. W. *Pure Appl. Chem.*, 1978, **50**, 1343–1362.

51. Nelson, C.; Gratzl, J. *Carbohydr. Res.*, 1978, **60**, 267–273.

52. Tajima, K. *Chem. Lett.*, 1987, 1319–1322.

53. Hustede, H.; Haberstroh, H.-J.; Schinzig, E. Gluconic acid, *Ullmann's Encyclopedia Industrial Chem.*, 5th Ed., Vol. A12, 1989, pp. 449–456.

54. Mehltretter, C. L. D-Glucaric acid, *Methods Carbohydr. Chem.*, 1963, **2**, 46–48.

55. Klingler, F. D. Oxocarboxylic acids, *Ullmann's Encyclopedia Industrial Chem.*, Electronic Release, 7th Ed., chap. 4, Wiley-VCH, Weinheim, 2007.

56. Hayes, D. J.; Ross, J.; Hayes, M. H. B.; Fitzpatrick, S. The biofine process, in *Biorefineries—Industrial Processes and Products*, Kamm, B.; Gruber, P.; Kamm, M. (Eds.), Wiley-VCH, New York, 2006, pp. 3–59.

57. DOE, Energy, Efficiency and Renewable Energy, Top value added chemicals from biomass. Screening for potential candidates from sugars: (a) Levulinic acid, pp. 45–48; (b) 3-Hydroxypropionic acid (3-HPA), pp. 29–31; (c) Four-carbon 1,4-diacids, pp. 22–25; (d) Itaconic acid, pp. 42–44; nrel.gov/docs/fy04osti/35523.pdf, 67 pp.

58. Huber, G. W.; Cortright, R. D., Dumesic, J. S. *Angew. Chem. Int. Ed.*, 2004, **43**, 1549–1551.

59. Huber, G. W.; Chheda, J. N.; Barret, C. J.; Dumesic, J. A. *Science*, 2005, **308**, 1446–1448.

60. For pertinent reviews on HMF, see (a) Lewkowski, J. Synthesis, chemistry and applications of 5-hydroxymethylfurfural and its derivatives, *ARKIVOC*, 2001, 17–54; (b) B. F. M. KUSTER, Manufacture of 5-hydroxymethylfurfural, *Starch/Stärke*, 1990, **42**, 314–321. Newer developments; (c) Bicker, M.; Hirth, J.; Vogel, H. Dehydration of fructose to 5-hydroxymethylfurfural in sub- and supercritical acetone, *Green Chem.*,

2003, **5**, 280–284.; (d) Lansalot-Matras, C.; Moreau, C. Dehydration of fructose to 5-hydroxymethylfurfural in ionic liquids, *Catalysis Commun.*, 2003, **4**, 517–520.

61. Wilson, E. K. Engineering cell-based factories, *Chem. Eng. News*, 2005, **83**(12), 41–44.

62. Riese, J., Bachmann, R. Industrial biotechnology: Turning the potential into profits, *Chem. Market Rep.*, 2004 (Dec.1.); mckinsey.com/clientservice/chemicals/potentialprofit.asp

63. For pertinent reviews, see (a) Lee, S. Y.; Hong, S. H.; Lee, S. H.; Park, S. J. Fermentative production of chemicals that can be used for polymer synthesis, *Macromol. Biosci.*, 2004, **4**, 157–164. (b) Tsao, G. T.; Cao, N. J.; Du, J.; Gong, C. S. Production of multifunctional organic acids from renewable resources, *Adv. Biochim. Eng. Biotechnol.*, 1999, **65**, 243–280.

64. (a) Cameron, D. C. 3-Hydroxypropionic acid: a new platform chemical from the biorefinery, 2003, ibeweb.org/meetings/2005/proceedings/abstracts/industry_abstracts. html; (b) Tullo, A. *Chem. Eng. News*, 2005, **83**(26), 11.

65. Lohbeck, K.; Haferkorn, H.; Fuhrmann, W. Maleic and fumaric acids, *Ullmanńs Encyclopedia Industrial Chem.*, 5th Ed., Vol. A16, 1990, pp. 53–62.

66. Carta, F. S.; Soccol, C. R.; Ramos, L. P.; Fontana, J. D. Production of fumaric acid by fermentation of enzymic hydrolyzates derived from cassava bagasse, *Bioresour. Technol.*, 1999, **68**, 23–28.

67. (a) Vemuri, G. N.; Eiteman, M. A.; Altman, E. Succinate production in dual-phase *E.coli* fermentations depends on the time of transition from aerobic to anaerobic conditions, *J. Ind. Microbiol. Biotechnol.*, 2002, **28**, 325–332; (b) Effects of growth mode and pyruvate carboxylase on succinic acid production by metabolically engineered *E.coli* strains, *Appl. Environ. Microbiol.*, 2002, **68**, 1715–1727; (c) Gokarn, R. R.; Eiteman, M. A.; Sridhar, J. Production of succinate by anaerobic microorganisms, in *Fuels and Chemicals from Biomass*, Saha, B. D.; Woodard, J. (Eds.), ACS Symposium Series No. 666, American Chemical Society, Washington, D.C., 1997, pp. 237–279.

68. Kane, J. H.; Finlay, A.; Amann, P. F. (Chas. Pfizer, Merck), Itaconic acid., U.S. Patent, 2,385,283 (1945); *Chem. Abstr.*, 1946, **40**, 995.

69. (a) Reddy, C. S.; Singh, R. P. Enhanced production of itaconic acid from corn starch and market refuse fruits by genetically manipulated *Aspergillus terreus* SKR10, *Bioresour. Technol.*, 2002, **85**, 69–71; (b) Willke, T.; Welter, K.; Vorlop, K. D. Biotechnical production of itaconic acid from sugar, *Appl. Microbiol. Biotechnol.*, 2001, **56**, 289–295.

70. DuPont, 2007:2.dupont.com/Sorona/en_US/index.html

71. Shell, 2007: shellchemicals.com/corterra/1,1098,280,00.html

72. Zhu, M. M.; Lawman, P. D.; Cameron, D. C. Improving 1,3-propanediol production from glycerol in a metabolically engineered *E.coli* by reducing accumulation of glycerol-3-phosphate, *Biotechnol. Prog.*, 2002, **18**, 694–699.

73. For an informative review, see Zeng, A. N.; Biebl, H. Bulk chemicals from biomass: Ease of 1,3-propanediol production and new trends, *Adv. Biochem. Eng./Biotechn.*, 2002, **74**, 239–259.

74. Tate & Lyle 2007, press release: 19.35.126.50/FeatureArticles/FeatureArticle218.asp

75. Altaras, N. E.; Cameron, D. C. Metabolic engineering of a 1,2-propanediol pathway in *E.coli*, *Appl. Environ. Microbiol.*, 1999, **65**, 1180–1185. (b) Enhanced production of

(R)-1,2-propanediol by metabolically engineered *E.coli, Biotechnol. Progr.*, 2000, **16**, 940–946. (c) Altaras, N. E.; Etzel, M. R.; Cameron, D. C. Conversion of sugars to 1,2-propanediol by *Thermoanaerobacterium thermosaccharolyticum, Biotechnol. Progr.*, 2001, **17**, 52–56.

76. Nishino, N.; Yochida, M.; Shiota, H.; Sakaguchi, E. Accommodation of 1,2-propane-diol and enhancement of aerobic stability in whole crop maize silage inocculated with *Lactobacillus buchneri, J. Appl. Microbiol.*, 2003, **94**, 800–807.

77. Wach, W. Fructose, *Ullmann's Encyclopedia Industrial Chem.*, Electronic Release, 7th Ed., Wiley-VCH, Weinheim, 2007.

78. Fontana, A.; Hermann, B.; Guirand, J. P., in *Inulin and Inulin-containing Crops*, A. Fuchs (Ed.), Elsevier Science Publishers, Amsterdam/London, 1993, pp. 251–258.

79. (a) Lichtenthaler, F. W. Towards improving the utility of ketoses as organic raw materials, *Carbohydr. Res.* 1998, **313**, 69–89.; (b) Schneider, B.; Lichtenthaler, F. W.; Steinle, G.; Schiweck, H. Distribution of furanoid and pyranoid tautomers of D-fructose in solution, *Liebigs Ann. Chem.*, 1985, 2454–2464.

80. Kanters, J. A.; Roelofson, G.; Alblas, B. P.; Meinders, I. The crystal structure of fruc-tose, *Acta Cryst.*, 1977, **B33**, 665–672.

81. Chan, J. Y. C.; Cheong, P. P. L.; Hough, L.; Richardson, A. C. *J. Chem. Soc., Perkin Trans, 1* 1985, 1447–1455.

82. Lichtenthaler, F. W.; Klotz, J.; Flath, F. J. Acylation and carbamoylation of D-fructose: acyclic, furanoid and pyranoid derivatives, *Liebigs Ann. Chem.*, 1995, 2069–2080.

83. Lichtenthaler, W. F.; Hahn, S.; Flath, F. J. Pyranoid *exo-* and *endo-*D-fructals and L-sorbals: practical routes for their acquisition and ensuing reactions, *Liebigs Ann. Chem.*, 1995, 2081–2088.

84. Boettcher, A.; Lichtenthaler, F. W. D-Fructose and L-sorbose-derived *exo-* and *endo-*hydroxyglycal esters and some of their chemistry, *Tetrahedron: Asymmetry*, 2004, **15**, 2693–2701.

85. Kang, J.; Lim, G. J.; Yoon, S. K.; Kim, M. Y. *J. Org. Chem.*, 1995, **60**, 564–577.

86. Brady, R. F., Jr. *Carbohydr. Res.*, 1970, **15**, 35–40.

87. Peters, S.; Lichtenthaler, F. W.; Lindner, H. J. *Tetrahedron: Asymmetry*, 2003, **14**, 2574–2579.

88. Weidenhagen, R.; Hermann, R. *Ber. Dtsch. Chem. Ges.*, 1937, **70**, 570–583; *Org. Synth. Coll.*, 1955, III 460–462.

89. (a) Brust, A.; Lichtenthaler, F. W., unpublished; (b) Streith, J.; Boiron, A.; Frankowski, A.; et al. A general one-pot synthesis of imidazolosugars, *Synthesis*, 1995, 944–946.

90. Rozanski, A.; Bielawski, K.; Boltryk, K.; Bartulewicz, D. *Akad. Med. Juiliana Marchlewskiego Bialymstoku*, 1991, 35–36, 57–63; *Chem. Abstr.*, 1992, **188**, 22471 m.

91. Ohle, H.; Hielscher, M. Tetraoxybutyl-chinoxalin, *Ber. Dtsch. Chem. Ges.*, 1941, **74**, 13–19.

92. Schiweck, H.; Munir, M.; Rapp, K.; Vogel, M. New developments in the use of sucrose as an industrial bulk chemical, in *Carbohydrates as Organic Raw Materials*, Lichtenthaler, F. W. (ed.), VCH, Weinheim, 1991, pp. 57–94; *Zuckerind. (Berlin)* 1990, **115**, 555–565.

93. El Haji, T.; Masroua, A.; Martin, J.-C.; Descotes, G. Synthesis of 5-(hydroxymethyl)-furan-2-carboxaldehyde and its derivatives by acid treatment of sugars on ion-exchange resins, *Bull. Soc. Chim. Fr.*, 1987, 855–860.

94. (a) Musau, R. M.; Munavu, R. M. Preparation of 5-hydroxymethyl-2-furaldehyde from D-fructose in the presence of DMSO, *Biomass* 1987, **13**, 67–74; (b) Larousse, C.; Rigal, L.; Gaset, A. Thermal degradation of 5-hydroxymethylfurfural, *J. Chem. Technol. Biotechnol.*, 1992, **53**, 111–116.

95. Schiavo, V.; Descotes, G.; Mentech, J. Hydrogenation of 5-hydroxymethylfurfural, in aqueous medium, *Bull. Soc. Chim. Fr.*, 1991, **128**, 704–711.

96. Elming, N.; Clauson-Kaas, N. 6-Methyl-3-pyridinol from 2-hydroxymethyl-5-aminomethyl-furan, *Acta Chem. Scand.*, 1956, **10**, 1603–1605.

97. (a) Leupold, E.; Wiesner, M.; Schlingmann, M.; Rapp, K. Catalytic oxidation of 5-hydroxymethylfurfural, Ger. Offen. DE 3 826 073 (1990); *Chem. Abstr.* 1990, *113*, 995; (b) Partenheimer, W.; Grushin, V. V. Synthesis of 2,5-diformylfuran and furan-2,5-dicarboxylic acid by catalytic air oxidation of HMF, *Adv. Synth. Catal.* 2001; **343**, 102–111; (c) Ribeiro, M. L.; Schuchardt, U. Cooperative effect of cobalt acetylacetonate and silica in the catalytic cyclization and oxidation of fructose to 2,5-furandicarboxylic acid, *Catal. Commun.*, 2003, **4**, 83–86.

98. Koch, H.; Pein, J. Condensations between 5-hydroxymethylfurfural, phenol, and formaldehyde, *Polym. Bull. (Berlin)*, 1985, *13*, 525–532; *Starch/Stärke*, 1983, **35**, 304–313.

99. Whistler, R. L.; BeMiller, J. N. α-D-Xylose, *Methods Carbohydr. Chem.*, 1962, **1**, 88–90.

100. Bär, A. Xylitol. *Ullmann's Encyclopedia Industrial Chem.*, 5th Ed., Vol. A25, 1994, pp. 416–417.

101. Olin, S. M. *Methods Carbohydr. Chem.*, 1962, **1**, 148–151.

102. Weygand, F. *Methods Carbohydr. Chem.*, 1962, **1**, 182–185.

103. Ferrier, R. J.; Sankey, G. H. *J. Chem. Soc. C*, 1966, 2339–2345.

104. Diehl, V.; Cuny, E.; Lichtenthaler, F. W. Conversion of D-xylose into hydrophilically functionalized pyrazoles, *Heterocycles*, 1998, **48**, 1193–1201.

105. Oikawa, M.; Müller, C.; Kunz, M.; Lichtenthaler, F. W. Hydrophilic pyrazoles from sugars, *Carbohydr. Res.*, 1998, **309**, 269–279.

106. Niu, W.; Molefe, M. N.; Frost, J. W. Microbial synthesis of the energetic material precursor 1,2,4-butanetriol, *J. Am. Chem. Soc.*, 2003, **125**, 12998–12999.

107. Brown, G. M.; Levy, H. A. *Acta Cryst.*, 1973, **B29**, 790–797; C. Hanson, J. C.; Sieker, L. C.; Jensen, L. H. *Acta Cryst.*, 1973, **B29**, 797–808.

108. Immel, S.; Lichtenthaler, F. W. The conformation of sucrose in water: a molecular dynamics approach, *Liebigs Ann. Chem.*, 1995, 1938–1947.

109. Lichtenthaler, F. W.; Immel, S. Computersimulation of chemical and biological properties of sucrose, *Internat. Sugar J.*, 1995, **97**, 12–22.

110. Lichtenthaler, F. W.; Pokinskyj, P.; Immel, S. Sucrose as a renewable organic raw material, *Zuckerind. (Berlin)*, 1996, **121**, 174–190.

111. Queneau, Y.; Fitremann, J.; Trombotto, S. The chemistry of unprotected sucrose: the selectivity issue, *Comptes Rend. Chim.*, 2004, **7**, 177–188.

112. Stoppok, E.; Matalla, K.; Buchholz, K. *Appl. Microbiol. Biotechnol.*, 1992, **36**, 604–610.

113. (a) Pietsch, M.; Walter, M.; Buchholz, K. Regioselective synthesis of new sucrose derivatives via 3-ketosucrose, *Carbohydr. Res.*, 1994, **254**, 183–194; (b) Simiand, C.; Samain, E.; Martin, O. R.; Driguez, H. Sucrose analogues modified at position 3, *Carbohydr. Res.* 1995, **267**, 1–15.

114. Kunz, M.; Puke, H.; Recker, C.; et al. (Südzucker AG), Process and apparatus for the preparation of mono-oxidized products from carbohydrates, Ger. Offen. DE 4 307 388 (1994); *Chem. Abstr.*, 1995, **122**, 56411.

115. Edye, L. A.; Meehan, G. V.; Richards, G. N. Platinum-catalyzed oxidation of sucrose, *J. Carbohydr. Chem.*, 1991, **10**, 11–23; 1994, **13**, 273–283.

116. Kunz, M.; Schwarz, A.; Kowalczyk, J. (Südzucker AG), Process for continuous manufacture of di- and higher-oxidized carboxylic acids from carbohydrates, Ger. Offen. DE 19 542 287 (1996); *Chem. Abstr.*, 1997, **127**, 52504.

117. Fritsche-Lang, W.; Leupold, E. I.; Schlingmann, M. Preparation of sucrose tricarboxylic acid, Ger. Offen. DE 3 535 720 (1987); *Chem. Abstr.*, 1987, **107**, 59408.

118. Lemoine, S.; Thomazeau, C.; Joannard, D.; et al. Sucrose tricarboxylate by sonocatalyzed TEMPO-mediated oxidation, *Carbohydr. Res.*, 2000, **326**, 176–184.

119. (a) Mondel, S. Practical approaches to surface-active 6,6-diamides and diesters of sucrose, Dissertation, Technical University Darmstadt, 1997; (b) Vlach, A. Preparation and polycondensation of sucrose-derived monomers, Dissertation, Technical University Darmstadt, 2001; (c) Cuny, E.; Mondel, S.; Lichtenthaler, F. W. Novel polyamides from disaccharide-derived dicarboxylic acids, *23. Int. Carbohydrate Symp., Whistler, Can.*, 2006, Abstr. TUE.PS 12; #292 at csi.chemie.tu-darmstadt.de/ak/fwlicht/publlist.html

120. Hurford, J. R., Surface-active agents derived from disaccharides, in *Developments in Food Carbohydrates*, Vol. 2, Lee, C. K. (Ed.), 1980, pp. 327–350, and literature cited therein.

121. Procter & Gamble, olean.com/

122. Mattson, F. H.; Volpenhein, R. A. (Procter & Gamble Co.), Low-caloric fat-containing food compositions, U.S. Patent 3,600,186 (1971); *Chem. Abstr.*, 1971, **75**, 139614.

123. Food & Drug Administration, Olestra and Other Fat Substitutes, 1995; fda.gov/opacom/backgrounders/olestra.html

124. Lichtenthaler, F. W.; Immel, S.; Pokinskyj, P. Selective 2-*O*-benzylation of sucrose. A facile entry into its 2-keto- and 2-deoxy derivatives and to sucrosamine, *Liebigs Ann. Chem.*, 1995, 1938–1947.

125. Gagnaire, J.; Cornet, A.; Bouchu, A.; et al. *Colloids and Surfaces A*, 2000, **172**, 125–138.

126. Danel, M.; Gagnaire, J.; Queneau, Y. *J. Mol. Catal. A*, 2002, **184**, 131–138.

127. (a) Kollonitsch, V. *Sucrose Chemicals*, The Int. Sugar Research Foundation, Washington, D.C., 1970; (b) Meath, A. R.; Booth, L. D. Sucrose and sucrose-modified polyols in urethane foams, in J. L. Hickson (Ed.), *Sucrochemistry*, ACS Symposium Series No. 41, American Chemical Society Washington, D.C., 1977, pp. 257–263; (c) Jhurry, D.; Deffieux, A. Sucrose-based polymers: polyurethanes with sucrose in the main chain, *Eur. Polym. J.*, 1997, **33**, 1577–1582.

128. Kunz, M. Isomaltulose, trehalulose and isomalt, *Ullmann's Encyclopedia Industrial Chem.*, 5th Ed., Vol. A25, 1994, pp. 426–429.

129. (a) isomalt.de/; (b) isomaltidex.com/html

130. Trombotto, S. Danel, M.; Fitremann, J.; et al. *J. Org. Chem.*, 2003, **68**, 6672–6678.

131. Lichtenthaler, F. W.; Klimesch, R.; Müller, V.; Kunz, M. Disaccharide-building blocks from isomaltulose: glucosyl-$\alpha(1\rightarrow5)$-D-arabinonic acid and ensuing products, *Liebigs Ann.*, 1993, 975–980.

132. Lichtenthaler, F. W.; Martin, D.; Weber, T.; Schiweck, H., 5-(α-D-Glucosyl-oxymethyl)-furfural: preparation from isomaltulose and exploration of its ensuing chemistry, (a) Ger. Offen.3 936 522 (1989); (b) *Liebigs Ann.*, 1993, 967–974.

133. Hanemann, T.; Haase, W.; Lichtenthaler, F. W. Disaccharide-derived liquid crystals, *Liquid Cryst.*, 1997, **22**, 47–50.

134. Lichtenthaler, F. W. Isomaltulose, *Carbohydr. Res.*, 1998, **313**, 81–89.

135. For pertinent reviews on synthetic glycopolymers, see (a) Varma, A. J.; Kennedy, J. F.; Galgali, P. Synthetic polymers functionalized by carbohydrates, *Carbohydr. Polym.*, 2004, **56**, 429–445; (b) Ladmiral, V.; Media, E.; Haddleton, D. M. Synthetic glycopolymers, *Eur. Polym. J.*, 2004, **40**, 431–449.

136. (a) Reppe, W.; Hecht, O. Ger. Patient 715, 268 (1936); U.S. Patient 2,157,347 (1939); *Chem. Abstr.*, 1939, **33**, 44456; (b) Mikhantév, B. I.; Lapenko, V. L. Vinylation of D-glucose and its acetone derivatives, *Zh. Obshch. Khim.*, 1957, **27**, 2840–2841.

137. Gruber, H.; Greber, G. Reactive sucrose derivatives, in *Carbohydrates as Organic Raw Materials*, Lichtenthaler, F. W. (Ed.), VCH, Weinheim, 1991, pp. 95–116.

138. (a) Jhurry, D.; Deffieux, A.; Fontanille, M.; et al. Linear polymers with sucrose side chains, *Makromol. Chem.*, 1992, **193**, 2997–3007.; (b) Fanton, E.; Fayet, C.; Gelas, J.; et al. Synthesis of 4-*O*- and 6-*O*-monoacryloyl derivatives of sucrose. Polymerization and copolymerization with styrene, *Carbohydr. Res.*, 1993, **240**, 143–152.

139. Ries, P.; Betorello, H. Surface modification of PVC with biodegradable polymers, *J. Appl. Polym. Sci.*, 1997, **64**, 1195–1201.

140. Mecking, S. Nature or petrochemistry?—Biologically degradable materials, *Angew. Chem.*, 2004, **116**, 1096–1104; *Angew. Chem. Int. Ed.*, 2004, **43**, 1078–1085.

141. ERRMA, errma.com

142. Searles, R. *Something in the Cellar: Wonderful World of Wine*, Souvenir Press Ltd., London, 1986.

3

PHOTOINITIATED SYNTHESIS: A USEFUL PERSPECTIVE IN GREEN CHEMISTRY

ANGELO ALBINI

Università degli Studi di Pavia, Pavia, Italy

INTRODUCTION

It is apparent that a photochemical reaction has the advantage of using a green reagent, that is, light, in many cases, even natural light.[1] Thus, promoting a reaction photochemically rather than thermally is "greener." As an example, rearranging an aryl ester to a hydroxyarylketone by shining light in a vessel sounds much more acceptable from the ecological point of view than carrying out the same reaction by adding an equimolecular amount of an unpleasant and waste-producing reagent, such as aluminium trichloride. However, many aspects should be considered before concluding that this is a useful perspective for industrial application. Let us first of all explore the potentiality of the method.

Photochemical reactions involve electronically excited states.[2] These are high energy, if extremely short-lived, species. Thus, only reactions involving very small activation energies occur before physical decay to the ground state. The situation is completely different from that of thermal chemistry. In the latter case, the effort is toward "activating" the substrate, for example, using a strong nucleophile to make a reluctant (weak) electrophile react at a reasonable rate. Excited states are very energetic and have a different electronic distribution with respect to ground states. This often results in a new reaction coordinate becoming accessible; usually, however, this statement holds only for a single path, which

Methods and Reagents for Green Chemistry: An Introduction, Edited by Pietro Tundo, Alvise Perosa, and Fulvio Zecchini

involves surmounting a low enough activation energy, as the other energies are incompatible with its lifetime. The combination of high energy and the short life-time makes reactions from the excited state both extremely fast and quite selective.

The photochemical reactions of a given molecule differ from those in the ground state and usually are much less affected by the medium, since they are much faster. During the last decades, a large body of knowledge has emerged, and now photochemical reactions are well organized and classified, as is done in thermal chemistry, according to the functionality involved in the organic molecule.[2] Thus, these reactions have become part of the palette chemists can use when planning a synthesis, as in the examples presented in this chapter.

Many new paths are open. Will these be used in industrially applied processes? The answer is probably negative in the near future. The problem mainly lies in having the reagent absorbing light in the desired way. In the laboratory, photo-chemical reactions are usually carried out in a very simple apparatus—a cylindrical vessel into which the lamp is inserted—and using diluted solutions in order that light absorption takes place homogeneously across the vessel. To make the process acceptable for industry, much more concentrated solutions should be used, which requires the use of a more elaborate reactor, for example, a falling film reactor, where a thin film of the solution is irradiated, thus minimizing side processes. In addition, the reactor must be engineered for maximal light absorption; otherwise, car-rying out the reaction will be too expensive in terms of the use of electricity. At any rate, an appropriate reactor needs to be built and this is obviously not economically reasonable as long as thermal alternatives are available, even when these involve some environmental disadvantage. If a sufficiently large series of processes for which no such alternative exists are to emerge, then photoinitiated processes will be applied, as already occurs for a (very limited) number of examples. In the meantime, and in view of the fact that more stringent regulations will foster the development of all fields of green chemistry, photochemistry should at least be considered something more than a mere laboratory curiosity.

3.2 PHOTOCHEMICAL REACTIONS

Chemical reactions from the excited state are not the only—and not necessarily the most important—way in which a reaction can be initiated by light absorption. A completely different approach involves the generation of some activated, but ground state, intermediate through the photochemical step, thus leading to a photoinitiated reaction, even if (some of) the chemical steps are actually thermal reactions. The first group, perhaps scientifically trivial, but quite useful in the laboratory and industry, involves the generation of an inorganic radical, for example, chlorine atoms from Cl_2.[3] Besides this, new methods are increasingly being developed for the generation of the reactive intermediates from organic molecules. These methods are based on a sensitization process. This process may involve atom abstraction, based on the fact that several excited states have a strong diradicalic character, and thus may abstract an atom from a substrate, generating a pair of radical as under mild conditions. Hydrogen abstraction (from alcohols, acetals, even alkanes) is typical. The thus-formed radicals can be used for the alkylation of electrophilic alkenes.

Conversely, while, as has been mentioned already, each excited function has its own chemistry, there are two properties that all excited states share: easy reduction (because these states can accept an electron in the half-filled highest occupied molecular orbit (HOMO) and oxidation (because they can donate an electron from the half-filled lowest unoccupied molecular orbit (LUMO). Thus, redox chemistry, practically nonexistent in ground-state organic chemistry, is ubiquitous in organic photochemistry.[4] Choosing a sensitizer that has no photochemistry of its own and easily enters redox processes, typically an aromatic, allows electron transfer sensitization to be carried out. The thus-formed radical ions have an interesting chemistry, which currently is being explored, including addition reactions and fragmentation reactions. The latter process is a further entry to radicals, which, along with the atom abstraction method just given, allows unconventional radical precursors (e.g., alkyl aromatics, silanes, ethers) to be used rather than the usual precursors, highly toxic and waste-producing stannanes.[5]

A key difference between photochemical and thermal reactions is that electronically excited states are situated at a much higher energy than the corresponding ground states (Figure 3.1). Conversely, excited states have a short lifetime (from microseconds to lower than nanoseconds for organic molecules), due to competition with physical deactivation. Thus, photochemical reactions take place on a high-lying potential surface, but can only overcome very small (a few kcal/mol) activation barriers, as opposed to the large activation energy of thermal reactions.

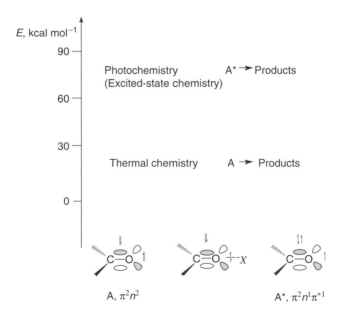

Figure 3.1 Energy profiles for photochemical and thermal reactions.

The other key difference is that the electronic structure is deeply changed in the excited states, and the energy of such states is 50 to 90 kcal M^{-1} for organic molecules, often making it possible that they react at rates typical of reactive intermediates, not saturated molecules. As an example, ketones are electrophiles (at the carbon atom) in the ground state. Actually these are weak electrophiles, and the use of a strong nucleophile or activation by a catalyst (an acid) is required in order to conveniently carry out a nucleophilic addition. Conversely, in the excited state ketones display a completely different reactivity, which can be likened to that of an alkoxy radical, both qualitatively (they undergo radical reactions at the oxygen atom) and quantitatively (e.g., hydrogen abstraction, from an alcohol, occurs with a rate constant around $10^6\ M^{-1}s^{-1}$)[2, 3] (Figure 3.2).

The different electron distribution in the excited state also may lead to other types of reactions. As an example, alkenes and polyenes display a low intermolecular reactivity, but undergo extremely fast rearrangements, since the π bonding character dramatically diminishes in the excited state. Thus, free rotation becomes feasible and, where appropriate, electrocyclic and sigmatropic processes take place[6] (Figure 3.3).

This is not to be taken as an indication that the π bonding character necessarily decreases in the excited state. Actually, there are molecules in which a π bond becomes stronger upon excitation. Indeed, oxygen has a weak diradicalic character in the (triplet) ground state, while it is a very strong π electrophile in the singlet excited state. It reacts in a way similar to electrophilic alkenes, but many orders of magnitude faster[7] (Figure 3.4).

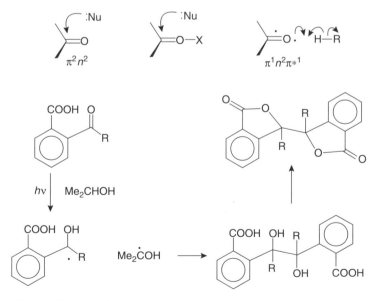

Figure 3.2   Reactions of an excited state: radical reactions of a ketone.

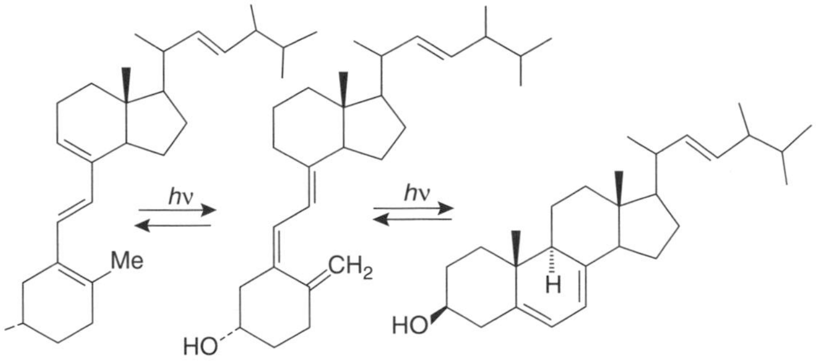

Figure 3.3 Reactions of an excited state: polyene isomerisation (previtamin D).

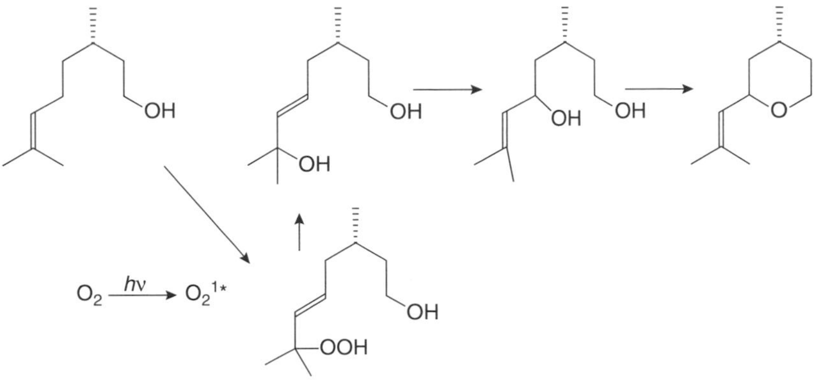

Figure 3.4 Reactions of an excited state: singlet oxygen (rose oxide).

As mentioned, electronically excited states are both easily reduced (since an electron can be accepted in the half vacant HOMO) and easily oxidized (since an electron can be donated from the half-filled LUMO) under much milder conditions than the corresponding ground states. As an example, reducing the ground state of benzophenone requires the use of a highly reactive reductant, such as sodium metal

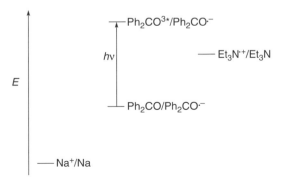

Figure 3.5 Reduction of benzophenone by sodium and of the excited state by Et$_3$N.

(to give the deep-blue radical anion, used as an indicator for anhydrous solvents). Conversely, a very mild reagent, such as triethylamine, is a good enough reducing agent for effecting the same reaction in the excited state (Figure 3.5).

3.3 PHOTOSENSITIZED REACTIONS

The previous examples involve direct photochemical reactions, namely, such that a molecule absorbs light and the thus formed excited-state reacts. In many cases, it may turn out to be advantageous to carry out a different experiment, in which another molecule absorbs the photon and then transfers electronic energy to the reagent (sensitization: the sensitizer does not undergo permanent chemical change and is recovered). In a more thorough alternative, the reaction does not proceed from the excited state, but rather the sensitizer activates the reagent by transforming it into a (ground state) highly reactive intermediate, such as a radical (via atom transfer sensitization) or a radical ion (electron transfer sensitization) (Figure 3.6).

A typical atom transfer sensitization process is shown in Figure 3.7: the alkylation of an unsaturated acid via a radical produced from an alcohol by hydrogen abstraction (in a green solvent, water).[8] This principle has been applied to a large series of radical alkylation reactions, where the radical precursor is an alcohol, an ether, or even an alkane.[9]

The hydrogen transfer photosensitization has been applied to the diastereoselective alkylation of chiral fumaric acid derivatives, where again the mild conditions of the photochemical method are advantageous (Figure 3.8).[10]

A convenient application uses an inorganic salt (e.g., a polyoxometallate)[11] or a semiconductor oxide[12] as the sensitizer. Such materials are often chemically more stable with respect to organic molecules, and again can be conveniently used for the generation of a radical from unconventional precursors.

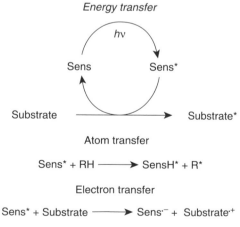

Energy transfer

hv

Sens   Sens*

Substrate   Substrate*

Atom transfer

Sens* + RH ⟶ SensH* + R*

Electron transfer

Sens* + Substrate ⟶ Sens·⁻ + Substrate·⁺

Figure 3.6 Various classes of photosensitization.

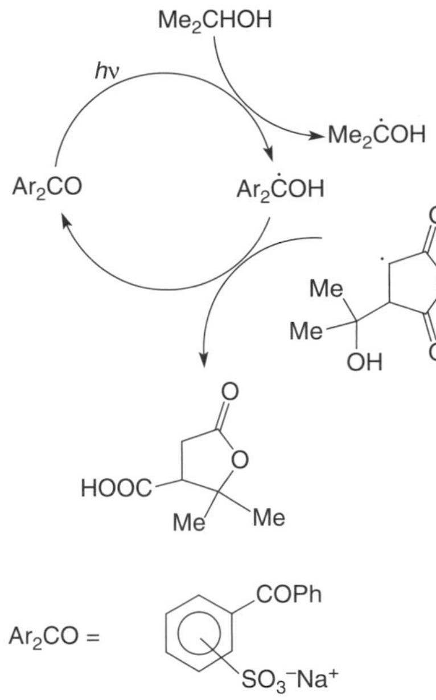

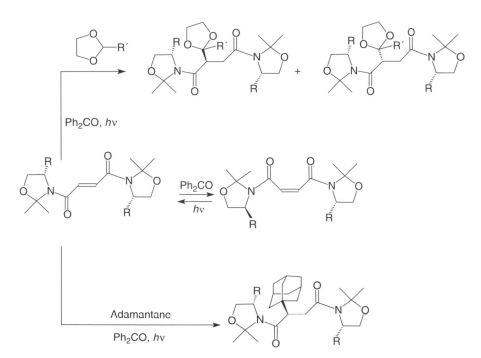

Figure 3.7 A typical atom transfer sensitization process.

Figure 3.8 Diastereoselective alkylation of chiral fumaric acid derivatives.

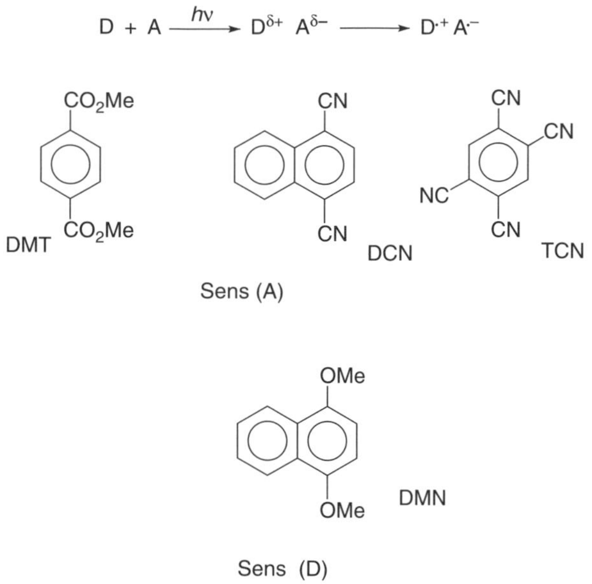

$$D + A \xrightarrow{h\nu} D^{\delta+} A^{\delta-} \longrightarrow D^{\cdot+} A^{\cdot-}$$

DMT

DCN

TCN

Sens (A)

DMN

Sens (D)

Figure 3.9   Photochemistry by electron transfer sensitization.

Several aromatic molecules undergo no efficient photochemistry of their own. Thus, these molecules are well suited for use as electron transfer sensitizers. According to the substrate and the conditions chosen, they may form either complexes (or tight radical ion pairs) or free, solvated radical ions (Figure 3.9).

Electron transfer sensitization allows either the radical cation or the radical anion of an aromatic alkene to form as desired, which finally results in nucleophile addition with Markovnikov and anti-Markovnikov regiochemistry. In an apolar solvent, the tight radical ion pair undergoes a stereoselective reaction when the electron-accepting sensitizer is chiral (Figure 3.10).[13]

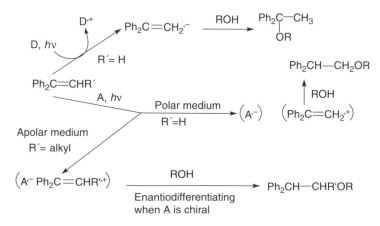

Figure 3.10   Regio- and stereoselective photosensitized addition via radical ions.

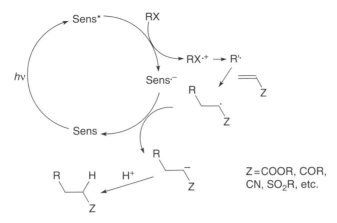

Figure 3.11 Alkylation via radicals generated through electron transfer sensitization: mechanism.

In other cases, the radical ion formed through electron transfer sensitization fragments (Figure 3.11). This is a new method for generating radicals from unconventional precursors under mild conditions.[5]

A range of examples of alkylation reactions via radicals generated through electron transfer sensitization is available in the literature, and a few of them are reported in Figure 3.12. Alkyl tin derivatives can be used as precursors, but in many cases these highly toxic reagents can be advantageously substituted by

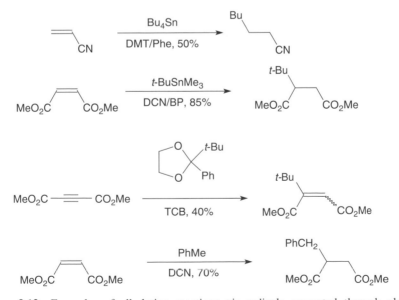

Figure 3.12 Examples of alkylation reactions via radicals generated through electron transfer sensitization.

silylated or by oxygenated precursors, such as alcohols, ethers, and acetals, as well as by alkylaromatics and alkylalkenes.[5,14]

The preceding examples give a hint of the synthetic versatility obtained in radical addition reactions by producing the key species through a photosensitized reaction. Several other classes of highly active reagents can be smoothly generated by a similar method, for example, carbenes or nitrenes, and the photochemical alternative is usually more versatile than the thermal counterpart, since the photochemically induced cleavage of the precursor is very fast and thus less affected by the experimental conditions. In other cases, photochemistry offers a convenient entry into a class of reactive intermediates otherwise inaccessible for all practical purposes, as recently illustrated by the case of aryl cations: once a photochemical method for their production has been found, a distinctive chemistry (selective addition to π nucleophiles) has been discovered.[15]

3.4 CONCLUSIONS

In conclusion, we reiterate that photoinitiated reactions have a dramatic advantage with respect to thermal reactions, since light is the green reagent par excellence. If we consider the possibility that such reactions will be applied in the foreseeable future, and in particular whether they will contribute to the development of green chemistry, we can conclude that there are both indisputable advantages and difficult to solve problems.

Pro: Photochemistry offers an inexhaustible source of new chemistry as follows:

- Distinctive chemical reactions of the excited state;
- A method for the mild generation of radicals (or other highly reactive intermediates) from unconventional (and green) precursors;
- A universal method for carrying out redox processes between organic molecules, which has no parallel in thermal chemistry;
- In every case, the photochemical step is not very dependent on conditions, affording a great deal of freedom of choice in arranging the experimental parameters (medium, additives), and thus allowing the resultant steps to be guided according to the plan.

Con: Significant limitations for an effective application are as follow:

- Photoinitiated reactions often only occur in a clean way at low concentrations, while it is difficult to have them this clean with a reduced amount of solvent;
- Light absorption must be maximized for industrial application, in order to avoid an unacceptable cost, and the knowledge of how to confront this point is not widespread;

• These reactions require an ad hoc apparatus, which make industry reluctant to invest in this direction unless large families of advantageous photochemical syntheses are developed.

REFERENCES

1. Albini, A.; Fagnoni, M.; Mella, M. *Pure Appl. Chem.*, 2000, **72**, 1321. Albini, A.; Fagnoni, M. *Green Chem.*, 2004, **6**, 1.

2. Horspool, W. H.; Song, P. S. (Eds.) *CRC Handbook of Organic Photochemistry and Photobiology*, CRC Press, Boca Raon, FL, 1995. Horspool, W. H.; Armesto, D. *Organic Photochemistry: A Comprehensive Treatment*, Ellis Horwood, New York, 1992.

3. Braun, A. M.; Maurette, M. T.; Oliveros, E. *Photochemical Technology*, Wiley, Chichester, UK, 1995.

4. Fox, M. A.; Chanon, M. (Eds.) *Photoinduced Electron Transfer*, Elsevier, Amsterdam, 1988. Mattay, J. (Ed.) *Top. Curr. Chem.*, 1990, 156; 1990, 158; 1991, 159; 1992, 163; 1993, 168.

5. Albini, A.; Mella, M.; Freccero, M. *Tetrahedron*, 1994, **50**, 575. Mella, M.; Fagnoni, M.; Freccero, M.; et al. *Chem. Soc. Rev.,* 1998, **27**, 81. Albini, A.; Fagnoni, M. *CRC Handbook Photochemistry and Photobiology*, CRC, Press, Boca Raton, FL, 2004, p. 4.

6. Jacobs, H. J. C.; Havinga, E. *Adv. Photochem.*, 1979, **11**, 305.

7. Ohloff, G.; Klein, E.; Schenck, G. O. *Angew. Chem*, 1961, **73**, 578. Rojahn, W.; Warnecke, H. U. *Dragoco Reps.*, 1980, **27** (6/7), 159.

8. Schenck, G. O.; Koltzenburg, G.; Grossmann, H. *Angew. Chem.*, 1957, **69**, 177.

9. Gonzalez-Cameno, A. Am.; Mella, M.; Fagnoni, M.; Albini, A. *J. Org. Chem.*, 2000, **65**, 297. Dondi, D.; Caprioli, I., Fagnoni, M.; Mella, M.; Albini, A. *Tetrahedron*, 2003, **59**, 947.

10. Campari, G.; Fagnoni, M.; Mella, M.; Albini, A. *Tetrahedron Asymmetry*, 2000, **11**, 1891. Marinkovic, S.; Hoffmann, N. *Eur. J. Org. Chem.*, 2004, 3102.

11. Hill, C. L. *Synlett*, 1995, 127. Dondi, D.; Fagnoni, M.; Molinari, A.; et al. *Chemistry*, 2004, **10**, 142.

12. Kisch, H. *Adv. Photochem.*, 2001, **26**, 93. Cermenati, L.; Richter, C.; Albini, A. *Chem. Commun.*, 1998, 805. Cermenati, L.; Dondi, D.; Albini, A. *Tetrahedron*, 2003, **59**, 6409.

13. Arnold, D. R.; Maroulis, A. J. *J. Am. Chem. Soc.*, 1976, **98**, 5931. Asaoka, S.; Kitazawa, T.; Wada, T.; Inoue, Y. *J. Am. Chem. Soc.*, 1999, **121**, 8486.

14. Fagnoni, M.; Mella, M.; Albini, A. *J. Org. Chem.*, 1998, **63**, 4026. Mella, M.; Fagnoni, M.; Albini, A. *Eur. J. Org. Chem.*, 1999, 2137.

15. Guizzardi, B.; Mella, M.; Fagnoni, M.; Freccero, M.; Albini, A. *J. Org. Chem.*, 2001, 66, 6344. Freccero, M.; Fagnoni, M.; Albini, A. *J. Am. Chem. Soc.*, 2003, **125**, 13181.

4

DIMETHYL CARBONATE AS A GREEN REAGENT

PIETRO TUNDO AND MAURIZIO SELVA

The Ca' Foscari University of Venice and National Interuniversity Consortium, "Chemistry for the Environment," Venice, Italy

INTRODUCTION

To overcome health and environmental problems at the source, the chemical industry must develop cleaner chemical processes by the design of innovative and environmentally benign chemical reactions. *Green chemistry* offers the tools for this approach.[1]

Green organic syntheses must meet, if not all, at least some of the following requirements: avoid waste,[2] be atom efficient,[3] avoid the use and production of toxic and dangerous chemicals, produce compounds that perform better or as well as existing ones and are biodegradable, avoid auxiliary substances (e.g., solvents) or use eco-compatible solvents (water or dense CO_2), reduce energy requirements, use renewable materials, and use catalysts rather than stoichiometric reagents.[4]

To focus on the more specific area of the replacement of harmful and undesirable compounds, a relevant case is given by methyl halides (CH_3X, $X = I$, Br, Cl), dimethylsulfate (DMS), and phosgene ($COCl_2$), which are reagents used for methylation and carbonylation reactions, respectively. For instance, Scheme 4.1 shows the methylation of phenol by CH_3X and DMS to give anisole, and the alkoxycarbonylation of an alcohol by $COCl_2$. These toxic and waste-producing reagents have a valuable green alternative in dimethylcarbonate (DMC).

Methods and Reagents for Green Chemistry: An Introduction, Edited by Pietro Tundo, Alvise Perosa, and Fulvio Zecchini
Copyright © 2007 John Wiley & Sons, Inc.

$$PhOH + (CH_3)_2SO_4 + NaOH \longrightarrow PhOCH_3 + NaCH_3SO_4 + H_2O$$

$$PhOH + CH_3I + NaOH \longrightarrow PhOCH_3 + NaI + H_2O$$

$$2ROH + COCl_2 + 2NaOH \longrightarrow ROCOOR + 2NaCl + 2H_2O$$

Scheme 4.1 Methylation and alkoxycarbonylation using DMS, CH_3I, and $COCl_2$.

$$PhOH + CH_3OCOOCH_3 \xrightarrow{\text{Cat. base}} PhOCH_3 + CO_2 + CH_3OH$$

$$ROH + CH_3OCOOCH_3 \xrightarrow{\text{Cat. base}} ROCOOCH_3 + CH_3OH$$

Scheme 4.2 Methylation and methoxycarbonylation using DMC.

Since 1980, with the development of gas liquid phase transfer catalysis (GL-PTC) as a continuous-flow (CF) method for organic syntheses,[5] our group has had a long-standing interest in the use of DMC as an environmentally friendly substitute for DMS and CH_3X in methylation reactions, and for phosgene in methoxycarbonylation reactions (Scheme 4.2). Among the specific advantages of DMC, and of alkyl carbonates in general, is that their building block is CO_2, an environmentally benign compound, which does not cause emissions of volatile organic compounds (VOC) in the atmosphere.

This chapter reports on the reactivity of organic carbonates as alkylating agents, with emphasis on the lightest term of the series, DMC. Under both CF and batch conditions, DMC can react with a number of nucleophilic substrates such as phenols, primary amines, sulfones, thiols, and methylene-active derivatives of aryl and aroxy-acetic acids. The mechanistic and synthetic aspects of these processes will be elucidated.

4.1 PROPERTIES OF DMC

Many of the properties of DMC make it a genuinely *green* reagent, particularly if compared to conventional alkylating agents such as methyl halides (CH_3X) and DMS or to phosgene used as a methoxycarbonylating reagent.

1. First of all, DMC is a nontoxic compound:[6] since the middle 1980s, in fact, it has not been produced from phosgene, but by catalytic oxidative carbonylation of methanol with oxygen through a process developed by Enichem (Italy)[7] and UBE Industries (Japan) (Scheme 4.3):[8]

In addition to improving procedural safety, this method of producing DMC avoids contamination from phosgene, and eliminates the need to dispose of by-product inorganic salts.

$$2CH_3OH + 1/2O_2 + CO \xrightarrow{\text{Cu salts}} CH_3OCOOCH_3 + H_2O$$

Scheme 4.3 Enichem synthesis of DMC.

An industrial procedure recently developed by Jiangsu Winpool Industrial Co., China (more than 10,000 t/y), is the cleavage of cyclic carbonates (Scheme 4.4). Importantly, syntheses like this do not use chlorine.[9]

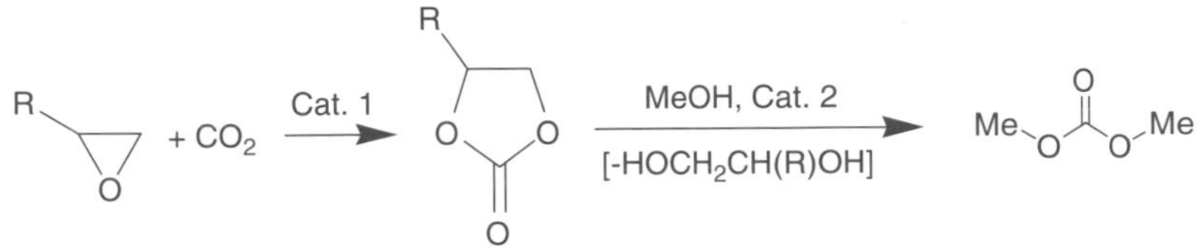

Scheme 4.4 Insertion of CO_2 into epoxides and cleavage of cyclic carbonates. Step 1. Catalyst: MgO, CaO. Step 2. Catalyst: zeolites exchanged with alkali and/or earth metal ions.

Some of the toxicological properties of DMC and phosgene and DMS are compared in Table 4.1.

TABLE 4.1 Comparison Between the Toxicological and Eco-Toxicological Properties of DMC, Phosgene, and DMS

Properties	DMC	Phosgene	DMS
Oral acute toxicity (rats)	LD_{50} 13.8 g/kg		LD_{50} 440 mg/kg
Acute toxicity per contact (cavy)	$LD_{50} > 2.5$ g/kg		
Acute toxicity per inhalation (rats)	LC_{50} 140 mg/L; (4 h)	LC_{50} 16 mg/m^3 (75 min)	LC_{50} 1.5 mg/L (4 h)
Mutagenic properties	None		Mutagenic
Irritating properties (rabbits, eyes, skin)	None	Corrosive	
Biodegradability (OECD 301 C)	>90% (28 days)	Rapid hydrolysis	Rapid hydrolysis
Acute toxicity (fish) (OECD 203)	NOEC 1000 mg/L		LC_{50} 10–100 mg/L (96 h)
Acute toxicity on aerobial bacteria of waste waters (OECD 209)	$EC_{50} > 1000$ mg/L		

Note: NOEC = Concentration that does not produce any effect.

TABLE 4.2   Some Physical and Thermodynamic
Properties of DMC

Melting point (°C)	4.6
Boiling point (°C)	90.3
Density (d_4^{20})	1.07
Viscosity (μ^{20}, cps)	0.625
Flashing point (°C, O.C.)	21.7
Dielectric constant (ε^{25})	3.087
Dipol moment (μ, D)	0.91
ΔH vap (kcal/kg)	88.2
Solubility H_2O (g/100 g)	13.9
Azeotropical mixtures	With water, alcohols, hydrocarbons

2. DMC is classified as a flammable liquid, smells like methanol, and does not have irritating or mutagenic effects, either by contact or inhalation. Therefore, it can be handled safely without the special precautions required for the poisonous and mutagenic methyl halides and DMS, and extremely toxic phosgene. Some physicochemical properties of DMC are listed in Table 4.2.

3. DMC exhibits a versatile and tunable chemical reactivity that depends on the experimental conditions. In the presence of a nucleophile (Y^-), DMC can react either as a methoxycarbonylating or as a methylating agent (Scheme 4.5).[10]

Although there is not always a clear cutoff between the two pathways of Scheme 4.5, it is generally observed that:

i. At the reflux temperature ($T = 90°C$), DMC acts primarily as a methoxycarbonylating agent by a $B_{AC}2$ (bimolecular, base-catalyzed, acyl cleavage, nucleophilic substitution) mechanism where the nucleophile attacks the

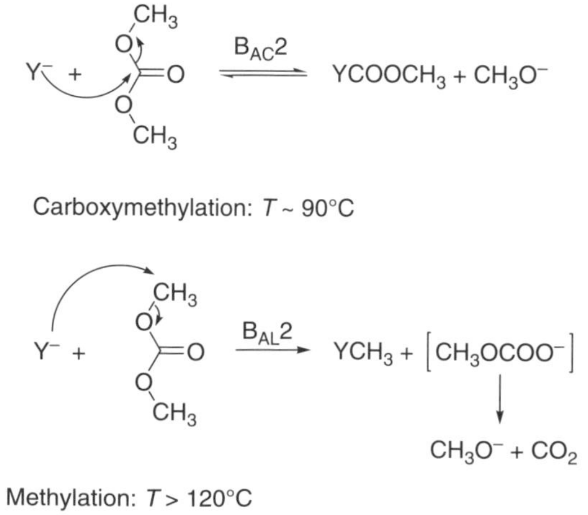

Carboxymethylation: $T \sim 90°C$

Methylation: $T > 120°C$

Scheme 4.5   Nucleophilic substitution on DMC by $B_{AC}2$ and $B_{AL}2$ mechanisms.

carbonyl carbon of DMC, giving the transesterification product. Under these conditions, DMC can replace phosgene;

ii. At higher temperatures (usually $T \geq 160°C$), DMC acts primarily as a methylating agent: a $B_{AL}2$ (bimolecular, base-catalyzed, alkyl cleavage, nucleophilic substitution) mechanism predominates where the nucleophile attacks the methyl group of DMC.

Of the two, only the methylation reaction is irreversible, because the CH_3OCO_2H that is formed decomposes to methanol and CO_2.

Since both methylation and methoxycarbonylation generate CH_3O^-, both reactions can be conducted in the presence of catalytic amounts of base. This avoids the formation of unwanted inorganic salts as by-products, and the related disposal problems. In principle, the methanol produced can be recycled for the production of DMC.[11] In contrast, methylation with methyl halides or DMS, and carbonylation with phosgene all generate stoichiometric amounts of inorganic salts.

4.2 REACTION CONDITIONS

The development of a new eco-friendly process is often associated with advanced reaction technologies. An aspect that sometimes imposes a careful balance between the environmentally benign character and the economic/safety feasibility of the process itself. The use of supercritical CO_2 (scCO_2) is an example: scCO_2 is among the most attractive green alternatives to replace conventional solvents, though its handling requires high-pressure operations (usually at $P > 130$ bar) which are energy-consuming and potentially dangerous.[12] This approach becomes a green solution only when real chemical benefits (higher selectivities, rates, yields) are achievable in the supercritical fluid.

Also, in the case of DMC, reaction conditions apparently are not green: the methylating ability of DMC can be exploited at a temperature $>160°C$ (above the boiling point of DMC itself: $90°C$), which implies an autogenic pressure (>3 bar) for batchwise processes. Such conditions are not prohibitive, however, especially according to the industrial practice, where pressures up to $20-30$ bar and temperatures up to $250°C$ are not a concern.

Moreover, from the environmental standpoint, advantages must be seen on the global balance: the DMC-mediated alkylation reactions are much safer than any other alkylation method known that uses conventional reagents. Not only the features of the reaction itself, but the peculiarity of the reagent(s), the base (truly catalytic), and the absence of wastes are also key aspects.

Discontinuous (batch) processes are carried out in pressure vessels (autoclaves) where DMC is maintained as liquid by autogenous pressure. Instead, CF reactions at atmospheric pressure require that both DMC and the reagent(s) in the vapor phase come into contact with a catalytic bed: a constraint that has spurred the development of new applications and alternative reaction engineering, namely, GL-PTC and the continuously fed stirred-tank reactor (CSTR).

Accordingly, under different conditions, DMC is used as a methylating reagent for a variety of substrates: phenols, thiols, thiophenols, aromatic amines, arylacetonitriles, arylacetoesters, aroxyacetonitriles, aroxyacetoesters, alkylarylsulfones, benzylarylsulfones, and lactones, either under CF or in batch conditions.

4.2.1 Continuous Flow: Plug-Flow and CSTR Reactors

Under GL-PTC conditions, a gaseous stream of reagent and DMC flows over a catalytic bed usually composed of a porous inorganic material (usually corundum in the form of a spherical extrudate of 1–3 mm of diameter), which acts as a support for both an inorganic base (an alkaline carbonate) and a phase-transfer (PT) agent such as phosphonium salts,[13] crown ethers,[14] and polyethylene glycols (PEGs). These latter in particular, although less efficient than other PT agents, are desirable because they are thermally stable, nontoxic, and inexpensive.[15]

In a typical configuration, the CF methylation reaction with DMC takes place in a plug-flow reactor made by a bed of K_2CO_3 coated with PEG 6000 (0.5–5% mol. eq.), and heated to 160–180°C.[5,16] A mixture of DMC and substrate (YH) is fed into the reactor where the base generates the reactive nucleophilic anion (Y^-) from the substrate. The role of the PT agent is to complex the alkaline metal cation, thereby increasing the basic strength of the solid carbonate. As shown in the Scheme of Figure 4.1, the immobilized PT agent is in the liquid phase throughout the reaction, and allows the continuous transfer of the products and reactants between the gas and liquid phases. The methylated product is then condensed and collected at the other end.

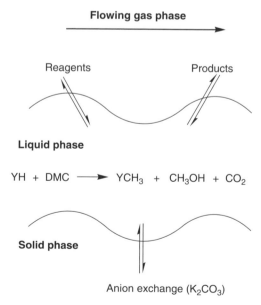

Figure 4.1 General picture of gas liquid phase transfer catalysis (GL-PTC).

TABLE 4.3   Reactions of DMC with Different Nucleophiles Under GL-PTC Conditions

Entry	Reagent	Product, Yield %
1	ArOH	ArOMe, 100
2	ArSH	ArSMe, 100
3	ROH	$ROCO_2Me + (RO)_2CO$
4	$PhCH_2CN$	$PhCH(Me)CN$, 98
5		

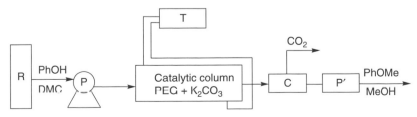

Ibuprofen precursor

Conditions: GL-PTC, plug-flow reactor, catalyst: K_2CO_3 coated with 0.5–5 mol % of PEG 6000, $T = 160–180°C$.

Quantitative conversions are obtained from all the substrates listed in Table 4.3.[4,16,17] Moreover, in the case of CH_2-active compounds, the reaction proceeds with a mono-methyl selectivity >99% (entries 4 and 5). An example reaction, shown in Figure 4.2, is the methylation of phenol under GL-PTC conditions.

In a typical experimental procedure, when a mixture of phenol (94 g, 1 mol) and DMC (2.0 mol) is made to flow over a 100-g catalytic bed composed of 95 wt % K_2CO_3 g and 5 wt % PEG 6000 at 180°C, pure anisole is recovered with a 100% yield in 1 h (residence time ~10 s).[17a,17b] Pyrocatecol and hydroquinone can also be selectively mono- or di-alkylated under CF conditions in a pilot plant scale.[10]

In Table 4.3, it should be noted, however, that hard alkoxide anions (RO^-) react with DMC via a $B_{Ac}2$ mechanism to yield exclusively transesterification products ($ROCO_2Me$) with no trace of methyl ethers (entry 3). Such a peculiar selectivity is currently under investigation.

An alternative CF methodology for DMC methylations was developed as well, by using a CSTR (Figure 4.3).[18] In this configuration, the catalyst fills the reactor in the form of a liquid slurry of the PT agent (usually PEG 1000) and K_2CO_3, and it is kept under vigorous stirring at the desired temperature (160–200°C). The mixture of DMC and the reagent is vaporized when it comes into contact with the

Figure 4.2   CF methylation of phenol in a plug-flow reactor under GL-PTC conditions. R: reagent's reservoir; P: metering pump; T: thermostat; C: condenser; P': product store.

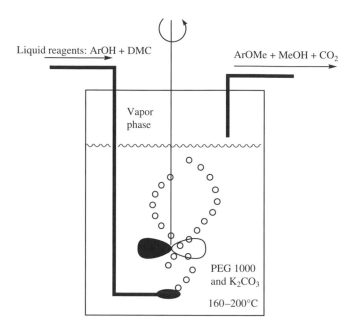

Figure 4.3 Schematic chart of a CSTR reactor for the *O*-methylation of phenols with DMC. Liquid reagents are vaporized by contact with the hot slurry (mechanically stirred) and bubbled through it. Reaction takes place instantaneously and anisoles are picked up from the vapor phase.

catalytic bed. When a suitable feeding rate is chosen, the apparatus works under gradientless conditions: the composition of the mixture collected at the outlet is equal to that present inside the reactor, and the reaction takes place instantaneously as far as reagents bubble through the bed. At atmospheric pressure, different phenols are transformed quantitatively into the corresponding anisoles with a weight hourly space velocity (WHSV) up to $9.5 \times 10^{-2} \, h^{-1}$.[10] The reaction could run without interruption for at least two weeks. Some results are listed in Table 4.4.

TABLE 4.4 The *O*-Methylation of Different Phenols (ArOH) with DMC in a CSTR[a]

ArOH, Ar	T (°C)	Flow Rate (mL/h)	Time of Flowing (h)	Substrate Converted (g)	Product, ArOMe (%)	WHSV × 100, g_{prod}/g_{bed} (h)
Ph	200	80	2.5	97.5	97	8.4
p-MeC$_6$H$_4$	160	80	4	128.0	98	9.5

[a]Reactions carried out over a catalytic bed of PEG 1000 (300 g) and K_2CO_3 (6 g). Molar ratios ArOH:DMC were 1:1.05 and 1:1.5 for PhOH and *p*-MeC$_6$H$_4$OH, respectively. WHSV: grams of ArOMe obtained hourly per gram of catalyst.

4.2.2 Batch Methylation Reactions with DMC

Under batch conditions, methylations with DMC must necessarily be run in sealed autoclaves, given its boiling point (90°C) and the reaction temperature (>160°C). Batch methylations with DMC can be performed on a number of different substrates and, under such conditions, the reaction mechanism can be conveniently investigated: in fact, the sampling of the reaction mixture at different conversions, and the identification of possible intermediates (see later) is easier with respect to CF-processes. For compounds that are susceptible to multiple methylation, the results are of special interest, since methylation with DMC totally inhibits multiple substitution in both N- and C-alkylation, for primary aromatic amines and for CH_2-active compounds, respectively.

The most interesting and best-studied reaction, particularly in view of its selectivity, is the mono-C-methylation of arylacetonitriles (Scheme 4.6).

In the presence of a weak base (usually K_2CO_3), these compounds can be

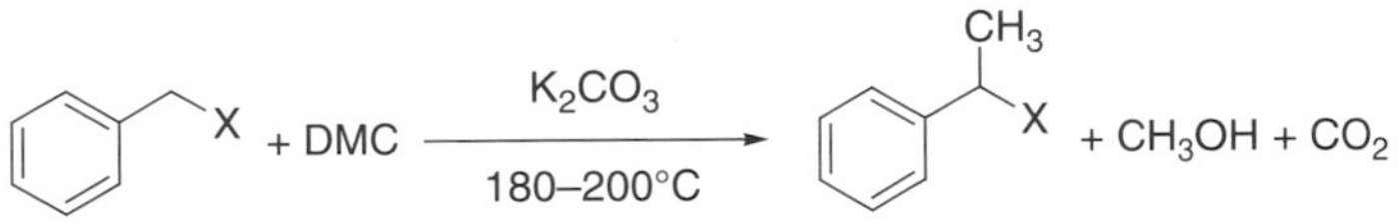

Scheme 4.6 Mono-C-methylation of arylacetonitriles (X = CN).

effectively mono-C-methylated with an unprecedented selectivity—greater than 99%—at complete conversions. For instance, when a mixture of $PhCH_2CN$, DMC, and K_2CO_3 in a 1:20:2 molar ratio, respectively, was allowed to react at 180°C for 3.75 h, 2-phenylpropionitrile was obtained in a 95% yield with a purity >99%.[19] For comparison, in the same reaction run under PTC conditions and using CH_3I, the mono- to dimethyl-derivative ratio never exceeded 2.4.[20] The very high monomethyl selectivity has turned out to be much interesting for the synthesis of precursors for anti-inflammatory drugs: some examples are listed in Table 4.5.

Similarly, in the presence of weak inorganic bases (K_2CO_3), the reactions of DMC with sulfones bearing α-methylenic groups (RCH_2SO_2R'; R = Alkyl, Aryl; R' = Aryl) afford the respective mono-C-methylated compounds [$RCH(CH_3)SO_2R'$] with >99% selectivity at complete conversions (Scheme 4.7).[21]

TABLE 4.5 Mono-C-Methylation of Arylacetonitriles

X	Ar	Conv. (%)	Selectivity in Mono-C-Methylation	Intermediate for
CN	4-Isobutylphenyl	99	99	Ibuprofen
CN	3-Carboxymethylphenyl	100	>99	Ketoprofen
COOCH$_3$	2-(6-Methoxynaphtyl)	100	>99	Naproxen

Conditions: Autoclave, substrate:DMC:K_2CO_3 = 1:18:2 molar ratio, T = 180–220°C.

$$RCH_2SO_2Ar + CH_3OCOOCH_3 \xrightarrow[180-210°C]{K_2CO_3} RCH(CH_3)SO_2Ar + CH_3OH + CO_2$$

Scheme 4.7 Mono-*C*-methylation of alkylarylsulfones.

It is noteworthy that in the case of methyl sulfones (ArSO$_2$Me), the reaction proceeds with the homologation of the methyl to an *i*-propyl group: that is, the methylation is still highly selective toward the substitution of only two of the three methyl protons. Aside from the synthetic results, this observation is relevant from the mechanistic viewpoint, as will be clarified in the discussion of Scheme 4.8.

Under the same conditions (batch or GL-PTC) discussed for CH$_2$-acidic compounds, primary aromatic amines also react with DMC. In this case, although the reaction yields selectively the mono-*N*-methylated amines with no dimethylated by-products, sizable amounts of methyl carbamates (ArNHCO$_2$Me) are formed.[10,17c] Much better results can be gathered in the presence of zeolites, particularly alkali metal exchanged Y and X faujasites. These aluminosilicates posses pseudospherical cavities (supercavities) of 11–8 Å in diameter, which can be accessed through channels whose size is 7.4 Å.[22]

In the presence of these solid catalysts, different anilines—even deactivated by both electronic and steric effects—yield the corresponding mono-*N*-methyl derivatives [ArNHMe] with selectivities of 93–98%, at conversions up to 95% (Scheme 4.8).[23]

A further added-value of the combined use of DMC and zeolites, is that derivatization (protection/deprotection) sequences may be avoided. In particular, in

$$XC_6H_4NH_2 + CH_3OCOOCH_3 \xrightarrow[130-150°C]{Y\text{-}Zeolite} XC_6H_4NHCH_3 + CH_3OH + CO_2$$

XC$_6$H$_4$NH$_2$, (X)	Catalyst	T (°C)	t (min)	Cat. : Substrate[a]	ArNHMe (Yield, %)	Selectivity, % (Mono/di)[b]
	NaY	130	195	1.2	84	98
H	K2CO3	180	220	1.8	13	87[c]
	–	150	200	–	2	100
p-NO$_2$	KY	150	600	3.3	79	93
p-CN	KY	150	270	3.3	83	98
o-CO$_2$Me	NaY	150	330	3.3	84	96
2,6(Me)$_2$	NaY	150	300	3.3	76	94

[a]Weight quotient between the catalyst and the amine.
[b]The % selectivity is calculated with the expression {[ArNHMe]/([ArNHMe] + [ArNMe$_2$])} × 100.
[c]Thirty-four percent of PhN(Me)CO$_2$Me was a by-product.

Scheme 4.8 Mono-*N*-methylation of aromatic amines.

the presence of sodium-exchanged Y-zeolite (NaY faujasite) as a catalyst, the reaction of DMC with ambient nucleophiles such as aminophenols, -benzylalcohols, -benzoic acids, and -benzamides not only shows a very high mono-*N*-methyl selectivity (up to 99%), but it proceeds with complete chemoselectivity toward the amino group (Scheme 4.9).[24]

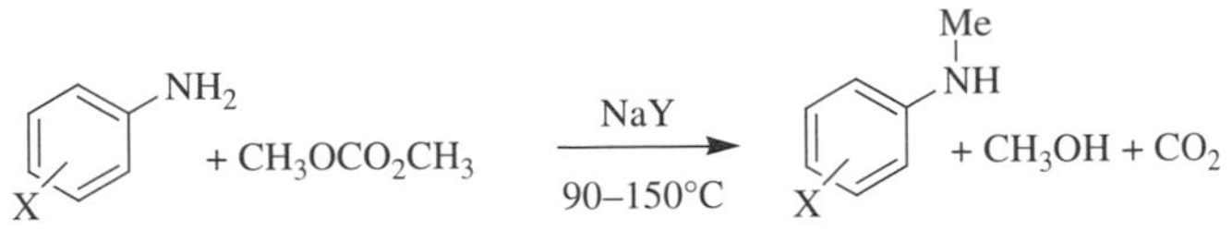

Substrate	T (°C)	Conv, (%)	ArNHMe, Isolated Yield, (%)
1a: X = *p*-OH	90	100	91
1b: X = *o*-OH	90	100	97
2a: X = *p*-CH$_2$OH	90	90	77
2b: X = *o*-CHOH$_2$	90	99	92
3a: X = *p*-CO$_2$H	130	100	74
3b: X = *o*-CO$_2$H	150	95	83
4a: X = *p*-CONH$_2$	90	96	86
4b: X = *o*-CONH$_2$	90	100	91

Scheme 4.9 The selective methylation of ambident aromatic amines.

The other nucleophilic functionalities (OH, CO$_2$H, CH$_2$OH, CONH$_2$) are fully preserved from alkylation and/or transesterification reactions.

In summary, all the nucleophiles indicated up to now are efficiently methylated (and monomethylated, where applicable) with DMC, both under CF and batch conditions.

4.3 MONOMETHYLSELECTIVITY: THE REACTION MECHANISM

4.3.1 CH$_2$-Acidic Compounds

The methylation of arylacetic acid derivatives is chosen as a model reaction for the mechanistic discussion. Experimental evidence of DMC-mediated alkylation of ArCH$_2$X (X = CN, CO$_2$Me) with DMC supports the hypothesis that the reaction does not proceed through a S_N2 displacement of the ArCH$^{(-)}$X nucleophile on DMC (B$_{Al}$2 mechanism).[19a] Rather, the selectivity arises from consecutive reactions involving two intermediate species observed during the reaction: $ArCH(CO_2Me)X$ (**3**) and $ArC(CH_3)(CO_2Me)X$ (**4**).

As an example, Figure 4.4 depicts the outcome of the mono-*C*-methylation of *o*-tolylacetonitrile with DMC: the two compounds o-$CH_3C_6H_4CH(CO_2Me)CN$ and o-$CH_3C_6H_4C(Me)(CO_2Me)CN$ (shown as **c** and **d**, respetively) are the reaction intermediates. In general, however, intermediates **3** are very reactive moieties (as confirmed by the kinetic investigation of the reaction, see later), whose identification in the reaction mixture very often fails. In the case of Figure 4.4, it is probably the steric hindrance of an *o*-Me substituent that affects the reactivity of compound **c**. The pattern for the reaction mechanism is outlined in Scheme 4.10.

Initially, the carbanion $[ArCH^{(-)}X]$ undergoes a methoxycarbonylation reaction by attacking the acyl carbon of DMC ($B_{Ac}2$ mechanism). The resulting intermediate $[ArCH(CO_2Me)X, (\mathbf{3})]$ reacts through its anion $[ArC^{(-)}(CO_2Me)X, (\mathbf{3^-})]$ with the alkyl carbon of DMC to yield the corresponding methyl derivative $[ArC(CH_3)(CO_2Me)X, (\mathbf{4}); B_{Al}2$ mechanism]. Finally, compound **4** is subjected to a demethoxycarbonylation reaction to produce the final product $[ArCH(CH_3)X]$.

This mechanism also applies to other CH_2-active compounds such as derivatives of aroxyacetic acid and benzylic sulfones, whose methylation reactions with

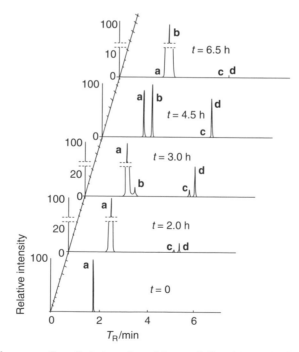

Figure 4.4 The mono-*C*-methylation of *o*-tolylacetonitrile with DMC. Gaschromatograms refer to different reaction times. (**a**) o-$CH_3C_6H_4CH_2CN$; (**b**) o-$CH_3C_6H_4CH(CH_3)CN$; (**c**) o-$CH_3C_6H_4CH(CO_2Me)CN$; (**d**) o-$CH_3C_6H_4C(Me)(CO_2Me)CN$.

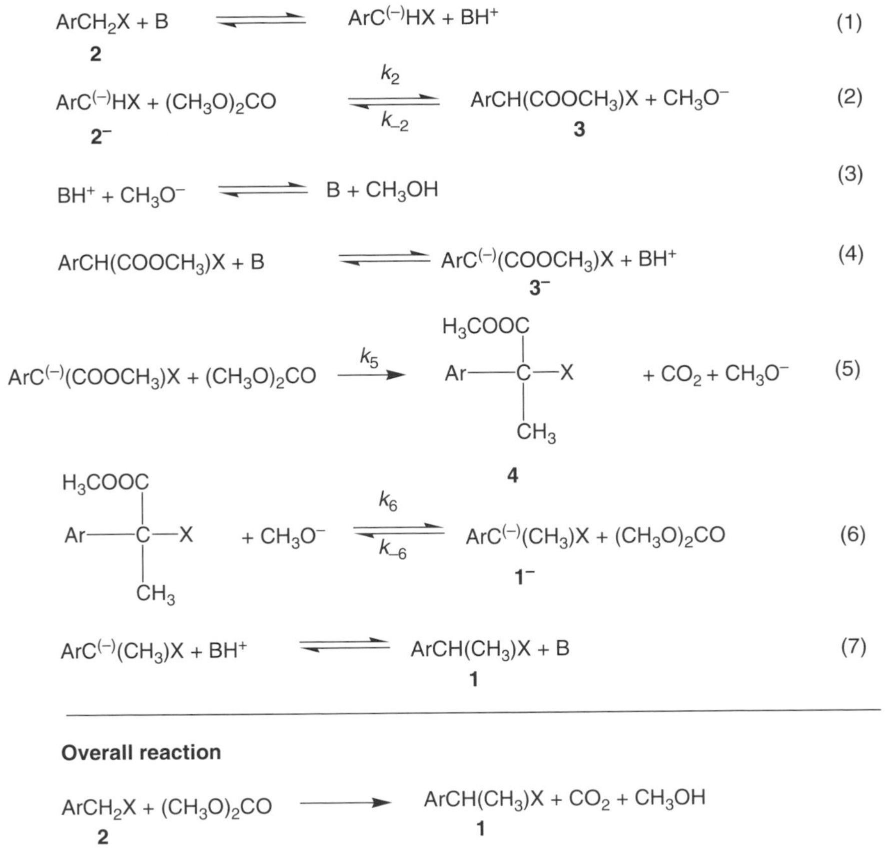

$$\text{ArCH}_2\text{X} + \text{B} \;\rightleftharpoons\; \text{ArC}^{(-)}\text{HX} + \text{BH}^+ \qquad (1)$$
$$\mathbf{2}$$

$$\text{ArC}^{(-)}\text{HX} + (\text{CH}_3\text{O})_2\text{CO} \;\underset{k_{-2}}{\overset{k_2}{\rightleftharpoons}}\; \text{ArCH(COOCH}_3)\text{X} + \text{CH}_3\text{O}^- \qquad (2)$$
$$\mathbf{2^-} \hspace{6cm} \mathbf{3}$$

$$\text{BH}^+ + \text{CH}_3\text{O}^- \;\rightleftharpoons\; \text{B} + \text{CH}_3\text{OH} \qquad (3)$$

$$\text{ArCH(COOCH}_3)\text{X} + \text{B} \;\rightleftharpoons\; \text{ArC}^{(-)}(\text{COOCH}_3)\text{X} + \text{BH}^+ \qquad (4)$$
$$\mathbf{3^-}$$

$$\text{ArC}^{(-)}(\text{COOCH}_3)\text{X} + (\text{CH}_3\text{O})_2\text{CO} \;\overset{k_5}{\longrightarrow}\; \underset{\underset{\text{CH}_3}{|}}{\overset{\overset{\text{H}_3\text{COOC}}{|}}{\text{Ar}-\text{C}-\text{X}}} + \text{CO}_2 + \text{CH}_3\text{O}^- \qquad (5)$$
$$\mathbf{4}$$

$$\underset{\underset{\text{CH}_3}{|}}{\overset{\overset{\text{H}_3\text{COOC}}{|}}{\text{Ar}-\text{C}-\text{X}}} + \text{CH}_3\text{O}^- \;\underset{k_{-6}}{\overset{k_6}{\rightleftharpoons}}\; \text{ArC}^{(-)}(\text{CH}_3)\text{X} + (\text{CH}_3\text{O})_2\text{CO} \qquad (6)$$
$$\mathbf{1^-}$$

$$\text{ArC}^{(-)}(\text{CH}_3)\text{X} + \text{BH}^+ \;\rightleftharpoons\; \text{ArCH(CH}_3)\text{X} + \text{B} \qquad (7)$$
$$\mathbf{1}$$

Overall reaction

$$\text{ArCH}_2\text{X} + (\text{CH}_3\text{O})_2\text{CO} \;\longrightarrow\; \text{ArCH(CH}_3)\text{X} + \text{CO}_2 + \text{CH}_3\text{OH}$$
$$\mathbf{2} \hspace{7cm} \mathbf{1}$$

Scheme 4.10 Mechanism of the mono-C-methylation of CH_2-active compounds ($X = CN$, CO_2CH_3) with DMC.

DMC proceed through the corresponding methoxycarbonylated and methyl methoxy carbonylated intermediates [$\text{WCH(CO}_2\text{Me)X}$ and $\text{WC(Me)(CO}_2\text{Me)X}$; $W = ArO$, $X = CN$, CO_2Me; $W = Ar$, $X = SO_2R$].

In the particular case of methyl sulfones (ArSO_2Me), the pathway of Scheme 4.10 also accounts for the previously mentioned homologation of the methyl to i-propyl group. In fact, once the methoxycarbonylated compound $\text{ArSO}_2\text{CH}_2\text{CO}_2\text{Me}$ is formed, only two protons remain available for the next methylation step.

In order to investigate in more depth the mechanism of Scheme 4.10, a detailed kinetic analysis has been performed by choosing the methylation of phenylacetonitrile (**2a**) and methyl phenylacetate (**2b**) with DMC as model reactions.[25] Some general considerations are the following.

In the case of compound **2a**, the rate-determining step of the overall transformation is the methoxycarbonylation reaction (step 2). The similarity of k_{-2} and k_6

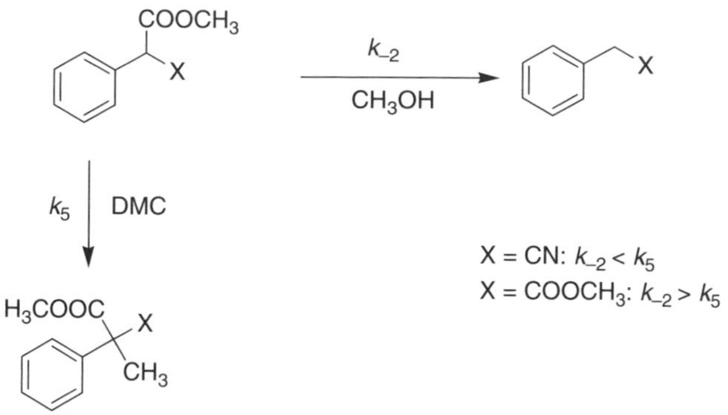

Scheme 4.11   Demethoxycarbonylation (k_2) vs. methylation (k_5) for CH_2-active compounds.

reveals that both the starting reagent 2a and its methyl derivative 1a undergo demethoxycarbonylation reactions at comparable rates while the methylation step of the intermediate 4a is the fastest reaction. In particular, at 140C, the pseudo-first-order rate constants of steps 6, 5, and the reverse of 2, namely, k_6, k_5, and k_2, of Scheme 4.8, are 2.910^3, 7.910^3, and 2.710^3 min^1.*

Overall, the comparison of the kinetic behavior of the investigated steps reveals that the nonequilibrium methylation reaction is crucial in driving the overall process to completion. In fact, the higher rate of step 5 allows both the rapid consumption of **3a** and the accumulation of **4a**, which serves as a reactant for step 6. In other words, both equilibria 2 and 6 are regulated by irreversible reaction 5. The mechanism shows the key action of the methoxycarbonyl group, which by increasing the acidity of **3**, acts as a promoter, significantly accelerating step 5.

At a high temperature, it is known that the lack of solvation may favor a $B_{Al}2$ mechanism with respect a $B_{Ac}2$ one.[26] In the case of DMC-mediated methylations, both reaction pathways occur in a definite sequence, which accounts for the high monomethyl selectivity observed in these reactions.

Finally, it should be noted that esters and nitriles in the demethoxycarbonylating step behave in a manner opposite to that in the methylating step. For nitriles, the methylation rate predominates over methoxycarbonylation; for esters, demethoxycarbonylation predominates (Scheme 4.11).

In order to clarify the different behavior of anion **2**$^-$ and **3**$^-$ (Scheme 4.10) toward DMC, various anions with different soft/hard character (aliphatic and aromatic amines, alcohoxydes, phenoxides, thiolates) were compared with regard to nucleophilic substitutions on DMC, using different reaction conditions.[27] Results were in good agreement with the hard−soft acid−base (HSAB) theory.[28] Accordingly, the high selectivity of monomethylation of CH_2 acidic compounds and primary aromatic amines with DMC can be explained by two different subsequent reactions, which are due to the double electrophilic character of DMC. The first

---

*DMC is used in large excess. (200 equiv. with respect to the substrate). Therefore, its concentration is assumed to be constant throughout the reaction.

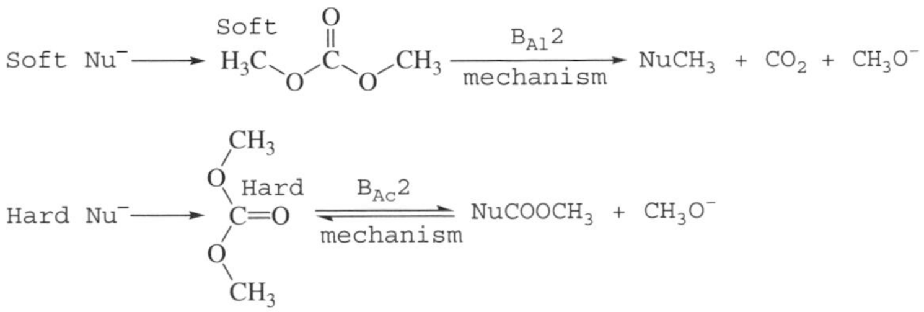

Figure 4.5 DMC as an ambident nucleophile.

step consists of a hard–hard reaction, and selectively produces a soft anion, which, in the second phase, selectively transforms into the final monomethylated product, via a soft–soft nucleophilic displacement (yield >99% at complete conversion, using DMC as solvent) (Fig. 4.5).

The combination of the dual electrophilic character of DMC with its reaction products allows two consecutive steps to occur in a selective way for what concerns both reaction sequence and yields: first, the hard–hard reaction occurs and produces a soft anion only; and second, a soft–soft nucleophilic displacement leads to the final product. Since hard–soft and soft–hard interactions are inhibited, double methylation and double carboxymethylation do not occur.

4.3.2 Amines

As far as the mechanism of DMC-mediated mono-N-methylation reactions is concerned, adsorption phenomena of reagents over zeolite catalysts have to be examined. Model cases of aniline and DMC are suitable to this investigation.

According to Czjiek et al.,[29] the steric requisites of aniline allows it to diffuse into the pores of NaY (supercages), where two sites of adsorption have been identified through diffraction techniques. Figure 4.6 describes the case of deuterated aniline. The first site (I) is in the proximity of a cation Na^+, which forms a π complex with $PhND_2$. The ion is perfectly centered in the aromatic ring of the aniline. In the second site (II), the amine is held by the interactions with the 12 oxygen atoms that form the supercavity of the zeolite.

Also, the adsorption of DMC on faujasites, has been described through two modes of interaction:[30, 31] IR experiments indicates that DMC acts as a base to form acid–base complexes with the Lewis acidic sites of the catalyst (Scheme 4.12).

It should be noted that the formation of both complexes (III and IV) implies a lengthening of the $O-CH_3$ bond of DMC, meaning that DMC undergoes an electrophilic activation within the pores of the solid.

Based on Figure 4.6 and Scheme 4.12, the proposed mechanism for the mono-N-methyl of aniline with DMC is described pictorially on Scheme 4.13. Once the reagents (amine and DMC) diffuse into the supercages of NaY, they can approach each other only according to the steric requisites of their adsorption

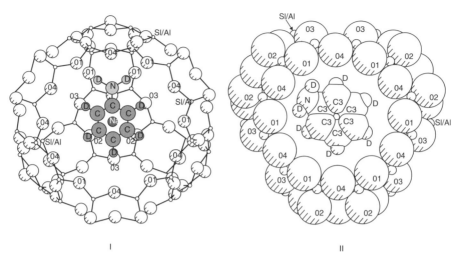

Figure 4.6 Two adsorption sites for PhND$_2$ in the NaY faujasite.

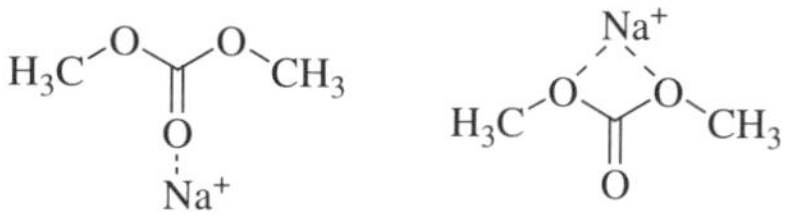

Scheme 4.12 Modes of adsorption (III and IV) of DMC over Na-exchanged faujasites.

patterns (for simplicity, Scheme 4.13 considers only one adsorption mode for both reagents).

Then, contrary to our previous hypothesis,[23] the reaction proceeds via a B$_{Al}$2 displacement of aniline on DMC. The product, mono-*N*-methyl aniline (PhNHMe), plausibly adsorbs into the zeolite in a different way with respect to aniline, because different H-bonds (*N*—H ···· O-zeolite) take place with the solid. As recently reported by Su et al.,[32] *N*-methyl amines also may interact with NaY by H-bonding between the protons of the methyl group and the oxygen atoms of the zeolite: this probably forces the molecule a bit far from the catalytic surface in a fashion less apt to meet DMC and react with it. This behavior can account for the mono-*N*-methyl selectivity observed, which is specific to the use of DMC: in the presence of alkali metal exchanged faujasites; in fact, the bis-*N*-methylation of primary aromatic amines occurs easily with conventional methylating agents (i.e., dimethyl sulfate).[33]

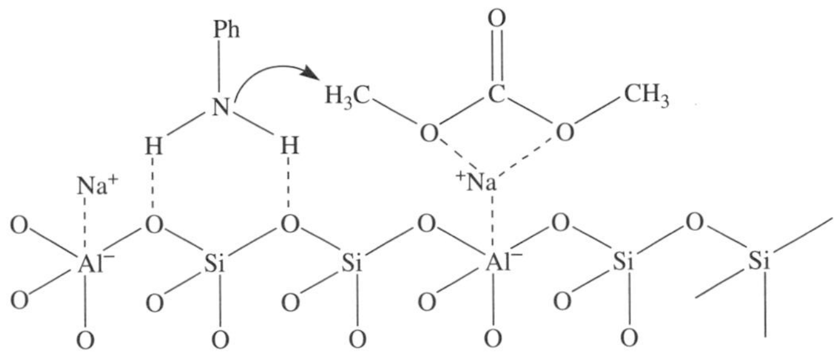

Scheme 4.13 $B_{Al}2$ displacement of aniline on DMC within the NaY supercage.

4.4 DMC AS A METHOXYCARBONYLATING AGENT

As shown in Scheme 4.4, selective methoxycarbonylation reactions with DMC are usually performed at relatively low temperatures. However, we have reported some interesting examples of this reactivity at higher temperatures that are dependent on the nucleophilicity of the substrates. These examples are discussed in the following section.

4.4.1 Oximes

At 180–190°C and in the presence of a weak base (usually K_2CO_3), ketoximes bearing an α-methylene react with DMC to yield N-methyl oxazolinones (Scheme 4.14).[34] This reaction is general for both linear and cyclic oximes, and exemplifies a selective methoxycarbonylation reaction at a high temperature. In fact, the first step is the $B_{Ac}2$ attack of the oxygen atom on the DMC carbonyl. Then, the reaction proceeds through an N-methylation and a final [3,3]-sigmatropic rearrangement, as shown in Scheme 4.15.

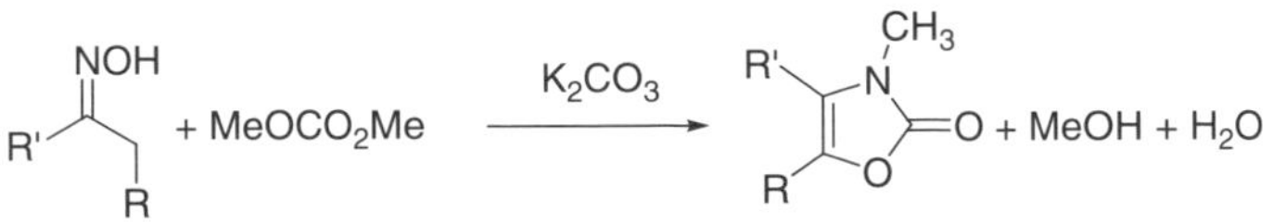

Scheme 4.14 N-Methyl oxazolinones from oximes.

Although oximes are ambident nucleophiles, a neat regioselectivity allows the formation of oxazolinones. The alpha effect, which is especially relevant for the substitution of $R_2C=NO^-$ anions at a carbonyl carbon,[35] may offer a good explanation for the $B_{Ac}2/B_{Al}2$ sequence (O-methoxycarbonylation and N-methylation

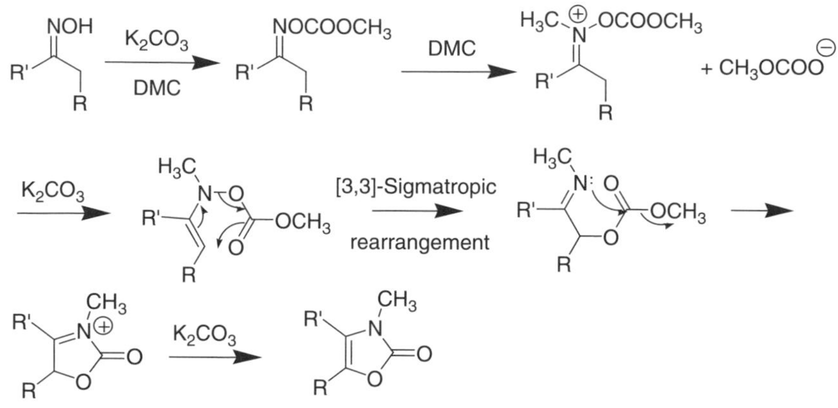

Scheme 4.15 Proposed mechanism for the synthesis of *N*-methyl oxazolinones from oximes.

reactions, respectively) in Scheme 4.15. This reaction is also the first ever reported example of a sigmatropic rearrangement involving DMC. We have gathered further evidence for this mechanism from the comparison of the reactivity of C_5–C_7 cyclic oximes, which is resumed in Scheme 4.16.

As can be seen, cyclohexanone- and cycloheptanone-oximes give the cyclic rearranged products, while cyclopentanone-oxime yields a complex reaction mixture with no traces of the corresponding oxazolinone. Since sigmatropic rearrangements proceed through a highly ordered transition state,[36] the required geometry is perhaps not achieved when a more strained C=C bond is involved, as occurs in the rigid cyclopentene ring (Scheme 4.15). Also, the reaction of allylimines with DMC give rearranged products. As an example, *N*-cyclohexylideneallylimine undergoes *N*-methylation by DMC followed by an aza-Claisen rearrangement (Scheme 4.17).[27]

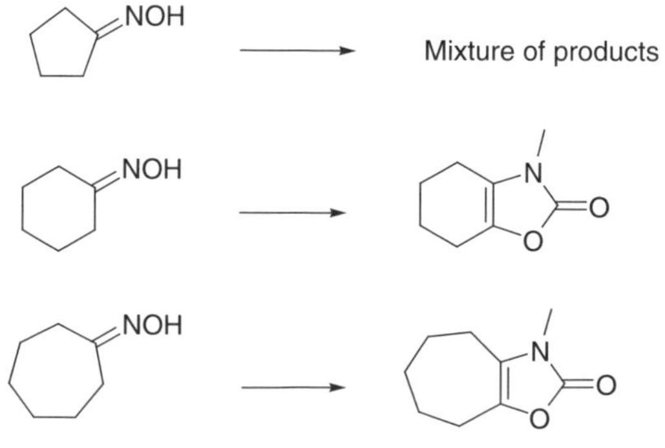

Scheme 4.16 The reactivity of C_5–C_7 cyclic oximes.

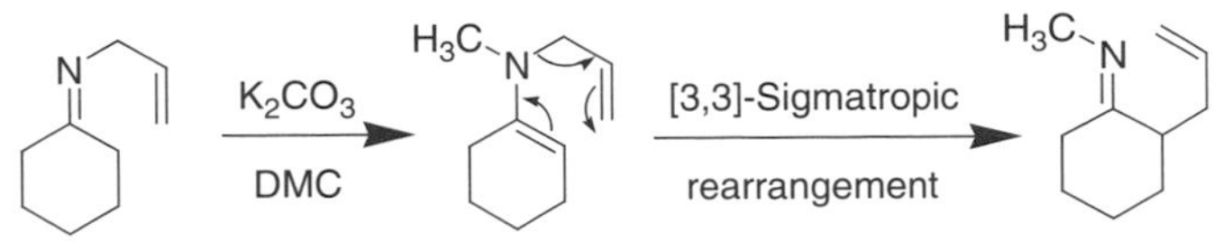

Scheme 4.17 Reaction of *N*-cyclohexylideneallylimine with DMC.

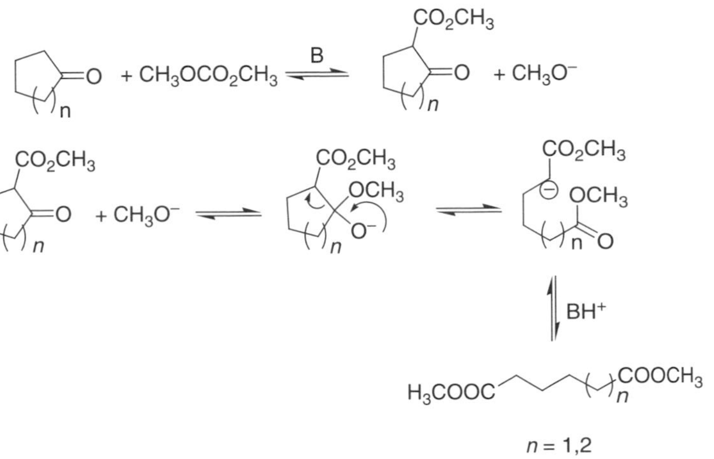

$$n = 1,2$$

Scheme 4.18 α,ω-Diesters from cyclic aliphatic ketones.

4.4.2 Ketones

At a high temperature (>200°C), the reaction with alicyclic ketones is a very attractive application of DMC as a methoxycarbonylating agent. In particular, cyclopentanone and cyclohexanone, whose reactions with DMC (or β = diethylaminoethyl cholride hydrochloride (DEC)] and a base (K₂CO₃) yield adipic and pimelic dimethyl (or diethyl) esters, respectively (Scheme 4.18).[37]

Industry, in fact, has a major interest in these diesters as building blocks for nylon 6,6 and nylon 7,7 in the production of polyesters and polyamides.[38] However, their present synthesis raises an environmental concern. For instance, the oxidation of cyclohexanone by nitric acid (for the preparation of adipic acid), accounts for more than 10% of the total yearly release of N₂O, which is among the main gases responsible for the greenhouse effect. The reaction of Scheme 4.14 represents an eco-friendly alternative synthesis of α,ω-diesters which uses green reagents and, relevantly, has a 100% atom economy.[3] The overall process is mechanistically described as a retro-Claisen condensation.

The reaction is even more general and applicable to ketones bearing an α-methylene. Yields up to 90% of the corresponding dimethyl esters are obtained in the case of benzylic substrates (Scheme 4.19).

$$ArCH_2COR + MeOCO_2Me \longrightarrow ArCH_2CO_2Me + RCO_2Me$$

Scheme 4.19   The reaction of benzylic ketones with DMC. Ar and R = Ph and *p*-tolyl.

4.4.3   Amines

The carbamation of primary amines with DMC is a known process, though only catalyzed reactions proceed with good yields and selectivity (Scheme 4.20).[39]

Our group has very recently reported that, when temperatures are 130–140°C, the reaction of Scheme 4.21 can proceed efficiently (yields up to 83% in methyl carbamates), without any catalyst in the presence of scCO$_2$ at 90 bar.[40] ScCO$_2$ plays a double role in this reaction: (1) it acts as a catalyst in the formation of the active nucleophiles (Scheme 4.21), and (2) at a pressure over 90 bar, it inhibits the formation of *N*-methylated carbamates [RN(Me)CO$_2$Me], which are possible by-products.

$$RNH_2 + MeOCO_2Me \xrightarrow{\text{catalyst}} RNHCO_2Me + MeOH$$

Scheme 4.20   Carbamation of amines with DMC. Catalyst: strong bases, Pb(II), alumina.

The mechanism of the reaction between amines with DMC has been investigated further.[41] While in the absence of bases they give methylation and carboxymethylation reactions without selectivity (B$_{Al}$2 and B$_{Ac}$2 mechanisms, respectively) (Table 4.6), in the presence of bases the B$_{Ac}$2 mechanism prevails. (Table 4.7) The carbamate already formed reacts further with DMC via B$_{Al}$2 mechanism to give the corresponding *N*-methyl derivative. These strong bases (potassium *tert*-butylate or sodium methoxide) catalyze the reaction of aliphatic

$$2\ RNH_2 + CO_2 \rightleftharpoons RNHCO_2^-\ RNH_3^+ \xrightarrow{\text{DMC}} RNHCO_2Me$$

Scheme 4.21   Carbamation of amines with DMC in the presence of CO$_2$.

and aromatic amines to give the corresponding carbamates quantitatively at 90°C. In these conditions high yields of carbamates are obtained in a few minutes (Table 4.7). Since aliphatic amines are harder nucleophiles than aromatic ones, they react faster with the carbonyl of DMC (entries 3–7). No *N,N*-dimethyl derivates were observed.

This pronounced double selectivity has been explained in terms of Pearson's HSAB theory.[28] According to this procedure, *N*-methylcarbamates have been prepared from primary aliphatic and aromatic amines, either at reflux temperature of

TABLE 4.6 Reaction of Amines and DMC in the Absence of a Base

Entry	Amine	$T\,(^\circ C)$	Time (h)	MNM (%)	DNM (%)	Carbamate (%)
1	Benzylamine	90	6	12	6	4
2	Aniline	200	24	15	37	9

Conditions: Amine/DMC molar ratio: 1/40. MNM = mono-*N*-methylation; DNM = *N,N*-dimethylation.

TABLE 4.7 Reaction of Amines and DMC in the Presence of a Base

Entry	Amine	Time (min)	Base	Carbamate (%)
1	*p*-Anisidine[a]	25	C_4H_9OK	77
2	*p*-Chloroaniline[b]	20	C_4H_9OK	86.5
3	Benzylamine[c]	1	C_4H_9OK	100
4	*n*-Decylamine[d]	3	CH_3ONa	100
5	Hexane-1,6-diamine	10	CH_3ONa	100
6	*n*-Octylamine	5	CH_3ONa	100
7	Phenethylamine	5	CH_3ONa	100

Conditions: Reflux temperature. 90°C; Amine/Base/DMC molar ratio: 1/1.2/40.
[a]After18h, 73% of *N*-methyl carbamate.
[b]After 18h, 87% of *N*-methyl carbamate.
[c]After 0.5h, 70% of *N*-methyl carbamate.
[d]After 6 h, 78% of *N*-methyl carbamate.

DMC (90°C) or at 230°C in autoclave. The reaction can be carried out in one step or through the isolation of the carbamate and the subsequent methylation reaction with DMC.

4.5 OTHER ORGANIC CARBONATES

Higher homologues of dialkyl carbonates exhibit alkylating/carboxyalkylating reactivity and selectivity similar to that of DMC. The main difference concerns the reaction rates that, as the alkyl chain of the carbonate increases, undergo a neat decrease according to the steric expectations for nucleophilic displacements. An exception is, of course, dibenzylcarbonate, owing to the resonance stabilizing effect of the benzyl group in the S_N2 transition state. In line with this observation, benzylation and more specifically, mono-*C*- and mono-*N*-benzylation reactions, take place with rates comparable to those of the corresponding methylation processes.

4.5.1 Dibenzylcarbonate

Because of the higher boiling point, dibenzylcarbonate (DBnC) allows much simpler conditions with respect to DMC: reactions can be performed at atmospheric pressure in normal glass apparatus. This fact, along with the peculiar high selectivity observed (at almost complete conversion) facilitates workup and separation of the mono-C-alkyl product. DBnC can be used to benzylate phenylacetonitrile, benzyl phenylacetate, and phenol, in refluxing DMF with a K_2CO_3 catalyst (Scheme 4.22).[42] The mechanism is analogous to the one sketched out for DMC (Scheme 4.10), and involves consecutive carboxybenzylation/benzylation steps.

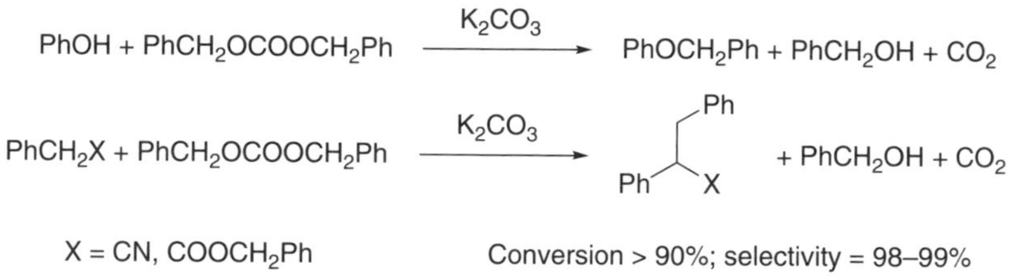

$$PhOH + PhCH_2OCOOCH_2Ph \xrightarrow{K_2CO_3} PhOCH_2Ph + PhCH_2OH + CO_2$$

$$PhCH_2X + PhCH_2OCOOCH_2Ph \xrightarrow{K_2CO_3} \underset{Ph\quad X}{\overset{Ph}{\diagdown}} + PhCH_2OH + CO_2$$

X = CN, COOCH$_2$Ph   Conversion > 90%; selectivity = 98–99%

Scheme 4.22   Benzylation of phenol and CH$_2$-active compounds with DBzlC.

4.5.2   Mixed Organic Carbonates

Asymmetrical alkyl methyl carbonates of the general formula $ROCO_2CH_3^\dagger$ can be conceived as selective methylating agents since their reactivity is discriminated by the structure and the length of the alkyl chain R. Also, it is the R group that imparts a practical advantage to the synthetic procedure: if heavy enough, it raises the boiling point of the carbonate, so that methylation reactions are allowed at ambient pressure. This overcomes one of the major operative drawbacks of batch methylations with DMC, that is, the need of pressure vessels. This idea was developed recently by our group in the investigation of the O-methylation of phenols (Scheme 4.23).[43] Some results are reported in Table 4.8.

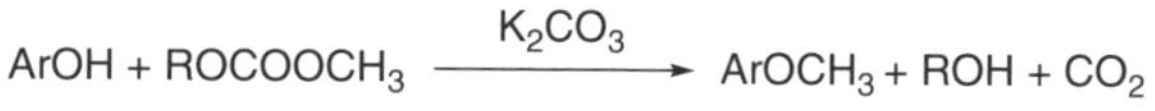

$$ArOH + ROCOOCH_3 \xrightarrow{K_2CO_3} ArOCH_3 + ROH + CO_2$$

Scheme 4.23   Methylation of phenols with mixed organic carbonates (ROCO$_2$CH$_3$).

As can be seen, asymmetrical carbonates give high chemoselective methylation reactions, provided that R has at least three carbon atoms ($R \geq n\text{-}C_3$, entries 2–4). Yet, in the case of reactive benzyl or allyl termini, the O-alkylation (forming PhOR) competes significantly with the formation of anisoles (entries 5–6).

Most satisfactory results can be obtained with the use of 2-(2-methoxyethoxy) ethyl carbonate [$CH_3O(CH_2)_2O(CH_2)_2OCOOCH_3$ (MEC) entry 4], which allows O-methyl selectivity up to 99% for different phenols (Table 4.9).

---

†These compounds can be easily obtained through the transesterification reaction of DMC with an alcohol ROH (see Ref. 38).

TABLE 4.8 **Reactions of Phenol with Different Alkyl Methyl Carbonates**[a]

Entry	R =	Time (h)[b]	Products (%)	
			PhOCH$_3$	PhOR
1	Et	15	90	·10
2	n-Pr	17	95	5
3	n-Bu	15	97	3
4	CH$_3$O(CH$_2$)$_2$O(CH$_2$)$_2$	20	>99	—
5	Bn	5	84	16
6	Allyl	21	83	17

[a]$T = 120°C$, phenol (3.3 mmol)/K$_2$CO$_3$,/**3** = 1:1.1:5. DMF (30 mL).
[b]Time for complete conversion of the substrate.

TABLE 4.9 **O-Methylation of Different Phenols by Methyl
2-(2-Methoxyethoxy)Ethyl Carbonate**[a]

Entry	Ar	Conv. (%)	Yield (%)[b]	Purity (%)
1	Ph	100	81	>99
2	p-MePh	100	79	>99
3	2-Naphthyl	100	83	>99

[a]$T = 140°C$, substrate/K$_2$CO$_3$/MEC = 1:1.1:5. Triglyme (50 mL).
[b]Isolated yields of O-methylated derivatives.

More recently, the use of MEC was also reported by us in the methylation of primary aromatic amines (p-XC$_6$H$_4$NH$_2$, X = H, Cl, NO$_2$) and of ambient anilines such as those listed in Scheme 4.9.[24–44] In the presence of a NaY faujasite and at atmospheric pressure, the reaction proceeds with a complete methyl chemoselectivity and even, most importantly, with a mono-N-methyl selectivity (90–97%) comparable to that achievable with DMC. As for DMC, selectivity arises from the synergistic effect of the reactivity of the carbonate and the amphoteric properties of the zeolite. In that case, however, a preliminary kinetic investigation was performed using alkyl- and alkoxy-substituted anilines, and it provides some general conclusions.[45a] This analysis indicates that the reaction selectivity toward methylated anilines (ArNHMe) does not depend on the polarity of the reaction solvent (when used), while a key role is played by the size of the zeolite cavities. In fact, as the bulkiness of the substituents grows, selectivity drops, because the diffusion of bigger molecules into the cavities is increasingly difficult at the point where it becomes obviated. For instance, from aniline to p-butylaniline, selectivity decreases from 99% to 90% accompanied by a decreased conversion from 100% to 9% (at comparable reaction times). Even more impressive, is the drop with 3,5-di-t-butylaniline whose size cannot fit the zeolite pores, and yields a 82% selectivity with 9% conversion.

The reactivity of MEC with amino-phenols, -benzylalcohols, -acids, and -benzamides (substrates of Scheme 4.9) is largely modified by the properties of aromatic substituents (X = OH, CO$_2$H, CH$_2$OH, CONH$_2$).[44b] Both steric and electronic effects on the reaction site (NH_2 group) and the establishment of direct acid–base interactions between substituents X and the catalyst must be considered to explain the observed scheme of rate constants. Weakly acid groups (CH_2OH, $CONH_2$ up to OH: pK_a of 16–17 up to 10) may help the adsorption over the NaY surface, and so favor the reaction. Aminobenzoic acids (pK_a of 4–5), however, are the least reactive substrates, presumably because carboxylic groups go through strong acid–base interactions with the catalyst.

These results represent the first ever reported evidence of strict cooperation between the steric requisites of the faujasite catalyst and the reactivity of an asymmetrical carbonate, in simultaneously inducing high methyl chemoselectivity and mono-*N*-methylselectivity for primary amines.

4.6   CONCLUDING REMARKS: THE GREEN CONTEXT

DMC and other dialkyl carbonates offer powerful perspectives for the development of alkylation/carboxyalkylation methods with low environmental impact. Moreover, these reactions are catalytic processes whose high selectivity allows minimization of the production of waste and of unwanted by-products as well.

Much more than in any previous period, mankind is living the binomial of "safeguarding the environment" and "implementation of the quality of life" where chemistry is the key science and, in particular, investigations in the area of green chemistry need to be a growing commitment for the chemical community. Educational programs in green chemistry, which are blossoming within IUPAC and other scientific organization, are establishing a new basis for communication between the chemical sciences and society.

ACKNOWLEDGMENTS

The Interuniversity Consortium Chemistry for the Environment, INCA, and the University of Ca' Foscari are gratefully acknowledged for the support to this work.

REFERENCES

1. (a) Tundo, P.; Anastas, P.; Black, D. StC.; et al. *Pure Appl. Chem.*, 2000, **72**, 1207; (b) *Green Chemistry, Theory and Practice*, Anastas, P. T.; Warner, J. C. (Eds.), Oxford University Press, Oxford, 1998.

2. Sheldon, R. A. *Pure Appl. Chem.*, 2000, **72**, 1233–1246.

3. Trost, B. M. *Science*, 1991, **254**, 1471–1477.

4. Anastas, P. T.; Williamson, T. in *Green Chemistry: Designing Chemistry for the Environment*, Anastas, P. T.; Williamson, T. (Eds.) ACS Symposium Series, No. 626, American Chemical Society, Washington D.C., 1996, pp. 1–17.

5. Tundo, P. *J. Org. Chem.*, 1979, **44**, 2048–1304.

6. Rivetti, F. *Dimethylcarbonate: An Answer to the Need for Safe Chemicals, in Green Chemistry: Challenging Perspectives*, Tundo, P.; Anastas, P. (Eds.) Oxford University Press, Oxford, 2000, pp. 201–219; *Registry of Toxic Effects of Chemical Substances*, Sweet, D. V. (Ed.) Vol. 2, Altanta, Geogia, pp. 186–1986.

7. Romano, U.; Rivetti, F.; Di Muzio, N. Dimethyl Carbonate, U.S. Patent 4,318,862, 1979. C.A. 1981, **95**, 80141w; Rivetti, F.; Romano, U.; Delledonne, D. Dimethylcarbonate and its production technology, in *Green Chemistry: Designing Chemistry for the Environment*, Anastas, P.; Williamson T. C. (Eds.) ACS Symposium Series, No. 626, American Chemical Society, Washington, D.C., 1996, pp. 70–80.

8. Nisihra, K.; Mizutare, K.; Tanaka, S. Process for preparing diester of carbonic acid, EP Patent Applied for 425, 197 (UBE Industries, Japan).

9. (a) Li, Y.; Zhao, X-; Wang, Y-J. *Appl. Catal. A: Gen.*, 2005, **279**, 205–208; (b) Kishimoto, Y.; Ogawa, I. *Ind. Eng. Chem. Res.*, 2004, **43**, 8155–8162.

10. Tundo, P. *Continuous Flow Methods in Organic Synthesis*, Horwood, Chichester, UK, 1991.

11. Delledonne, D.; Rivetti, F.; Romano, U. O. *J. Organomet. Chem.*, 1995, **448**, C15–C19.

12. Jessop, P. G.; Leitner, W. (Eds.) Wiley-VCH, Weinheim, 1999, pp. 259–413.

13. Starks, C. M. *J. Am. Chem. Soc.*, 1971, **93**, 195–199.

14. Cinquini, M.; Tundo, P. *Synthesis*, 1976, 516–519.

15. (a) Lee, D.; Chang, V. *J. Org. Chem.*, 1978, **43**, 1532–1536; (b) Shirai, M.; Smod, J. Decarboxylation reactions: reactivity of a free carbonate anion in ethereal solvents, *J. Am. Chem. Soc.*, 1980, **102**, 2863–2865.

16. Tundo, P.; Selva, M. *Chemtech*, 1995, **25**(5), 31–35.

17. (a) Tundo, P.; Trotta, F.; Moraglio, G.; Ligorati, F. *Ind. Eng. Chem. Res.*, 1988, **27**, 1565–1571; (b) Tundo, P.; Trotta, F.; Moraglio, G.; Ligorati, F. *Ind. Eng. Chem. Res.*, 1989, **28**, 881–890; (c) Tundo, P.; Trotta, F.; Moraglio, G. *J. Chem. Soc. Perkin Trans. I*, 1989, 1070–1071.

18. Bomben, A.; Selva, M.; Tundo, P.; Valli, L. *Ind. Eng. Chem. Res.*, 1999, **38**, 2075–2079.

19. (a) Selva, M.; Marques, C. A.; Tundo, P. *J. Chem. Soc. Perkin Trans. I*, 1994, 1323–1328; (b) Loosen, P.; Tundo, P.; Selva, M. U.S. Patent 5, 278, 533, 1994; (c) Tundo, P.; Selva, M. *Green Chemistry: Designing Chemistry for the Environment*, Anastas, P. T.; Williamson, T. C. (Eds.) American Chemical Society Symposium Series, No. 626, American Chemical Society, Washington, D.C., 1996, pp. 81–91.

20. Mikolajczyk, M.; Grzejszczak, S.; Zatorski, A.; et al. *Tetrahedron Lett.*, 1975, 3757–3760.

21. Bomben, A.; Selva, M.; Tundo, P. *J. Chem. Res. (S)*, 1997, 448–449.

22. Schwochow, F.; Puppe, L. *Angew. Chem. Int. Ed.*, 1975, **14**, 620–628.

23. Selva, M.; Bomben, A.; Tundo, P. *J. Chem. Soc. Perkin Trans. I*, 1997, 1041–1045.

24. Selva, M.; Tundo, P.; Perosa, A. *J. Org. Chem.*, 2003, **68**, 7374–7378.

25. Tundo, P.; Selva, M.; Perosa, A.; Memoli, S. *J. Org. Chem.*, 2002, **67**, 1071–1077.

26. Takashima, K.; Josè, S. M.; do Amaral, A. T.; Riveros, J. M. *J. Chem. Soc., Chem. Commun.*, 1983, 1255–1256.

27. Tundo, P.; Rossi, L.; Loris, A. *J. Org. Chem.*, 2005, **70**, 2219–2224.

28. (a) Pearson, R. G. *J. Am. Chem. Soc.*, 1963, **85**, 3533; (b) Pearson, R. G.; Songstad, J. *J. Am. Chem. Soc.*, 1967, **89**, 1827; (c) Gazquez, J. L.; Mèndez, F. J. *J. Phys. Chem.*, 1994, **98**, 4049.

29. Czjzek, M.; Vogt, T.; Fuess, H. Aniline in Yb, NaY, *Zeolites*, 1991, **11**, 832–836.

30. Bonino, F.; Damin, A.; Bordiga, S.; et al. *Angew. Chemie, Engl. Int. Ed.*, 2005, **44**, 4774–4777.

31. Beutel, T. *J. Chem. Soc., Faraday Trans.*, 1998, **94**, 985.

32. Docquir, F.; Toufar, H.; Su, B. L. *Langmuir*, 2001, **17**, 6282–6288.

33. Onaka, M.; Ishikawa, K.; Izumi, Y. *Chem Lett.*, 1982, 1783.

34. Marques, C. A.; Selva, M.; Tundo, P.; Montanari, F. *J. Org. Chem.*, 1993, **58**, 5765–5770.

35. Kice, J. L.; Legan, E. *J. Am. Chem. Soc.*, 1973, **95**, 3912–3917.

36. (a) Brown, A.; Dewar, M. J. S.; Schoeller, W. MINDO/2 *J. Am. Chem. Soc.*, 1970, **92**, 5516–5517; (b) Shea, K. J.; Phillips, R. B. *J. Am. Chem. Soc.*, 1980, **102**, 3156–3162.

37. Selva, M.; Marques, C. A.; Tundo, P. *Gazz. Chim. It.*, 1993, **123**, 515–518.

38. Tundo, P.; Memoli, S.; Selva, M. Synthesis of α,ω-diesters, European patent pending PCT/EP01/09241.

39. (a) Bortnick, N.; Luskin, L. S.; Hurwitz, M. D.; Rytina, A. W. *t*-Carbinamines, RR'R"CNH$_2$. III. *J. Am. Chem. Soc.*, 1956, **78**, 4358–4361; (b) Fu, Z-H.; Ono, Y. *J. Mol. Catal.*, 1994, **91**, 399–405; (c) Vauthey, I.; Valot, F.; Gozzi, C.; et al. *Tetrahedron Lett.*, 2000, 6347–6350.

40. Selva, M.; Tundo, P.; Perosa, A. *Tetrahedron Lett.*, 2002, **43**, 1217–1219.

41. Tundo, P.; Bressanello, S.; Loris, A.; Sathicq, G. *Pure Appl. Chem.*, 2005, **77**(10), 1719–1725.

42. Selva, M.; Marques, C. A.; Tundo, P. *J. Chem. Soc. Perkin Trans. I*, 1995, 1889–1893.

43. Perosa, A.; Selva, M.; Tundo, P.; Zordan, F. *Synlett*, 2000, 272–274.

44. Selva, M.; Tundo, P.; Perosa, A. *J. Org. Chem.*, 2001, **66**, 677–680.

45. (a) Selva, M.; Tundo, P.; Perosa, A. *J. Org. Chem.*, 2002, **67**, 9238–9247; (b) Selva, M.; Tundo, P., Foccardi, T. *J. Org. Chem.*, 2005, **70**, 2476–2485.

PART 2

ALTERNATIVE REACTION CONDITIONS

5

IONIC LIQUIDS: "DESIGNER" SOLVENTS FOR GREEN CHEMISTRY

Natalia V. Plechkova and Kenneth R. Seddon

QUILL Centre, The Queen's University of Belfast, Northern Ireland, United Kingdom

INTRODUCTION

Ionic liquids are now widely recognized as an important component of green chemistry. This chapter, after exploring the field of green chemistry, explains some of the unique features of these neoteric solvents, and suggests that they should be part of the arsenal of solvents used by all synthetic chemists.

5.1 PROBLEM? WHAT PROBLEM?

To prevent readers from immediately dashing to their library, let us briefly introduce the concept of green chemistry, prior to examining the *rôle* of ionic liquids. To illustrate its importance, and to explain what green chemistry is, and why it is needed, we start by showing you what green chemistry is not. Figure 5.1 is a photograph taken in Central China in 2000, in a beautiful mountainous area frighteningly close to the main panda breeding grounds. Those of you with sharp eyes will spot a factory in the lower right quadrant, swallowed in its own pollution. This is *not* green chemistry; this is what green chemistry is meant to prevent.

This image clearly represents a grossly undesirable situation; in most modern minds, pollution of this magnitude is considered objectionable. However, this has

Methods and Reagents for Green Chemistry: An Introduction, Edited by Pietro Tundo, Alvise Perosa, and Fulvio Zecchini
Copyright © 2007 John Wiley & Sons, Inc.

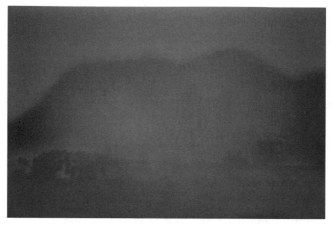

Figure 5.1   Environmental pollution in China. (Photo: K. R. Seddon.)

not always been the case. In 1854, Charles Dickens published *Hard Times*, a won-derful novel set in Coketown, a fictional town located in industrialised Northern England. The town is described as follows:

> It was a town of red brick, or of brick that would have been red if the smoke and ashes had allowed it; but as matters stood it was a town of unnatural red and black like the painted face of a savage. It was a town of machinery and tall chimneys, out of which interminable serpents of smoke trailed themselves for ever and ever, and never got uncoiled. It had a black canal in it, and a river that ran purple with ill-smelling dye.

Half a century later, things had become even worse. In a pioneering book, *The Soul of London*, widely considered one of the first "modern" novels, Ford Madox Ford wrote in 1905:

> Let him [a Londoner] go to one of the larger towns well outside his Home Counties, and he will have it forced in on him that he has no municipal buildings costing well-nigh a million, that he has no ship-canals, that his atmosphere has not half the cor-rosive properties that it should have to betoken the last word of wealth, of progress, and of commercial energy.

To repeat: "his atmosphere has not half the corrosive properties that it should have to betoken the last word of wealth, of progress, and of commercial energy." Here is our inheritance from the industrial revolution; this is the founding princi-pal upon which industrial chemistry developed worldwide. The basis of the indus-try that we built is the simple equation:

Pollution = Wealth

One would like to think, as we enter the new millennium, that our thinking would have advanced since 1905; that industrialists and politicians had moved beyond this. It is incredible to find, then, in March 2001, President George W. Bush making the following statement: "We'll be working with our allies to reduce greenhouse gases, but I will not accept a plan that will harm our economy and hurt American workers." A similar statement could have fallen from the mouth of Josiah Bounderby, the power behind Coketown: self-interest before global good. Not content to leave matters ambiguous, however, we can turn to the White House Web site for a clearer statement in 2005:[1]

> I've asked my advisors to consider approaches to reduce greenhouse gas emissions, including those that tap the power of markets, help realize the promise of technology and ensure the widest possible global participation.... Our actions should be measured as we learn more from science and build on it. Our approach must be flexible to adjust to new information and take advantage of new technology. We must always act to ensure continued economic growth and prosperity for our citizens and for citizens throughout the world.

It is notable that the (future) actions are for economic growth and prosperity, and not to protect the global environment of our planet! And how little attitudes have changed (for a more detailed and eloquent suite of arguments, see Ref. 2). The United States, with 5% of the world's population, emits nearly one-third of the world's carbon dioxide. It promised to cut emissions by 7% over 1990 levels by 2012 at the latest, but its emissions in fact rose by more than 10% between 1990 and 2000 (see Figure 5.2).[3] While the Kyoto Protocol[4] is a deeply flawed document (and do read it, not the distorted accounts in the press), it was the best that we had.

From the Environmental Protection Agency (EPA)[5] in 2000, we note that:

> Fossil fuels burned to run cars and trucks, heat homes and businesses, and power factories are responsible for about 98% of U.S. carbon dioxide emissions, 24% of methane emissions, and 18% of nitrous oxide emissions. Increased agriculture,

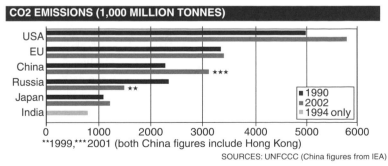

Figure 5.2   Data on carbon dioxide emissions.[3]

deforestation, landfills, industrial production, and mining also contribute a significant share of emissions. In 1997, the United States emitted about one-fifth of total global greenhouse gases. Estimating future emissions is difficult, because it depends on demographic, economic, technological, policy, and institutional developments. Several emissions scenarios have been developed based on differing projections of these underlying factors. For example, by 2100, in the absence of emissions control policies, carbon dioxide concentrations are projected to be 30–150% higher than today's levels.

From the same source:[5]

Global mean surface temperatures have increased 0.5–1.0°F since the late 19th Century [see Figure 5.3]. The 20th Century's 10 warmest years all occurred in the last 15 years of the century. Of these, 1998 was the warmest year on record. The snow cover in the Northern Hemisphere and floating ice in the Arctic Ocean have decreased. Globally, sea level has risen 4–8 inches over the past century. Worldwide precipitation over land has increased by about one percent. The frequency of extreme rainfall events has increased throughout much of the United States. Increasing concentrations of greenhouse gases are likely to accelerate the rate of climate change. Scientists expect that the average global surface temperature could rise 1–4.5°F (0.6–2.5°C) in the next fifty years, and 2.2–10°F (1.4–5.8°C) in the next century, with significant regional variation. Evaporation will increase as the climate warms, which will increase average global precipitation. Soil moisture is likely to decline in many regions, and intense rainstorms are likely to become more frequent. *Sea level is likely to rise two feet along most of the U.S. coast [emphasis added]*.

This last sentence is especially poignant and prophetic when we remember the recent disaster in New Orleans.

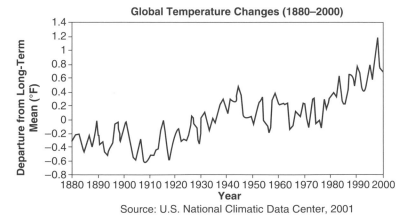

Figure 5.3 The hard data demonstrating global warming.[5]

But, the EPA has been even more specific: in a Report to Congress in December 1989, entitled *The Potential Effects of Global Climate Change on the United States*, they unambiguously stated:[6]

> A rise in sea level would further accelerate the rate of land loss in coastal Louisiana. Even a 50-centimeter rise in sea level (in combination with land subsidence) would inundate almost all of the delta and would leave New Orleans, most of which is below sea level and only protected with earthen levees, vulnerable to a hurricane.

The evidence for global warming is now incontravertible (see, e.g., Figure 5.3).*

5.2 GREEN CHEMISTRY

So, having seen the background, and hopefully appreciating the magnitude and importance of the problem, let us turn to green chemistry. The father of green chemistry, and its leading proponent, is Prof. Paul Anastas. To summarize a decade of definition and development in one sentence, green chemistry is the *design* of chemical products and processes that reduce or eliminate the use and/or generation of hazardous substances. There are twelve basic principles of green chemistry,[7] which we have condensed into what we consider the thirteenth principle: "If you do what you have always done, then you will get what you have always got." Green chemistry is the conscience of chemistry; we ignore it at our peril. However, the term green chemistry is all-inclusive. It is a multidisciplinary and cross-sectorial activity, and centrally involves the following disciplines, *inter alia*:

- Economics
- Engineering
- Political science
- Ethics and psychology
- Environmental science
- Chemistry
- Biology
- Toxicology

There is no doubt (notice economics at the top of the list) that processes have to be profitable; no industrialist, no matter how green his or her heart, will ever introduce a green process to replace a "black" process if it will lose money. However, green processes are intrinsically more economic than black processes,

---

*Two major recent reports ["The Stern Review: The Economics of Climate Change" from the UK government (http://www.hm-treasury.gov.uk/independent_reviews/stern_review_economics_climate_change/stern_review_report.cfm) and "Climate Change 2007: The Physical Science Basis" from the Intergovernmental Panel on Climate Change (IPCC; http://www.ipcc.ch)] have amplified and emphasised the points made here.

as the fundamental principles of green chemistry include atom efficiency, 100% selectivity and conversion, and zero waste. It is now up to chemists and chemical engineers, working as a close-knit team, to develop exciting new green technologies to allow two new overriding equations to be established:

$$\text{Pollution} = \text{Economic disaster}$$
$$\text{Green chemistry} = \text{Wealth}$$

But a word of caution to academics; your work is not green because you choose to call it green and publish it in *Green Chemistry*! Improving a yield, eliminating a toxic reagent, increasing a selectivity, are all admirable, but do not (in themselves) constitute green chemistry, just a green attitude. The term green chemistry has to be applied to a total process, not to an individual step, and involves commercialization. In other words, it is the result of the efforts of a large number of people with disparate skills, including chemists, economists, engineers, material scientists, industrialists, academics, and lawyers. If it is just a reaction reported in a paper, it may one day contribute toward a green process, but it is not green chemistry per se. To be truly green, it has to be used; otherwise, it is just a pipedream—a lonely paper with three citations, and a green flight of fancy!

5.3 SOLVENTS

One of the key components of any green industrial process has to be the selection of the solvent, since working with conventional solvents results in the emission of volatile organic compounds (VOCs), a major source of environmental pollution (see Figure 5.4).[8,9] There are four[†] principal strategies to avoid conventional organic solvents:[10,11]

1. No solvent (heterogeneous catalysis)
2. Water
3. Supercritical fluids
4. Ionic liquids

The first of these, the solventless option, is the best established,[12] and is central to the petrochemical industry, the greenest chemical sector.[13] The second, to use water as the solvent, can be an ideal approach,[14] but suffers from the difficulty of dissolving many organic compounds in water, undemonstrated scale-up, and the cost of the cleanup of organic contaminated water. Supercritical fluids represent a

---

[†]It is conventional to include perfluorocarbons (fluorous phases) in this list. However, these have extremely high global warming potentials: they are from 140–23,900 times more potent than CO_2 in their abilities to trap heat in the atmosphere over a century. Moreover, they remain in the atmosphere almost indefinitely, and so will accumulate as long as they are used on a significant scale.[9] We do not consider their use as a green option.

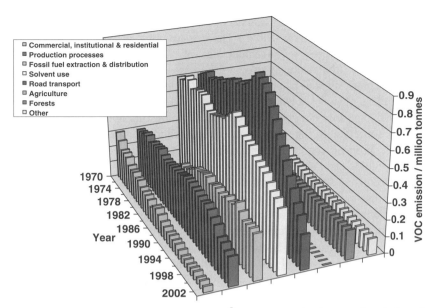

Figure 5.4 UK VOC emissions (1970–2002).[8] Note the predominance of emissions from solvent usage in the latest data.

new phase of matter for chemical synthesis, and wonderful developments have been described elsewhere,[15] including the recent commercialization of this technology by Thomas Swan, in a newly commissioned plant in Consett (County Durham, United Kingdom).[16] The final choice, ionic liquids (which, like supercritical fluids, are neoteric solvents, but unlike supercritical fluids, have immeasurably low vapor pressures at room temperature) are the subject of this overview. If must be stressed, however, even to the ad nauseum limit, that these methodologies are not in competition with each other, they are complementary. They form a toolbox of alternative solvent strategies (sometimes referred to as alternative reaction media or green solvents), giving a significant choice to the industrialist looking for the best option for his or her green process.

5.4 RATIONALE AND DEFINITIONS: WHY USE IONIC LIQUIDS?

Ionic liquids are, quite simply, liquids that are composed entirely of ions.[17–22] Thus, molten sodium chloride is an ionic liquid: a solution of sodium chloride in water (a molecular solvent) is an ionic solution. The term *ionic liquids* was selected with care, as it is our belief that the more commonly used phrase *molten salts* (or simply *melts*) is referential, and invokes a flawed image of these solvents as being high-temperature, corrosive, viscous media (cf. molten cryolite). The reality is that room-temperature ionic liquids can be liquid at temperatures as low as $-96°C$, and are typically colorless, fluid, and easily handled. To use the term molten salts to describe these novel systems is as archaic as describing a car as a horseless carriage. Moreover, in the patent and recent academic literature, ionic liquids are now taken as being liquids that are composed entirely of ions, which are fluid at, around, or below 100°C. There is nothing sacred about the temperature of 100°C, it is merely a convenient marker.

In this chapter, I have already used a term new to chemistry: *neoteric solvents.*[10] The word neoteric is well established in the English language.[§] It is used here to indicate a class of novel solvents that have remarkable new properties, that "break new ground," and that offer a huge potential for industrial application. It is applied not only to ionic liquids, but also to supercritical fluids, another type of solvent that shows huge promise for green synthesis. Thus the term neoteric solvents covers both ionic liquids and supercritical fluids, and brings together under one banner the two most promising solvent systems for cleaning up the modern chemical industry.

Perhaps the most important question to be answered here is: Why would any sane chemist wish to use a room-temperature ionic liquid as a solvent for the study of industrially relevant catalytic process? To address this question, it is worth considering the history of reactive chemistry. For two millennia, chemistry has been studied in water, the solvent that covers 70% of the surface of our planet. In the nineteenth century, the range of known chemistry was dramatically expanded by the use of a wider range of solvents: organic solvents (e.g., alcohols, chlorinated hydrocarbons, arenes, and nitriles) caused a blossoming in the known organic chemistry; liquid ammonia opened up low-oxidation-state chemistry; nitrogen(IV) oxide and bromine(III) fluoride allowed the exploration of high-oxidation-state chemistry; and solvents such as sulfuric acid and liquid hydrogen chloride combined high solvation with acid catalysis. However, for all the huge chemical variety provided by this wide range of chemically disparate solvent systems, they all have one thing in common—they are all *molecular* solvents. There has been no learning experience from studying reactive chemistry in *ionic* solvents. Thus, our basic understanding of reactive chemistry is potentially flawed, as it is derived from a biased data set. Will apparently well-understood, conventional reaction types, such as the Friedel–Crafts reaction, follow the mechanisms that we teach undergraduates when studied in an ionic environment? Will the short-lived reaction intermediates found in conventional solvents be long-lived stable intermediates in an ionic liquid? Will changes in the relative stabilities of reaction intermediates change the available reaction pathways? Will the expected changes in reaction kinetics shift the balance between thermodynamic and kinetic control, even for well-known, simple reaction types, and hence produce different products? The answers to these questions will only be discovered by extensive experimental studies of these systems, which are now well under way in many laboratories around the world. Of one thing we can be sure, both the thermodynamics and the kinetics of any reaction will be different in an ionic fluid than in a molecular solvent, and so the outcome of the reaction may well be different too, with the choice of ionic liquid dictating the outcome (*vide infra*).

[§]Neoteric is defined in the Oxford English Dictionary as meaning: recent, new, modern.

5.5 IONIC LIQUIDS FOR GREEN CHEMISTRY

Room-temperature ionic liquids have been developed over the past decade as green solvents for industrial applications,[18–20,22] ranging from the petrochemical industry, via heavy chemicals, fine chemicals, agrochemicals, and pharmaceuticals,[23] to the nuclear industry,[24] solar cells,[25] and electrochemistry.[17] Recent independent reports,[26,27] and many reviews,[28,29] have highlighted ionic liquids as representing a state-of-the-art, innovative approach to green chemistry, and they are even suitable media for preparing and studying nanoclusters.[30] The potential of ionic liquids to act as solvents for a broad spectrum of chemical processes is attracting increasing attention from industry, because these processes promise significant environmental benefits.[18–20,31] Prof. Paul Anastas, former director of the ACS Green Chemistry Institute,[32] ex-chief of the Industrial Chemistry Branch at the Environmental Protection Agency in Washington, D.C., and advisor to both President Clinton and President Bush, is convinced that ionic liquids can contribute significantly to sustainable chemistry and the development of green technology. He was reported in *Chemical & Engineering News* as stating:[33]

> Green chemistry aims to design the hazards out of chemical products and processes, including solvents. With ionic liquids, you do not have the same concerns as you have with, for example, volatile organic solvents, which can contribute to air pollution. Ionic liquid chemistry is a very new area that is not only extremely interesting from a fundamental chemistry point of view, but could also have a very large impact on industry.

Thus, room-temperature ionic liquids have the potential to provide environmentally friendly solvents for the chemical and pharmaceutical industries. The ionic liquid environment is very different from normal polar and nonpolar organic solvents; both the thermodynamics and the kinetics of chemical reactions are different, and so the outcome of a reaction may also be different. Organic reactions that have been successfully studied in ionic liquids include Friedel–Crafts,[34] Diels–Alder,[35] Heck catalysis,[36] chlorination,[37] enzyme catalysis,[38,39] polymerization,[40] cracking,[41] oxidation,[42] and hydrogenation.[43] Indeed, a comprehensive review of Heck chemistry[44] concluded that "the ionic liquid process appears as one of the cleanest recyclable procedures so far described for the Heck reaction."

One of the suggestions to arise out of this deluge of new chemistry is that ionic liquids are "designer" solvents.[33] For this to be true, there is an implication (implicit rather than explicit) that for every chemical reaction of interest, it should be possible to design, or tailor, a solvent to optimize that reaction. Given that the chemical industry currently uses only about 600 molecular solvents, the claim of designer solvents for ionic liquids must imply that many, many more ionic liquids could be prepared, by orders of magnitude, than are currently available, or indeed imaginable, for conventional solvents. In this chapter, we examine the question, How many ionic liquid systems could there be?

5.6 THE CATIONS AND THE ANIONS

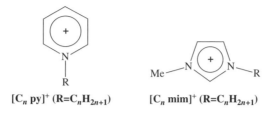

$[C_n \, py]^+ \, (R=C_nH_{2n+1})$ $[C_n \, mim]^+ \, (R=C_nH_{2n+1})$

Despite the discovery of room-temperature ionic liquids in 1914, by Paul Walden,[45] no attention was paid to them until 1948,[46] when Hurley and Wier used fluids based on the N-ethylpyridinium cation, $[N$Etpy$]^{+}$.[47,¶] However, it was not until Osteryoung investigated the N-butylpyridinium cation, $[N$Bupy$]^+$, that the field was really born.[48] A tremendous boost to this area was achieved when Hussey and co-workers[49] discovered that the 1-ethyl-3-methylimidazolium cation, [emim]$^+$, conferred even better properties on the systems under study, which were then mostly based upon the tetrachloroaluminate(III) anion. The next step in expanding the area came when Wilkes and Zaworotko discovered that there were many more ionic liquid systems possible with the [emim]$^+$ cation, if the anion were to be changed to, say, tetrafluoroborate or ethanoate.[50] Moreover, these new ionic liquids had the added advantages over the traditional tetrachloroaluminate(III) systems that they were air- and water-stable. In our own laboratories, we were investigating the effects of modifying the cation

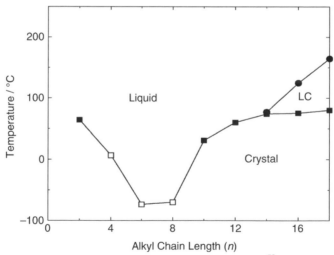

Figure 5.5   Phase diagram for the ionic liquids [C_n mim][PF$_6$].[52] Open squares are glass transitions, closed squares are melting points, and circles are clearing transitions; LC is a liquid crystalline phase.

¶Two parallel, and equally useful, systems of abbreviations exist: $[N$Etpy$]^+$ is also referred to as $[C_2$ py$]^+$, $[N$Bupy$]^+$ as $[C_4$ py$]^+$, [emim]$^+$ as $[C_2$ mim]$^+$, and so forth.

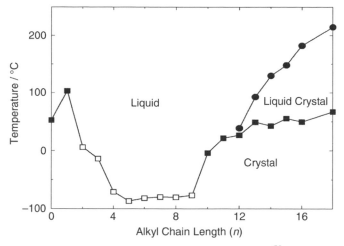

Figure 5.6 Phase diagram for the ionic liquids [C$_n$ mim][BF$_4$],[53] key as for Figure 5.5.

(by replacing the *N*-alkyl substituents with longer chains) to form [C$_n$ mim]$^+$ and [C$_n$ py]$^+$ ($n = 2–18$),[51] the effects of changing the anion,[52,53] the effects of hydrogen bonding,[54] the nature of the anionic equilibria,[55,56] purification techniques,[55] and the properties of binary and ternary mixtures of ionic liquids.[57]

These changes in anion and cation were not merely a case of methyl, ethyl, propyl, butyl, and then futile. The change of anion dramatically affects the chemical behavior and stability of the ionic liquid; the change of cation has a profound effect on the physical properties, such as melting point, viscosity, and density.[52,53,58] This readily can be seen by examining the phase diagrams for the hexafluorophosphate[52] and tetrafluoroborate[53] salts (see Figures 5.5 and 5.6, respectively).

5.7 HOW DO I PREPARE THEE? LET ME COUNT THE WAYS

So, the observations in the preceding section raised the obvious question: Just how many ionic liquids are there? For each cation series that we consider, a matrix of salts can be generated with the following common anions:

n	Cl$^-$	[Al$_2$Cl$_7$]$^-$	[BF$_4$]$^-$	[PF$_6$]$^-$	[CF$_3$SO$_3$]$^-$	[CH$_3$CO$_2$]$^-$	[RCO$_2$]$^-$	[COCl$_4$]$^{2-}$	[NiCl$_4$]$^{2-}$
2									
4									
6									
8									
10									
12									
14									
16									
18									

This therefore gives us 81 ionic liquids for each cation type (N.B.: we only consider the even chain lengths, as Figure 5.6 shows us that there is no advantage to be gained in using the more expensive odd chain lengths). Taking only the previously mentioned imidazolium skeleton (and hence ignoring the dozens of combinations that could be generated with 2-, 4-, and 5-substitution patterns), pyridinium, 4-methylpyridinium, and 3-methylpyridinium, there are 324 ionic liquids already defined here. Well, actually 612, since we should consider all the carboxylates, not just ethanoate! Adding only four more anions (e.g., nitrate, hydrogentartrate, lactate, and hydrogensulfate) and four more series of cations (e.g., $[PPh_3R]^+$, $[PPh_2MeR]^+$, $[NPh_3R]^+$, and $[NPh_2MeR]^+$) generates a possible maximum of 1512 ionic liquids. If we now consider all tetra-alkylammonium and tetra-alkylphosphonium salts (and there are 13,122 cations, without considering either branched chains or enantiomers), the total possible number of ionic liquids is now 277,074. It only takes a little imagination to select a few more anions and cations (and we have not yet considered optical isomers, or any aromatic skeleton other than pyridine or imidazole), and it can be seen that it is not impossible that there will be at least one million (10^6) simple ionic liquids that can be easily prepared in the laboratory. But that total is just for *simple* systems. If there are one million possible simple systems, then there are one billion (10^{12}) binary combinations of these, and one trillion (10^{18})$^{||}$ ternary systems possible![59] And if this seems too far-fetched, it is entirely plausible that to generate the required combination of reactivity, solubility, and viscosity, industrial applications will need to work with ternary systems.[57] A simple, but viable as a zeroth-order approximation, view is that the anion defines the chemistry that can be performed, and the cation controls the physical properties. Thus the simple ionic liquids are good, but unoptimized, first choices. The binary systems can be used to fine-tune the physical properties (such as viscosity, density, and miscibility), by adding (for example) a second ionic liquid with a common anion but different cation. The ternary systems can be used to minimize cost, by replacing an expensive component with a less expensive one, to the point where the desired chemistry ceases to be as effective.

Thus, it possible to consider an artist's palette of anions and cations (see Figure 5.7), being used to create a vast range of ionic liquids, the anions being chosen to control the chemistry, and the cations to engineer the physical properties. In reality, then, the descriptor of designer solvents is justified and apt. And the recent work coming from Grätzel's and MacFarlane's groups is rapidly increasing the ranges of both anions[60] and cations,[61] while Ohno's elegant work has expanded ionic liquids into amino acid derivatives.[62] Moreover, the use of microwave synthesis has speeded up ionic liquid synthesis considerably, as well as making their synthesis a model of green chemistry, being stoicheiometric, atom

---

$^{||}$American readers should note, *en passant*, that the original (European) definition of a trillion is a million million millions, that the definition of a billion is a million millions, and that what they refer to as a billion is correctly referred to as a milliard (how the desire to be a billionaire can distort language!).

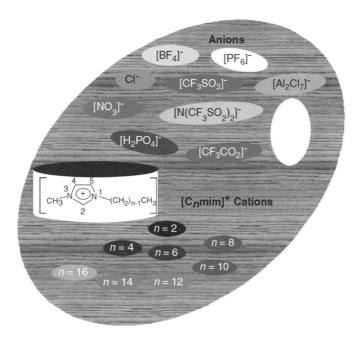

Figure 5.7 The artist palette of anions and cations. (Design: M. J. Torres.)

economic, and quantitative.[63] In addition, new routes to purer ionic liquids, not starting with haloalkanes, has increased the scope of the possible applications.[64]

5.8 IONIC LIQUID PROPERTIES: ARE IONIC LIQUIDS "GREEN"?

There is not enough space in an overview of this length to do justice to this important topic, covered at length elsewhere,[21] but the following bullet points highlight some the key features of ionic liquids (as a class of materials—not all ionic liquids possess all of these properties, but all the properties are exhibited by some ionic liquids):

- Liquid range of $300°C$ ($-96-+200°C$)
- Excellent solvents for organic, inorganic, and polymeric materials
- Acidic compositions are superacids ($pK_a \approx -20$)
- Some are very water sensitive and must be used in a dry box; others are hydrophobic and air stable
- Thermally stable under conditions up to $200°C$
- *Now* — easy to buy and simple to prepare
- *No* measurable vapor pressure at room temperature
- Nonflammable (but some are explosive!)

- Exhibit Brønsted, Lewis, Franklin, and "super" acidity
- *Highly* solvating—therefore low volumes used, thus process intensification
- Catalysts as well as solvents
- Highly selective reactions
- *New chemistry*

However, and this must be clearly and unambiguously stated, as it is frequently misquoted, ionic liquids are *not* intrinsically green—we could design noxious ionic liquids with the greatest ease (such as, for example, ionic liquids containing alkaloid-based cations and cyanide as the anion). We frequently ask students (and indeed faculty) why they consider their work to be green, and all too often receive the answer: "Because I am using ionic liquids" (usually delivered to us as though we must be mentally retarded not to realize this!). This is quite simply *wrong*, and shows ignorance of the whole field. Nevertheless, ionic liquids can be *designed* to be green, and important work on the toxicity of ionic liquids is continuing both at the University of Bremen[65] and the University of Notre Dame.[66] In addition, the nonflammability of ionic liquids adds an invaluable safety bonus; solvents that will not burn are intrinsically safer to use industrially than flammable solvents. This should make a very interesting case study under the U.S. Green Chemistry Research and Development Act of 2005, S.1270,[67] which (to the best of our knowledge) is the first time the principles of green chemistry have been enshrined in law. John Marburger [Office of Science and Technology Policy (OSTP)] has made a strong case for using green chemical approaches as a creative response to terrorism,[68] and the use of ionic liquids in industrial processes will make them intrinsically safer than equivalent processes using conventional solvents, as many ionic liquids do not burn.

5.9   IONIC LIQUID APPLICATIONS

There now is a proven extensive organic chemistry in ionic liquids, of which a few key reactions are listed below:[21,26,28]

- Friedel–Crafts
- Heck and Suzuki coupling
- Oxidation (with air and/or dioxygen)
- Reduction
- Chiral hydrogenation
- Oligomerization
- Polymerization
- Sulfonation
- Nitration

- Halogenation
- Diazotization
- Diels–Alder
- N-Alkylation and O-alkylation
- Aldol condensation

It is clear that there are few restrictions to the possible reactions that can be studied. Of particular interest is an article demonstrating that the nature of the ionic liquid can dramatically influence the outcome of a reaction.[69] In a study of the reactions between toluene and nitric acid, three completely different products were obtained by using three different ionic liquids (see Figure 5.8). Despite the thousands of articles demonstrating the synthetic utility of ionic liquids, however, there are only, as yet, a few on mechanism, the most notable being from Welton's group.[70] In addition, three key articles,[39,71] published in 2000, were the motivation for the creation of the whole field of enzymatic catalysis in ionic liquids.[38]

Despite the large number of academic publications in this area, perhaps the most significant advance was the public announcement, in 2003, that BASF was operating the first commercial process employing ionic liquids.[31,72] This new multiton process is called the BASIL™ (*b*iphasic *a*cid *s*cavenging utilizing *i*onic *l*iquids) process, and was developed by the BASF team (Matthias Maase, Rüdiger Büttner, and Holger Ganz) under the leadership of Dr. Klemens Massonne, from

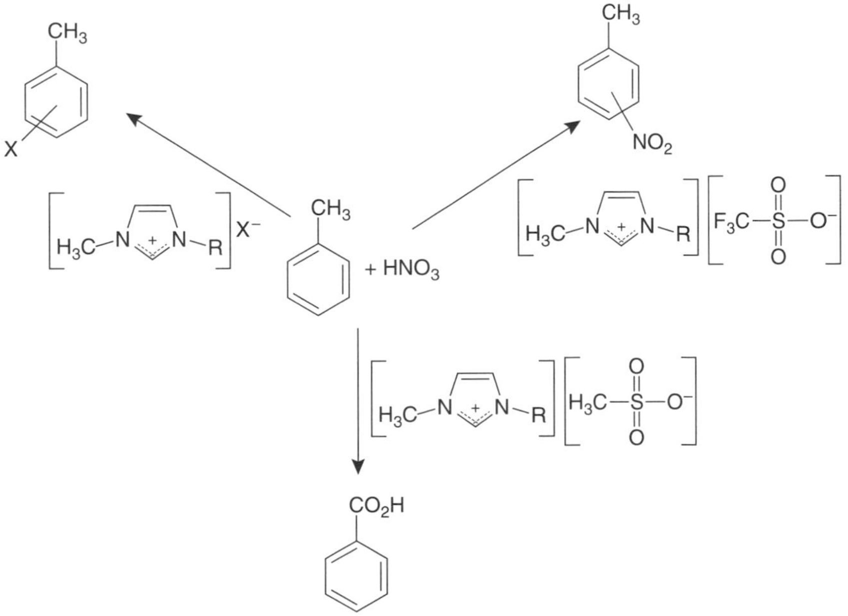

Figure 5.8 The reactions between toluene and nitric acid in (a) a halide-based ionic liquid, (b) a triflate-based ionic liquid, and (c) a mesylate-based ionic liquid.

Figure 5.9 The BASIL reactor showing the reaction mixture, the upper phase being the solvent-free pure product, and the lower the ionic liquid. (Photo: BASF.)

Ludwigshafen, Germany. BASF generated the ionic liquid in situ, to remove waste acidic hydrogen chloride. The key reaction involved is of a well-known generic type:

$$PhP(R)Cl + R'OH \rightarrow PhP(R)(OR') + HCl$$

The product, an alkoxyphenylphosphine, is used as an important precursor for the synthesis of photoinitiators that are themselves used in the manufacture of printing inks, as well as glass fiber and wood coatings. Conventionally, the waste HCl is removed by adding a simple base (such as triethylamine), and filtering off the waste solid chloride salt that is formed. The ingenuity of the BASF team was to use the more expensive 1-methylimidazole, mim, as a base, to form the salt 1-methylimidazolium chloride, [Hmim]Cl, which melts above 75°C. Above this temperature, this salt is a colorless, dense ionic liquid, immiscible with the reaction mixture, which separates as a discrete layer (Figure 5.9). This is then removed by gravity separation (much easier and cheaper than filtration), and the 1-methylimidazole is regenerated from the ionic liquid, and recycled.[73] The space–time yield for the formation of the product alkoxyphenylphosphine was thus increased from $8 \text{ kg m}^{-3} \text{ h}^{-1}$ using NEt_3 to $690,000 \text{ kg m}^{-3} \text{ h}^{-1}$ using 1-methylimidazole, a factor of 80,000 increased productivity! In early 2002, BASIL was established in regular production for the synthesis of alkoxyphenyl-phosphines and has been in regular operation since then.

Although BASIL was the first publicly recognized industrial ionic liquid process, Stephen Falling, of the Eastman Chemical Company, announced at the first International Congress on Ionic Liquids (COIL; Salzburg, June 2005)[74] that they had, in fact, been running an ionic liquid process since 1996, for the production of 2,5-dihydrofuran (1400 tonnes per annum).[75] At the same meeting, Dan Tempel, from Air Products, announced that they had developed a new subatmospheric technology for storing (and delivering) Lewis basic gases (such as arsine and phosphine) and Lewis acidic gases (such as boron(III) fluoride) based on using Lewis acidic ionic liquids (such as $[C_n\text{mim}][CuCl_2]$) and Lewis basic ionic liquids (such as $[C_n\text{mim}][BF_4]$), respectively.[75] This new technology appears to be safer and more efficient than the current technology (based on solids adsorption technology). Also announced at COIL was that Degussa is using ionic liquids as paint additives and is investigating them for use in lithium ion batteries, and that IoLiTec has developed a high-tech process for the cleaning of nozzles.[74] Other companies with advanced research and/or pilot plants for ionic liquid applications include: IFP, Arkema, SASOL, Chevron, and Eli Lilly.[74] A review of these industrial applications will appear elsewhere,[76] and QUILL has published a map for the industrialization of ionic liquids,[77] it is clear that ionic liquids are now well established in the industrial consciousness, and will not disappear.

5.10 THE QUILL CENTRE

Quill was founded in April 1999 as an industrial consortium, with members from all sectors of the chemical industry.[78] It is based on the well-proven industry/university cooperative research center (IUCRC) concept developed by the U.S. National Science Foundation and is only the second IUCRC in Europe. There were 17 founding industrial members of the Quill consortium, and the current membership includes (listed alphabetically) bp, Chevron, Cytec, DuPont, Eastman Chemicals, ICI, Invista, Merck, Novartis, Procter and Gamble, SACHEM, SASOL, Shell, Strata, and UOP. Research carried out between QUB and individual companies, or by QUILL itself, has generated more than 20 patent applications, many of which have now been published, from as diverse a range of industries as BNFL, BP Chemicals, Cytec, ICI, Quest International, and Unichema Chemie BV. In a recent report in *Nature*,[79] the need for collaboration between government, industry, and academic institutions to form sustainable chemistry centers was stressed as vital in order to rethink traditional chemistry processes to be not only beneficial to the environment but also to make economic sense for industry. Quill, under the codirection of Professors Kenneth R. Seddon and Jim Swindall OBE, is one of these chemistry centers, and is the first (and only) in the world to focus on ionic liquids. It also has been recently awarded Marie Curie status by the European Union (EU), as a training center for ionic liquid technology, and received the 2006 Queen's Anniversary Prize for Higher and Further Education.

One of the important spin-offs of QUILL is that there are now many commercial sources of ionic liquids, up to and including the 50-ton scale, including Dupont (dupont.com/fluorointermediates/pdf/K15303.pdf), BASF (http://www2.basf.de/en/intermed/nbd/products/ionic_liquids/products/?id=Ce3wVA 2tgbw2.1-) Cytec (cytec.com), Merck (ionicliquids-merck.de), and SACHEM (sacheminc.com/othermarkets/group/ionicliquids.php). In addition, Acros Organics (fisher.co.uk/whats-new/newprods/chems/ionic.htm) markets ionic liquids supplied by QUILL, and Strata Technology (stratatec.co.uk/) Accelergy (accelergy.com) manufacture engineering equipment, including autoclaves and high-speed mixers, for use with ionic liquids. It is our belief that the commercial availability of these fluids and engineering equipment is a key contributor to the rapid (greater than exponential) increase in articles concerning ionic liquids being published in the open literature, and their high citation impact.[80] Other indicators of the growth in this field are the number of international meetings primarily concerned with ionic liquids, including a NATO Advanced Workshop,[20] three symposia, each of 10 sessions, concerning the green applications of ionic liquids at the ACS National Meetings in spring 2001 in San Diego,[19] autumn 2002 in Boston,[18] and autumn 2003 in New York.[22] DeChema hosted a meeting on green solvents for catalysis in Bruchsal, Germany, in October 2002, and the articles were published in *Green Chemistry*.[81] In addition there was a special issue dedicated to ionic liquids,[82] a second meeting on green solvents for synthesis, held two years later,[83] and a meeting on green solvents for processing was held in 2006.

5.11 CONCLUSION

> There are more things in heaven and earth, Horatio,
> Than are dreamt of in your philosophy.
>
>   W. Shakespeare, *Hamlet*, Act I, Scene V

> Now, my own suspicion is that the universe is not only queerer than we suppose, but queerer than we *can* suppose.
>
>   J. B. S. Haldane, *Possible Worlds*, 1927

In conclusion, we must concur with both Shakespeare and Haldane. There are more things (a trillion more) in heaven and earth than are dreamt of in our philosophy, and, yes, the universe is not only queerer than we suppose, but queerer than we *can* suppose. To place the preceding observations in context, there are only about 600 molecular solvents commonly in use in industry today. So the reasons for discussing ionic liquids as designer solvents are valid: with a trillion solvents to select from, an optimum solvent for any given reaction will exist. The only problem will be finding it. Thus, we believe that the future of ionic liquids will, to some extent, be guided by the principles of combinatorial chemistry and high-throughput experimentation. The next few years will be the true test of ionic liquid technology, and they look to be very exciting years indeed!

ACKNOWLEDGMENTS

K.R.S. thanks both Prof. Pietro Tundo and Dr. Alvise Perosa for their support, friendship, and enthusiasm over the years of the Venice Green Chemistry Summer School, and for believing in ionic liquids when many scientists thought they were all hype with no substance. Their *joie de vivre* is infectious, and transfers itself to teachers and students alike. They are the reason for the success of the project. We would both like to express our gratitude for the continuing support of the QUILL Industrial Advisory Board, bp, and the EPSRC (Portfolio Partnership Scheme, grant no. EP/D09538/1).

REFERENCES

1. Bush, G. W. The White House, 2005; whitehouse.gov/news/releases/2005/05/20050518-4.html

2. Moore, M. *Stupid White Men*, Penguin, London, 2002.

3. BBC, *BBC News Online*, 2005; http://news.bbc.co.uk/1/hi/sci/tech/3143798.stm

4. United Nations Framework Convention on Climate Change, *Kyoto Protocol*, 1997; http://unfccc.int/resource/docs/convkp/kpeng.html

5. U.S. Environmental Protection Agency, *Global Warming—Climate*, 2000; http://yosemite.epa.gov/oar/globalwarming.nsf/content/climate.html

6. U.S. Environmental Protection Agency, *The Potential Effects of Global Climate Change on The United States*, 1989; http://yosemite.epa.gov/oar/globalwarming.nsf/UniqueKeyLookup/RAMR5CKNNG/$File/potential_effects.pdf

7. Anastas, P. T.; Warner, J. *Green Chemistry: Theory and Practice*, Oxford University Press, Oxford, 1998.

8. Environment Agency (UK), *Greenhouse Gas Emissions*, 2005; environment-agency.gov.uk/yourenv/432430/432434/432446/435321/?lang = _e

9. U.S. Environmental Protection Agency, *High Global Warming Potential (GWP) Gases*, 2004; epa.gov/highgwp/

10. Seddon, K. R., Room-temperature ionic liquids—neoteric solvents for clean catalysis, *Kinet. Catal.*, 1996, **37**(5), 693–697; Seddon, K. R. Room-temperature ionic liquids—neoteric solvents for clean catalysis, *Kinet. Katal.*, 1996, **37**(5), 743–748.

11. Seddon, K. R. Ionic liquids for clean technology, *J. Chem. Technol. Biotechnol.*, 1997, **68**(4), 351–356; Seddon, K. R. Ionic liquids for clean technology: an update, in *Molten Salt Forum: Proceedings of the 5th International Conference on Molten Salt Chemistry and Technology*, Wendt, H. (Ed.), Vol. 5–6, 1998, pp. 53–62.

12. Ertl, G.; Knözinger, H.; Weitkamp, J. (Eds.) *Handbook of Heterogeneous Catalysis*, 5 Volume Set VCH, Weinheim, 1997.

13. Sheldon, R. A., Consider the environmental quotient, *Chemtech*, 1994, **24**(3), 38–47; Sheldon, R. A., The role of catalysis in waste minimization, in *Precision Process*

Technology: Perspectives for Pollution Prevention, Weijnen M. P. C.; Drinkenburg A. A. H. (Eds.), Kluwer, Dordrecht, The Netherlands, 1993, pp. 125–138.

14. Kolb, H. C.; Finn, M. G.; Sharpless, K. B. Click chemistry: diverse chemical function from a few good reactions, *Angew. Chem., Int. Ed.*, 2001, **40**, 2004–2021; Klijn, J. E. Engberts, J. B. F. N., Organic chemistry: fast reactions "on water", *Nature*, 2005, **435**, 746–747, 900; Narayan, S.; Muldoon, J.; Finn, M. G.; et al. "On water": unique reactivity of organic compounds in aqueous suspension, *Angew. Chem., Int. Ed.*, 2005, **44**(21), 3275–3279; Li, C.-J.; Chan, T.-H. *Organic Reactions in Aqueous Media*, Wiley, New York, 1997; Grieco, P. A. (Ed.) *Organic Synthesis in Water*, Blackie, London, 1998.

15. Poliakoff, M.; Fitzpatrick, J. M.; Farren, T. R.; Anastas, P. T. Green chemistry: science and politics of change, *Science*, 2002, **297**, 807–810; Freemantle, M. Green process uses ionic liquid and CO_2, *Chem. Eng. News*, 1999, **77**(19), 9.

16. Thomas Swan & Co., 2005, thomas-swan.co.uk/

17. Ohno, H. (Ed.) *Electrochemical Aspects of Ionic Liquids*, Wiley-Interscience, Hoboken, NJ, 2005; Ohno, H. (Ed.) *Ionic Liquids: The Front and Future of Material Development*, CMC Press, Tokyo, 2003.

18. Rogers, R. D., Seddon, K. R. (Eds.) *Ionic Liquids as Green Solvents: Progress and Prospects*, ACS Symposium Series, No. 856, American Chemical Society, Washington D.C., 2003.

19. Rogers, R. D.; Seddon, K. R. (Eds.) *Ionic Liquids: Industrial Applications for Green Chemistry*, ACS Symposium Series, No. 818, American Chemical Society, Washington D.C., 2002.

20. Rogers, R. D.; Seddon, K. R.; Volkov, S. (Eds.) *Green Industrial Applications of Ionic Liquids*, NATO Science Series II: Mathematics, Physics and Chemistry, Vol. 92, Kluwer, Dordrecht, The Netherlands, 2002.

21. Wasserscheid, P.; Welton, T. (Eds.) *Ionic Liquids in Synthesis*, Wiley-VCH, Weinheim, 2003.

22. Rogers, R. D.; Seddon, K. R. (Eds.) *Ionic Liquids IIIB: Fundamentals, Progress, Challenges, and Opportunities—Transformations and Processes*, ACS Symposium Series, No. 902, American Chemical Society, Washington D.C., 2005; Rogers, R. D.; Seddon, K. R. (Eds.) *Ionic Liquids IIIA: Fundamentals, Progress, Challenges, and Opportunities—Properties and Structure*, ACS Symposium Series, No. 901, American Chemical Society, Washington D.C., 2005.

23. Earle, M. J.; Seddon, K. R.; McCormac, P. B. The first high yield green route to a pharmaceutical in a room temperature ionic liquid, *Green Chem.*, 2000, **2**(6), 261–262.

24. Allen, D.; Baston, G.; Bradley, A. E.; et al. An investigation of the radiochemical stability of ionic liquids, *Green Chem.*, 2002, **4**(2), 152–158.

25. Grätzel, M. Highly efficient nanocrystalline photovoltaic devices, *Plat. Met. Rev.*, 1994, **38**, 151–159; Grätzel, M. Dye-sensitized solar cells, *J. Photochem. Photobiol. C-Photochem. Rev.*, 2003, **4**(2), 145–153.

26. Kitazume, T. Organic synthesis in ionic liquids, *Kagaku Kogyo*, 2000, **51**(6), 437–444.

27. Hagiwara, R., Ito, Y. Room temperature ionic liquids of alkylimidazolium cations and fluoroanions, *J. Fluorine Chem.*, **105**(2), 2000, 221–227; Welton, T. Room-temperature Ionic Liquids. Solvents for Synthesis and Catalysis, *Chem. Rev.*, 1999, **99**, 2071–2084;

Olivier, H. Recent developments in the use of non-aqueous ionic liquids for two-phase catalysis, *J. Mol. Catal. A: Chem.*, 1999, **146**(1–2), 285–289.

28. Earle, M. J., Seddon, K. R. Ionic liquids. Green solvents for the future, *Pure Appl. Chem.*, 2000 **72**(7), 1391–1398; Earle, M. J.; Seddon, K. R. Ionic liquids: green solvents for the future, in *Clean Solvents: Alternative Media for Chemical Reactions and Processing*, Abraham M.; Moens, L. (Eds.) ACS Symposium Series, No. 819, American Chemical Society, Washington, D.C., 2002), pp. 10–25; Earle, M. J. Ionic liquids: solvents for the twenty-first century, in *Ionic Liquids: Industrial Applications for Green Chemistry*, Rogers R. D., Seddon, K. R. (Eds) ACS Symposium Series, No. 818, American Chemical Society, Washington D.C., 2002, pp. 90–105.

29. Rooney, D. W., Seddon, K. R. Ionic liquids, in *Handbook of Solvents*, G. Wypych (Ed.), ChemTech Publishing, Toronto, 2001, pp. 1459–1484.

30. Ott, L. S.; Cline, M. L.; Deetlefs, M. et al. Nanoclusters in ionic liquids: evidence for *N*-heterocyclic carbene formation from imidazolium-based ionic liquids detected by H-2 NMR, *J. Am. Chem. Soc.*, 2005, **127**(16), 5758–5759; Hamill, N. A., Hardacre, C.; McMath, S. E. J. In situ XAFS investigation of palladium species present during the Heck reaction in room temperature ionic liquids, *Green Chem.*, 2002, **4**(2), 139–142.

31. Rogers, R. D.; Seddon, K. R. Ionic liquids—solvents of the future? *Science*, 2003, **302**, 792–793; Seddon, K. R. Ionic liquids: a taste of the future, *Nature Mater.*, 2003, **2**, 363–364.

32. American Chemical Society, *ACS Green Chemistry Institute*, 2005; chemistry.org/greenchemistryinstitute

33. Freemantle, M. Designer solvents—ionic liquids may boost clean technology development, *Chem. Eng. News*, 1998, **76**(13), 32–37.

34. Seddon, K. R.; Hardacre, C.; McCauley, B. J. Catalyst comprising indium salt and organic ionic liquid and process for Friedel-Crafts reactions, World Patent WO 03, 028883 (2003); Earle, M. J.; McAuley, B. J.; Ramani, A.; et al. Chemical reactions catalyzed by fluoroalkylsulfonated compounds, preferably bis-triflimide compounds, World Patent WO 02, 072519 (2002); Csihony, S.; Mehdi, H.; Homonnay, Z.; et al. In situ spectroscopic studies related to the mechanism of the Friedel-Crafts acetylation of benzene in ionic liquids using $AlCl_3$ and $FeCl_3$, *J. Chem. Soc., Dalton Trans.*, 680–685; 2002, Valkenberg, M. H.; Sauvage, E.; De Castro-Moreira, C. P.; Hoelderich, W. F. Immobilized ionic liquids, World Patent WO 0132308 (2001); Adams, C. J.; Earle, M. J.; Roberts, G.; Seddon, K. R. Friedel-Crafts reactions in room temperature ionic liquids, *Chem. Commun.*, 1998, 2097–2098; Boon, J. A., Levisky, J. A., Pflug, J. L.; Wilkes, J. S. Friedel-Crafts reactions in ambient-temperature molten salts, *J. Org. Chem.*, 1986, **51**(4), 480–483; Earle, M. J.; Hakala, U.; Hardacre, C.; et al. Chloroindate(III) ionic liquids: recyclable media for Friedel-Crafts acylation reactions, *Chem. Commun.*, 2005, 903–905; Earle, M. J.; Hakala, U.; McAuley, B. J.; et al. Metal bis{(trifluoromethyl)sulfonyl}amide complexes: highly efficient Friedel-Crafts acylation catalysts, *Chem. Commun.*, 2004, 1368–1369.

35. Earle, M. J., McCormac, P. B.; Seddon, K. R. Diels-Alder reactions in ionic liquids: a safe recyclable green alternative to lithium perchlorate-diethyl ether mixtures, *Green Chem.*, 1999, **1**(1), 23–25; Doherty, S.; Goodrich, P.; Hardacre, C.; et al. Marked enantioselectivity enhancements for Diels-Alder reactions in ionic liquids catalysed by platinum diphosphine complexes, *Green Chem.*, 2004, **6**(1), 63–67.

36. Kobayashi, S.; Jørgensen, K. A. (Eds.) *Cycloaddition Reactions in Organic Synthesis*, Wiley-VCH, Weinheim, Germany, 2002; Carmichael, A. J., Earle, M. J., Holbrey, J. D.; et al. The Heck reaction in ionic liquids: a multiphasic catalyst system, *Org. Lett.*, 1999, **1**, 997–1000; Forsyth, S. A., Gunaratne, H. Q. N.; Hardacre, C.; et al. Utilisation of ionic liquid solvents for the synthesis of Lily-of-the-Valley fragrance {beta-Lilial (R), 3-(4-t-butylphenyl)- 2-methylpropanal}, *J. Mol. Catal. A-Chem.*, 2005, **231**(1–2), 61–66.

37. Winterton, N.; Seddon, K. R.; Patell, Y. Halogenation of unsaturated hydrocarbons in ionic liquids, World Patent WO 0037400 (2000).

38. Van Rantwijk, F.; Lau, R. M., Sheldon, R. A. Biocatalytic transformations in ionic liquids, *Trends Biotechnol.*, 2003, **21**(3), 131–138; Sheldon, R. A.; Lau, R. M.; Sorgedrager, M. J.; et al., Biocatalysis in ionic liquids, *Green Chem.*, 2002, **4**(2), 147–151; Sheldon, R. A. Catalytic reactions in ionic liquids, *Chem. Commun.*, 2001, 2399–2407.

39. Cull, S. G.; Holbrey, J. D.; Vargas-Mora, V.; et al. Room-temperature ionic liquids as replacements for organic solvents in multiphase bioprocess operations, *Biotechnol. Bioeng.*, 2000, **69**(2), 227–233; Lau, R. M.; van Rantwijk, F.; Seddon, K. R.; Sheldon, R. A. Lipase-catalyzed reactions in ionic liquids, *Org. Lett.*, 2000, **2**(26), 4189–4191.

40. Hardacre, C.; Holbrey, J. D.; Katdare, S. P.; Seddon, K. R. Alternating copolymerisation of styrene and carbon monoxide in ionic liquids, *Green Chem.*, 2002, **4**(2), 143–146; Carmichael, A. J.; Haddleton, D. M.; Bon, S. A. F.; Seddon, K. R. Copper(I) mediated living radical polymerisation in an ionic liquid, *Chem. Commun.*, 2000, 1237–1238.

41. Adams, C. J.; Earle, M. J.; Seddon, K. R. Catalytic cracking reactions of polyethylene to light alkanes, *Green Chem.*, 2000, **2**(1), 21–24.

42. Seddon, K. R.; Stark, A. Selective catalytic oxidation of benzyl alcohol and alkylbenzenes in ionic liquids, *Green Chem.*, 2002, **4**(2), 119–123.

43. Anderson, K.; Goodrich, P.; Hardacre, C.; McCath, S. E. J. Hydrogenation processes performed in ionic liquids, World Patent, WO 02, 094740 (2002); Dyson, P. J.; Ellis, D. J.; Parker, D. G.; Welton, T. Arene hydrogenation in a room-temperature ionic liquid using a ruthenium cluster catalyst, *Chem. Commun.*, 1999, 25–26; Adams, C. J.; Earle, M. J.; Seddon, K. R. Stereoselective hydrogenation reactions in chloroaluminate(III) ionic liquids: a new method for the reduction of aromatic compounds, *Chem. Commun.*, 1999, 1043–1044.

44. Beletskaya, I. P.; Cheprakov, A. V. The Heck reaction as a sharpening stone of palladium catalysis, *Chem. Rev.*, 2000, **100**, 3009–3066.

45. Von Walden, P. Über die Molekulargrösse und elektrische Leitfähigkeit einiger geseh-molzenen Salze, *Bull. Acad. Imp. Sci. St. Petersburg*, 1914, **8**, 405–422.

46. Wier, T. P. Jr., U.S. Patent 4,446,350 (1948); Wier, T. P. Jr., Hurley, F. H. U.S. Patent 4,446,349, 1948; Hurley, F. H. U.S. Patent 4,446,331 (1948).

47. Hurley, F. H.; Wier, T. P. The electrodeposition of aluminium from nonaqueous solutions at room temperature, *J. Electrochem. Soc.*, 1951, **98**, 207–212; Hurley, F. H.; Wier, T. P. *J. Electrochem. Soc.*, 1951, **98**, 203–206.

48. Chum, H. L.; Koch, V. R.; Miller, L. L.; Osteryoung, R. A. An electrochemical scrutiny of organometallic iron complexes and hexamethylbenzene in a room temperature molten salt, *J. Am. Chem. Soc.*, 1975, **97**, 3264–3265.

49. Wilkes, J. S.; Levisky, J. A.; Wilson, R. A.; Hussey, C. L. Dialkylimidazolium chloroaluminate melts—A new class of room-temperature ionic liquids for electrochemistry, spectroscopy, and synthesis, *Inorg. Chem.*, 1982, **21**(3), 1263–1264.

50. Wilkes, J. S.; Zaworotko, M. J. Air and water stable 1-ethyl-3-methylimidazolium based ionic liquids, *J. Chem. Soc., Chem. Commun.*, 1992, 965–967.

51. Abdul-Sada, A. K.; Atkins, M. P.; Ellis, B.; et al. Process and catalysts for the alkylation of aromatic hydrocarbons, World Patent WO 95, 21806 (1995); Abdul-Sada, A. K.; Ambler, P. W.; Hodgson, P. K. G.; et al. Ionic liquids of imidazolium halide for oligomerization or polymerization of olefins, World Patent WO 95, 21871 (1995); Bowlas, C. J.; Bruce, D. W.; Seddon, K. R. Liquid-crystalline ionic liquids, *Chem. Commun.*, 1996, 1625–1626.

52. Gordon, C. M., Holbrey, J. D., Kennedy, A. R.; Seddon, K. R. Ionic liquid crystals: hexafluorophosphate salts, *J. Mater. Chem.*, 1998, **8**(12), 2627–2636.

53. Holbrey, J. D.; Seddon, K. R. The phase behaviour of 1-alkyl-3-methylimidazolium tetrafluoroborates: ionic liquids and ionic liquid crystals, *J. Chem. Soc., Dalton Trans.*, 1999, 2133–2139.

54. Abdul-Sada, A. K.; Greenway, A. M.; Hitchcock, P. B.; et al. Structure of room temperature haloaluminate ionic liquids, *J. Chem. Soc., Chem. Commun.*, 1986, 1753–1754; Abdul-Sada, A. K.; Al-Juaid, S.; Greenway, A. M.; et al. Upon the hydrogen-bonding ability of the H(4) and H(5) protons of the imidazolium cation, *Struct. Chem.*, 1990, **1**, 391–394; Abdul-Sada, A. K.; Elaiwi, A. E.; Greenway, A. M.; Seddon, K. R. Evidence for the clustering of substituted imidazolium salts *via* hydrogen bonding under the conditions of fast atom bombardment mass spectrometry, *Eur. Mass Spectrom.*, 1997, **3**(3), 245–247; Hitchcock, P. B.; Seddon, K. R.; Welton, T. Hydrogen-bond acceptor abilities of tetrachlorometallate(II) complexes in ionic liquids, *J. Chem. Soc., Dalton Trans.*, 1993, 2639–2643; Avent, A. G.; Chaloner, P. A.; Day, M. P.; et al. Evidence for hydrogen-bonding in solutions of 1-ethyl-3-methylimidazolium halides, and its implications for room-temperature halogenoaluminate(III) ionic liquids, *J. Chem. Soc., Dalton Trans.*, 1994, 3405–3413; Elaiwi, A.; Hitchcock, P. B.; Seddon, K. R.; et al. Hydrogen bonding in imidazolium salts and its implications for ambient-temperature halogenoaluminate(III) ionic liquids, *J. Chem. Soc., Dalton Trans.*, 1995, 3467–3472; Avent, A. G.; Chaloner, P. A.; Day, M. P.; et al. Evidence for hydrogen-bonding in solutions of 1-methyl-3-ethyl-imidazolium halides, as determined by ^{1}H, ^{35}Cl and ^{127}I N.M.R. spectroscopy, and its implications for room- temperature halogenoaluminate(III) ionic liquids, in *Proceedings of the Seventh International Symposium on Molten Salts*, vol. PV 90-17. Hussey, C. L.; Wilkes, J. S.; Flengas, S. N.; Ito, Y. (Eds.) The Electrochemical Society Inc., Pennington, N.J., 1990, pp. 98–133.

55. Abdul-Sada, A. K.; Avent, A. G.; Parkington, M. J.; et al. The removal of oxide impurities from room-temperature halogenoaluminate ionic liquids, *J. Chem. Soc., Chem. Commun.*, 1987, 1643–1644; Abdul-Sada, A. K.; Avent, A. G.; Parkington, M. J.; et al. Removal of oxide contamination from ambient-temperature chloroaluminate(III) ionic liquids, *J. Chem. Soc., Dalton Trans.*, 1993, 3283–3286.

56. Abdul-Sada, A. K.; Greenway, A. M.; Seddon, K. R.; Welton, T. Upon the existence of $[Al_3Cl_{10}]^-$ in room-temperature chloroaluminate ionic liquids, *Org. Mass Spectrom.*, 1989, **24**(10), 917–918; Dent, A. J.; Seddon, K. R.; Welton, T. The structure of halogenometallate complexes dissolved in both basic and acidic room-temperature

halogenoaluminate(III) ionic liquids, as determined by EXAFS, *J. Chem. Soc., Chem. Commun.*, 1990, 315–316.

57. Abdul-Sada, A. K.; Seddon, K. R.; Stewart, N. J. Ionic liquids of ternary melts, World Patent WO 95 21872 (1995).

58. Seddon, K. R.; Stark, A.; Torres, M.-J. Influence of chloride, water, and organic solvents on the physical properties of ionic liquids, *Pure Appl. Chem.*, 2000, **72**(12), 2275–2287; Seddon, K. R.; Stark, A.; Torres, M. J. Viscosity and density of 1-alkyl-3-methylimidazolium ionic liquids, in *Clean Solvents: Alternative Media for Chemical Reactions and Processing*, Eds. Abraham, M.; Moens, L. ACS Symposium Series, Vol. 819, American Chemical Society, Washington D.C., 2002, pp. 34–49.

59. Seddon, K. R. Ionic liquids: designer solvents? in *The International George Papatheodorou Symposium: Proceedings*, Boghosian, S.; Dracopoulos, V.; Kontoyannis, C. G.; and Voyiatzis, G. A. (Eds). Institute of Chemical Engineering and High Temperature Chemical Processes, Patras, Greece, 1999, pp. 131–135.

60. MacFarlane, D. R.; Forsyth, S. A.; Golding, J.; Deacon, G. B. Ionic liquids based on imidazolium, ammonium and pyrrolidinium salts of the dicyanamide anion, *Green Chem.*, 2002, **4**(5), 444–448; Golding, J.; Forsyth, S.; MacFarlane, D. R.; et al. Methanesulfonate and *p*-toluenesulfonate salts of the *N*-methyl-*N*-alkylpyrrolidinium and quaternary ammonium cations: novel low cost ionic liquids, *Green Chem.*, 2002, **4**(3), 223–229; Bônhote, P.; Dias, A. P.; Papageorgiou, N.; et al. Hydrophobic, highly conductive ambient-temperature molten salts, 1996, *Inorg. Chem.*, **35**(5), 1168–1178; Papageorgiou, N.; Athanassov, Y.; Armand, M.; et al. The performance and stability of ambient-temperature molten salts for solar cell applications, *J. Electrochem. Soc.*, 1996, **143**(10), 3099–3108.

61. Sun, J.; MacFarlane, D. R.; Forsyth, M. A new family of ionic liquids based on the 1-alkyl-2-methylpyrrolinium cation, *Electrochim. Acta*, 2003, **48**(12), 1707–1711; Forsyth, S.; Golding, J.; MacFarlane, D. R.; Forsyth, M. *N*-methyl-*N*-alkylpyrrolidinium tetrafluoroborate salts: ionic solvents and solid electrolytes, *Electrochim. Acta*, 2001, **46**(10–11), 1753–1757; Golding, J. J.; Macfarlane, D. R.; Spiccia, L.; et al. Weak intermolecular interactions in sulfonamide salts: structure of 1-ethyl-2-methyl-3-benzylimidazolium bis[(trifluoromethyl)sulfonyl]amide, *Chem. Commun.*, 1998, 1593–1594.

62. Fukumoto, K.; Yoshizawa, M.; Ohno, H. Room temperature ionic liquids from 20 natural amino acids, *J. Am. Chem. Soc.*, 2005, **127**(8), 2398–2399.

63. Deetlefs, M.; Seddon, K. R. Improved preparations of ionic liquids using microwave irradiation, *Green Chem.*, 2003, **5**(2), 181–186.

64. Carmichael, A. J.; Deetlefs, M.; Earle, M. J.; et al. Ionic liquids: Improved syntheses and new products, in *Ionic Liquids as Green Solvents: Progress and Prospects*, Rogers, R. D.; Seddon, K. R. (Eds.) ACS Symposium Series, Vol. 856, American Chemical Society, Washington D.C., 2003, pp. 14–31; Holbrey, J. D.; Reichert, W. M.; Swatloski, R. P.; et al. Efficient, halide free synthesis of new, low cost ionic liquids: 1,3-dialkylimidazolium salts containing methyl- and ethyl-sulfate anions, *Green Chem.*, **4**(5), 2002, 407–413; Seddon, K. R.; Carmichael, A. J.; Earle, M. J. Process for preparing ambient temperature ionic liquids, World Patent WO 01, 40146 (2001); Earle, M. J.; Seddon, K. R. Preparation of imidazole carbenes and the use thereof for the synthesis of ionic liquids, World Patent, WO 01, 77081 (2001).

65. Jastorff, B., Störmann, R., Ranke, J.; et al. How hazardous are ionic liquids? Structure–activity relationships and biological testing as important elements for sustainability evaluation, *Green Chem.*, 2003, **5**, 136–142; Ranke, J.; Molter, K.; Stock, F.; et al. Biological effects of imidazolium ionic liquids with varying chain lengths in acute Vibrio fischeri and WST-1 cell viability assays (vol. 58, p. 396, 2004), *Ecotoxicol. Environ. Saf.*, 2005 **60**(3), 350–350; Stock, F.; Hoffmann, J.; Ranke, J.; et al. Effects of ionic liquids on the acetylcholinesterase—A structure-activity relationship consideration, *Green Chem.*, 2004, **6**(6), 286–290; Stepnowski, P.; Olafsson, G.; Helgason, H.; Jastorff, B. Preliminary study on chemical and physical principles of astaxanthin sorption to fish scales towards applicability in fisheries waste management, *Aquaculture*, 2004, **232**(1–4), 293–303; Ranke, J.; Molter, K.; Stock, F.; et al. Biological effects of imidazolium ionic liquids with varying chain lengths in acute Vibrio fischeri and WST-1 cell viability assays, *Ecotoxicol. Environ. Saf.*, 2004, **58**(3), 396–404.

66. Bernot, R. J.; Brueseke, M. A.; Evans-White, M. A.; Lamberti, G. A. Acute and chronic toxicity of imidazolium-based ionic liquids on *daphnia magna*, *Environ. Toxicol. Chem.*, 2005, **24**(1), 87–92.

67. United States Senate, *Green Chemistry Research and Development Act of 2005*, 2005, http://frwebgate.access.gpo.gov/cgi-bin/
getdoc.cgi?dbname = 109_cong_bills&docid = f:s1270is.txt.pdf

68. Marburger, J., *Green Chemistry and Homeland Security*, 2003, acs.org/portal/a/
ContentMgmtService/resources?id = c373e9f7d880b6688f6a4fd8fe800100

69. Earle, M. J.; Katdare, S. P.; Seddon, K. R. Paradigm confirmed: the first use of ionic liquids to dramatically influence the outcome of chemical reactions, *Org. Lett.*, 2004, **6**(5), 707–710.

70. Dyson, P. J.; Laurenczy, G.; Ohlin, C. A.; et al. Determination of hydrogen concentration in ionic liquids and the effect (or lack of) on rates of hydrogenation, *Chem. Commun.*, 2003, 2418–2419; Crowhurst, L., Mawdsley, P. R.; Perez-Arlandis, J. M.; et al., Solvent-solute interactions in ionic liquids, *Phys. Chem. Chem. Phys.*, 2003, **5**(13), 2790–2794; Lancaster, N. L.; Salter, P. A.; Welton, T.; Young, G. B. Nucleophilicity in ionic liquids. 2. Cation effects on halide nucleophilicity in a series of bis(trifluoromethylsulfonyl)imide ionic liquids, *J. Org. Chem.*, 2002, **67**(25), 8855–8861; Anderson, J. L.; Ding, J.; Welton, T.; Armstrong, D. W. Characterizing ionic liquids on the basis of multiple solvation interactions, *J. Am. Chem. Soc.*, 2002, **124**(47), 14247–14254; Aggarwal, A.; Lancaster, N. L.; Sethi, A. R.; Welton, T. The role of hydrogen bonding in controlling the selectivity of Diels-Alder reactions in room-temperature ionic liquids, *Green Chem.*, 2002, **4**(5), 517–520; Lancaster, N. L.; Welton, T.; Young, G. B. A study of halide nucleophilicity in ionic liquids, *J. Chem. Soc., Perkin Trans. II*, 2001, 2267–2270; Cammarata, L.; Kazarian, S. G.; Salter, P. A.; Welton, T. Molecular states of water in room temperature ionic liquids, *Phys. Chem. Chem. Phys.*, 2001, **3**(23), 5192–5200.

71. Erbeldinger, M.; Mesiano, A. J.; Russell, A. J. Enzymatic catalysis of formation of Z-aspartame in ionic liquid—An alternative to enzymatic catalysis in organic solvents, *Biotechnol. Prog.*, 2000, **16**(6), 1129–1131.

72. Freemantle, M. BASF's smart ionic liquid, *Chem. Eng. News*, 2003, **81**(13), 9.

73. Maase, M.; Massonne, K.; Halbritter, K.; et al. Method for the separation of acids from chemical reaction mixtures by means of ionic fluids, World Patent WO 03, 062171 (2003).

74. Seddon, K. R., *COIL*, 2005, http://events.dechema.de/COIL.html

75. Freemantle, M. Ionic liquids make splash in industry, *Chem. Eng. News*, 2005, **83**(31), 33–38.

76. Plechkova, N. V.; Seddon, K. R. *Chem. Soc. Rev.*, to be submitted.

77. Atkins, M. P.; Davey, P.; Fitzwater, G.; et al. *Ionic Liquids: A Map for Industrial Innovation* Q001, January 2004, QUILL, Belfast, 2004.

78. Seddon, K. R. QUILL rewrites the future of industrial solvents, *Green Chem.*, 1999, **1**, G58–G59.

79. Horton, B. Green chemistry puts down roots, *Nature*, 1999, **400**(6746), 797–799.

80. ISI Essential Science Indicators, *Ionic Liquids*, 2004, esi-topics.com/ionic-liquids/index.html.

81. Leitner, W.; Seddon, K. R.; Wasserscheid, P. Green solvents for catalysis (Editorial), *Green Chem.*, 2003, **5**(2), G28.

82. Seddon, K. R. Ionic liquids (Editorial), *Green Chem.*, 2002, **4**(2), G25–G26.

83. Leitner, W.; Seddon, K. R.; Wasserscheid, P. Green solvents for synthesis meeting, Bruchsal, Germany, October 2004, *Green Chem.*, 2005, **7**(5), 253.

6

SUPPORTED LIQUID-PHASE SYSTEMS IN TRANSITION METAL CATALYSIS

ALVISE PEROSA AND SERGEI ZINOVYEV

The Ca' Foscari University of Venice and National Interuniversity Consortium, "Chemistry for the Environment," Venice, Italy

INTRODUCTION

Liquid multiphasic systems, where one of the phases is catalyst-philic, are attractive for organic transformation, as they provide built-in methods of catalyst separation and product recovery, as well as advantages of catalytic efficiency. The present chapter focuses on recent developments of catalyst-philic phases used in conjunction with heterogeneous catalysts. Interest in this field is fueled by the desire to combine the high catalytic efficiency typical of homogeneous catalysis with the easy product–catalyst separation features provided by heterogeneous catalysis and in situ phase separations.

Multiphasic Systems

A multiphasic system for a chemical reaction can be constituted by any combination of gaseous, liquid, and solid phases. If a catalyst is present, it can be homogeneous or heterogeneous, thereby adding further phases—and degrees of freedom—to the system. Extra phases add new variables to a reaction, and it is therefore necessary that this be done for an advantage, such as an easier separation of the products, improved rates and selectivity, improved catalyst stability,

Methods and Reagents for Green Chemistry: An Introduction, Edited by Pietro Tundo, Alvise Perosa, and Fulvio Zecchini better catalytic efficiency, or improved environmental performance. It is just as obvious that there is a large number of cases where added phases generate advantages, as is testified by the growing number of articles in this area.

The most frequent multiphasic systems in the literature are biphasic systems. Industrially, the most relevant are gas–solid (G–S) systems where gaseous reactants are fluxed over a solid catalyst, generating products that are collected at the outlet. The synthesis of ammonia is an obvious example.

In a gas–liquid (G–L) system, a reagent gas is brought into contact with a liquid solution where reactant and homogeneous catalyst are dissolved. A typical case is that of homogeneous catalytic hydrogenation.

In a liquid–solid (L–S) system, a heterogeneous (or heterogenized) catalyst is used to promote a reaction in solution. This can be run in batch or in continuous-flow, and there are numerous examples of reactions done in an L–S system.

Gas–liquid–solid (G–L–S) where a reagent gas is brought into contact with a liquid solution where the reactant is dissolved and where a heterogeneous catalyst is suspended, that are triphasic systems, can be considered biphasic L–S systems in the context of this review.

Conversely, liquid–liquid (L–L) systems appear a little less obvious as systems for a chemical reactions, because partitioning of the species between the different phases becomes a critical issue. For example, two reactants can be in separate phases, and may need to be brought together by a phase-transfer catalyst (PTC),[1,2] or by a surfactant. The advantage here lies in the possibility of carrying out a reaction between two species with opposite polarity, without the need for a solvent such as acetone, dimethyl sulfoxide (DMSO), or dimethyl formamide (DMF). The phase-transfer-catalyzed halogen exchange reaction is an example. Recently, more work has been done on L–L biphasic systems that involve "neoteric" solvents such as dense carbon dioxide, polyethylene glycols (PEGs), and ionic liquids, which are catalyst-phylic. These solvents can often aid in catalyst separation and product recovery by phase separation of the two.[3]

Solid–solid (S–S) systems are now being investigated in view of eliminating solvents from chemical reactions. Here the paradigm is "the best solvent is no solvent." Just mixing two solids can often lead efficiently and cleanly to a product; however, there are limitations that are mainly due to the choice of reagents and to mass and heat transport.[4,5]

It should be pointed out that many biphasic systems have found their way into the chemical industry, starting from PTC and continuous flow (CF) processes. The reasons are that efficiency can be increased (rates, selectivity, energy requirements, reaction intensification), making them more economic and often more environmentally compatible, in short, more sustainable.

What we highlight here are some new recent multiphasic reaction systems for catalysis. The systems described here have in common a catalyst-philic phase, which contains, or coats a catalyst (mainly heterogeneous), or in some instances is the catalyst itself (PTC). There are two or three separate phases, and a general composition that can be summarized as: liquid-liquid-solid (L–L–S), or liquid–liquid–liquid–solid (L–L–L–S). One of the Ls indicates the liquid-ionic/hydrophilic phase (water, PEG, PT agent, ionic liquid, etc.) that is rich in the active catalytic species.

Before the 1990s there was little in the literature on multiphasic L–L–S and L–L–L–S systems used for chemical reactions. There is, however, a relatively large volume of work done on other types of multiphasic systems related to the present topic: supported liquid-phase catalysis (SL-PC), and gas liquid phase transfer Catalysis (GL-PTC).[6] The common denominator in both cases is the presence of an interfacial liquid layer of a hydrophilic compound between the catalyst and the bulk of the reaction.

In SL-PC, a catalyst is supported on a solid matrix in the form of the film of a nonvolatile liquid phase adsorbed on the solid. The catalytic film can be, for example, a molten salt or a molten oxide (e.g., Deacon's catalyst ($CuCl_2/KCl$) used to oxidize HCl with oxygen for the chlorination of ethylene in the synthesis of vinyl chloride, Figure 6.1; V_2O_5 for the oxidation of sulphurous to sulphuric anhydride). Alternately, it can be a liquid phase (e.g., ethylene glycol, PPh_3, butyl benzyl phthalate, etc.) that contains a soluble catalytic species such as a metal complex.

The reagents flow through in the gaseous phase (if they flowed in the liquid state, the catalytic species would be washed away), and the product diffuses in the gaseous stream and is collected at the outlet. The main drawback is that only relatively light compounds can be reacted, since they have to be in the gas phase, and low boiling products (and by-products) must be formed so that they can be easily recovered.

SL-PC was developed in view of industrial applications, since CF methods are largely preferred in that context. As an example, the hydroformylation of light olefins (up to C6), propylene in particular, was thoroughly studied by Sholten and co-workers to the pilot plant stage and to the calculations for a large-scale plant (20,000 ton/y). The catalyst was hydridocarbonyltris-(triphenylphosphine)-rhodium(I) dissolved in liquid PPh_3 as the stationary phase. It is noteworthy that the problems of Rh leaching and stability seemed to be resolved by operating under SL-PC conditions and by using the correct CF parameters. The limitation, for the industrial

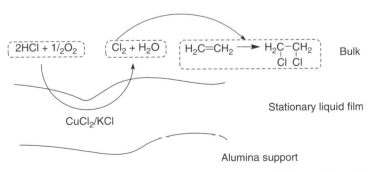

Figure 6.1 SL-PC for the synthesis of vinyl chloride with Deacon's catalyst.

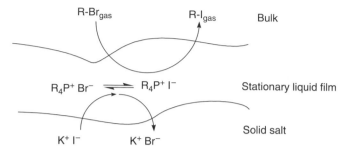

Figure 6.2   GL-PTC for the continuous-flow Finkelstein reaction in the gas phase.

application, was the low conversion per pass necessary to avoid the aldol condensation side reaction that ensues when the space velocity is low.

The situation changes if the reaction is not truly catalytic, that is, when the
reagent bed promotes the reaction, but is also consumed, such as in the case of a
halogen exchange reaction carried out under GL-PTC conditions (Figure 6.2).[7] In
this case, for example, a gaseous alkyl bromide can be fluxed over a reagent bed
made by a quaternary phosphonium bromide impregnated on silica gel, and potassium iodide. At the reaction temperature (150°C), the phosphonium salt is
molten, and the reaction takes place by diffusion of the alkyl halide into the
liquid stationary phase, followed by halide exchange and diffusion of the product
back to the gaseous stream. Iodide is replenished by exchange with the solid salt.

While the onium salt remains globally unchanged, the iodide is consumed
during the reaction. GL-PTC was developed, among other thing, for reactions
between ionic nucleophiles and gaseous electrophilic substrates, and for base
mediated reactions. The PT immobilized catalyst is a liquid at the operating temperature; it dissolves the salt, it activates the anion, and promotes anion exchange
with the solid bed.

Conceptually, GL-PTC and SL-PC are closely related. In fact, both involve the
presence of a stationary liquid interfacial layer, between the flowing gas phase
and the support, where the reaction takes place. Various examples of the preceding techniques have been reported over the years, and many are collected and
cited in the monograph mentioned earlier.[6]

Target

The present chapter targets multiphasic catalytic systems that can be represented
in general as L–L–S and L–L–L–S systems (Figure 6.3). The liquid phases are
two or three, and separate at ambient conditions. One of the Ls is a catalyst-philic
liquid phase that can be either ionic or hydrophilic, the equivalent to the supported liquid film described in the previous section. Figure 6.3 shows the two
different arrangements of the multiphasic systems that is considered here.

The underlying idea is that what is sometimes a separate liquid phase
that holds the catalyst, can, in other conditions, be considered a supported

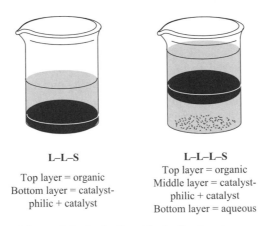

L–L–S

Top layer = organic
Bottom layer = catalyst-
philic + catalyst

L–L–L–S

Top layer = organic
Middle layer = catalyst-
philic + catalyst
Bottom layer = aqueous

Figure 6.3 L–L–S and L–L–L–S systems.

liquid film. The perspective depends on the amount of this phase: if the catalytic amount of liquid quat (or ionic liquid) is used, it forms a membrane that adheres to the catalyst; while if an excess is present, it forms a third—catalyst-philic—liquid phase.

The nature of this ionic/hydrophilic liquid phase can be quite diverse: it can be made by an onium salt (e.g., ammonium or phosphonium), by an ionic liquid (e.g., imidazolium salts), by polyethyleneglycols, and even water. What is required is that the catalyst-philic phase is not miscible with the other phases (organic or aqueous phases).

Interest in multiphasic systems with a catalyst-philic phase derives from the following nine factors:

1. The ability to form three separate stable liquid phases is intriguing and peculiar.
2. The catalyst can be "immobilized."
3. The catalyst can be stabilized.
4. The catalytic activity can be modified.
5. Separation of products can be simplified.
6. Phase transfer (catalysis) becomes an issue.
7. The solvents used can be advantageous from an environmental standpoint.
8. There may be process advantages (intensification, separation, environmental, energy requirements, etc.).
9. Both homogeneous and heterogeneous catalysts can benefit from the new conditions.

Each multiphasic system is discussed separately, by first describing its composition and behavior, then by describing the reactions its has been used for, and finally by rating its efficiency, advantages, drawbacks, and so forth.

6.1 LIQUID–LIQUID–SOLID MULTIPHASIC SYSTEMS

This section describes catalytic systems made by a heterogeneous catalyst (e.g., a supported metal, dispersed metals, immobilized organometallic complexes, supported acid–base catalysts, modified zeolites) that is immobilized in a hydrophilic or ionic liquid catalyst-philic phase, and in the presence of a second liquid phase—immiscible in the first phase—made, for example, by an organic solvent. The rationale for this multiphasic system is usually ease in product separation, since it can be removed with the organic phase, and ease in catalyst recovery and reuse; because the latter remains immobilized in the catalyst-philic phase, it can be filtered away, and it does not contaminate the product. These systems often show improved rates as well as selectivities, along with catalyst stabilization.

6.1.1 Supported Liquid Phase

Research in this field started in the wake of the reports of SL-PC. Consisting of a catalyst-containing supported liquid layer for CF reactions in the gas phase, the concept was transferred to batch reactions, using a catalyst dissolved in a supported aqueous phase. This was first referred to as supported aqueous-phase catalysis (SAPC) by Davis in an article published in *Nature* in 1989.[8] Later, the concept was extended, using a variety of names, but the essence has remained the same: a supported catalyst-philic phase.

In SAPC, a hydrophilic support such as silica is contacted with a water-soluble organometallic complex by aqueous-phase impregnation. After evacuation of the water phase, the organometallic complex becomes distributed on the support. Exposure to water vapor for a fixed time allows precise amounts of water to condense on the solid surface. The solid, coated by an aqueous film of catalyst, is placed in an immiscible organic phase that contains the reagents. The reactants diffuse from the organic phase into the porous solid, where they react at the water-organic interface, and the products diffuse back to the bulk organic phase. Along with the advantage of immobilization of the organometallic species, SAPC offers a high surface area for support, which translates into a high interfacial area, and the possibility of selectivity variations from bulk equilibrium product distribution through the effect of the interface.[8,9] One of the differences between SAPC and SL-PC is that, while SL-PC is designed for gas-phase reagents, SAPC is very efficient for liquid-phase reagents. Figure 6.4 shows schematically an SAPC.

The prototype reaction was the hydroformylation of oleyl alcohol (water insoluble) with a water-soluble rhodium complex, $HRh(CO)[P(m\text{-}C_6H_4SO_3Na)_3]_3$ (Figure 6.5). Oleyl alcohol was converted to the aldehyde (yield = 97%) using 2 mol % Rh with respect to the substrate and cyclohexane as the solvent, at 50 atmospheres CO/H_2, and 100°C. The SAPCs were shown to be stable upon recycling, and extensive work proved that Rh is not leached into the organic phase. Since neither oleyl alcohol nor its products are water soluble, the reaction must take place at the aqueous–organic interface where Rh must be immobilized. Also, if the metal catalyst was supported on various controlled pore glasses with

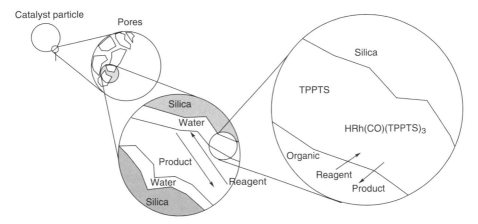

Figure 6.4 Schematic diagram of a SAPC. (From Ref. 9.)

different surface areas, the resulting conversions would be proportional to the interface area. It is noteworthy that the reaction did not proceed in liquid water as solvent with an unsupported catalyst. But, if the support was added, the reaction started, implying that the components self-assembled to yield the active SAPC. Rates were higher than with the corresponding homogeneous system.[10,11]

Analogously, the SAPC catalyzed hydroformylation reaction was carried out using other water-soluble metal complexes of Pt and Co. Pt complexes in the presence of an Sn co-catalyst underwent hydrolysis of the Pt–Sn bond, which led to lower reaction selectivity.[12] With the corresponding Co catalyst, good hydroformylation selectivities and conversions could be achieved, provided excess phosphine was used.[13] Other authors performed hydrogenation of α,β-unsaturated aldehydes using SAPC, and Ru and Ir water-soluble complexes.[14]

SAPC was then applied to asymmetric hydrogenation catalysis using the chiral Ru catalyst [Ru(BINAP-4SO$_3$Na)(C$_6$H$_6$)Cl]Cl.[15–17] The immobilization technique was the same as that used earlier. In this case, however, the presence of water caused the cleavage of the Rh–Cl bond, whose presence is critical for asymmetric induction. Therefore, a different nonvolatile hydrophilic liquid—inert with respect to the metal–halogen bond—was needed to replace water. The highly polar ethylene glycol was chosen as the catalyst-philic phase, coupled with a nonpolar mixture of cyclohexane and chloroform as the hydrophobic organic phase. The asymmetric reduction shown in Figure 6.6 proceeded with 96% enantiometric excess (ee) at 100% conversion in the system—now called supported liquid-phase

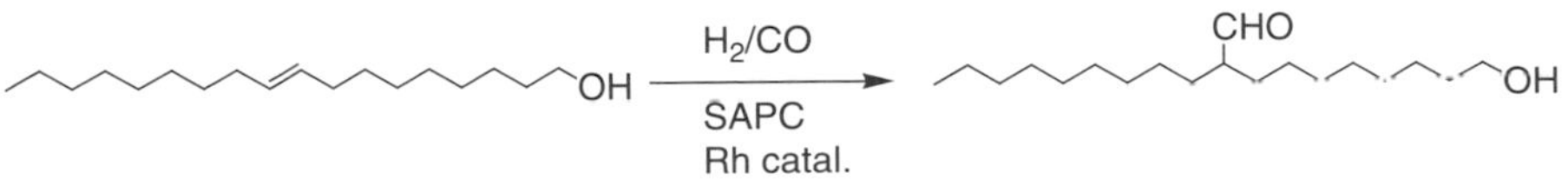

Figure 6.5 SAPC hydroformylation of oleyl alcohol.

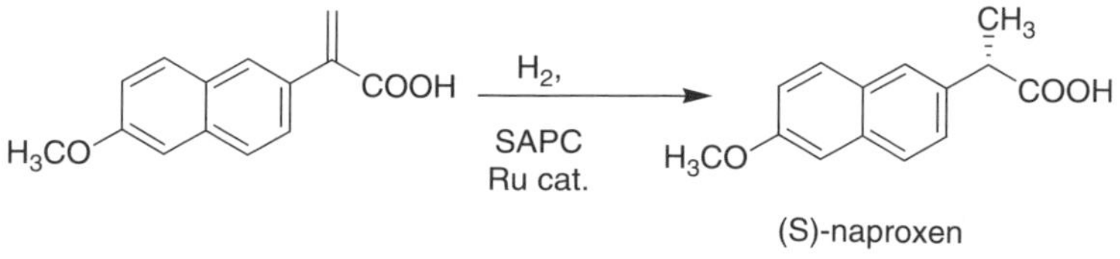

Figure 6.6   Asymmetric SAPC reduction.

catalysis (**SL-PC**)—without loss of Ru at a detection limit of 32 ppb. The SLP-Ru-BINAP-4SO$_3$Na was found to be at least 50 times more active than its two-phase (EtOAc/H$_2$O) analog, and only slightly less active than the homogeneous counterpart. In the absence of the silica support conversion did not exceed 2%. In this case, the stability of the SLPC catalyst also was confirmed by adding all the individual components of the SL-PC system separately, and by observing that they self-assemble and are more stable in the heterogeneous configuration than separated, which implies that the reverse process (i.e., separation of the SL-PC components) is unlikely.

A very similar kind of system was also described by Naughton and Drago shortly after.[18] They described supported homogeneous film catalysts (**SHFCS**), where water-soluble Rh catalysts were dissolved in PEG catalyst-philic phases immobilized on a silica gel support, and used for hydroformylation reactions. What was different in this case was that the catalyst-philic phase was modified by inserting a nonionic surfactant—Surfynol 485—that enhanced catalytic activity, presumably by solubilizing the substrate in the catalyst-philic PEG phase, and increasing the concentration of the alkene available for hydroformylation. Other hydrophilic polymers, such as polyvinylpyrrolidinone, polyethelene oxide, and polyvinyl alcohol, were also tested as catalytic films. Finally, high-boiling polar compounds such as formamide and glycerol were also employed; however, all showed lower activity.

Analogously, over the years, Arai and co-workers have investigated silica-supported ethylene glycol as a catalyst-philic phase, which contained a metal precursor, for C–C bond-forming reactions, such as the Heck reaction. They describe a multiphasic system with an organic phase (solvent) that contains only reactants and products without any catalyst. The products could be recovered by simple filtration, and the catalyst recycled many times without deactivation, since it did not precipitate, thus making the catalytic system stable and reusable (Figure 6.7).[19,20]

The method of catalyst immobilization is one for the reasons for the success of the SAPC approach. Rather than covalently linking an organometallic complex to a support—which usually leads to loss of catalytic efficiency and leaching of the metal—it is the catalyst-philic phase that is immobilized.

Horvath recognized that SAPC solved the problem posed by the solubility of lyophilic substrates in aqueous biphasic catalysis with water-soluble homogeneous catalysts.[21] He compared biphasic aqueous–organic catalysis with SAPC, in order to clarify whether in SAPC the catalyst remained dissolved in the

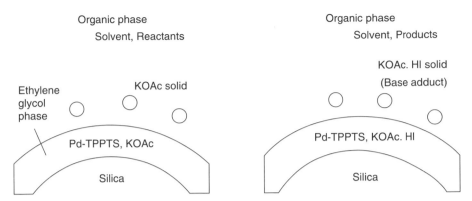

Figure 6.7 Representative ethylene glycol-based SAPC catalyst.

aqueous phase, or if it works at the aqueous–organic interface. High-pressure infrared (IR) studies indicated that water acts as an immobilization agent rather than a solvent. This was apparent from the fact that in SAPC water (not the catalyst) leached from the support in an amount that left only two monolayers of water on the hydrophilic support. This led to the theory that the water layer holds the water-soluble phosphines by hydrogen bonding the hydrated Na-sulphonate groups to the surface (Figure 6.8). The metal coordinated by the phosphines is therefore found precisely at the interface between the supported aqueous phase and the bulk organic phase (Figure 6.9), and is readily available for hydrophobic substrates.

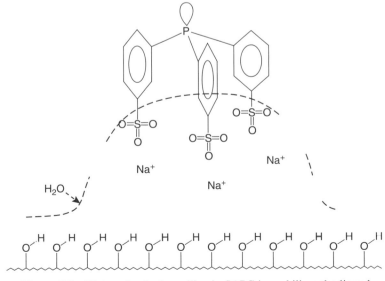

Figure 6.8 Water adsorbed on silica in SAPC immobilizes the ligand.

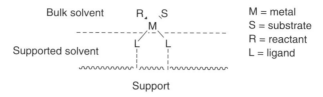

Figure 6.9 Schematic coordination of water-soluble phosphines with a metal at the water–organic interface.

Since these first reports on the use of SAPC, the concept has been applied to a large number of reactions, with different metals and ligands. It is peculiar that the technology was renamed "glass bead technology" in a review on the topic.[22] The investigated reactions range from hydroformylation, to hydrogenation, to Wacker oxidation, to Heck couplings,[23] to Suzuki couplings, to allylic substitution; using Rh, Pd, Co, Pt, and in the presence of supported phases, water, PEGs, ethylene glycol.

A recent technological modification of SLPC and SAPC introduced supercritical carbon dioxide (scCO$_2$) as the "organic" phase.[24] In a SAPC–scCO$_2$ system, using Ru[P(m-C$_6$H$_4$SO$_3$Na)$_3$]$_3$, it was possible to efficiently reduce cinnamaldehyde to the corresponding unsaturated alcohol with high selectivity (96%). Here the high solubility of reactant gases in scCO$_2$ overcomes G–L–L mass-transfer limitations.

Recently the concept has been reformulated and applied to a new class of solvents: ionic liquids. These were supported on silica—covalently anchored or adsorbed—and used as catalyst-philic phases for metal complexes in hydroformylation and hydrogenation reactions. The concept is identical to that of SAPC, but the acronym was modified to SILC (supported ionic liquid catalysis). The first type of reaction that was reported was the hydroformylation of 1-hexene, using Rh as catalyst.[25] The ionic liquid phase was made by butylmethylimidazolium hexafluorophosphate ([bmim][PF$_6$]) supported on silica, modified by covalently anchoring ionic liquid fragments (Figure 6.10). What was obtained was a system where [bmim][PF$_6$] was supported on silica and contained the active catalytic species HRh(CO)[P(m-C$_6$H$_4$SO$_3$Na)$_3$]$_3$ plus an excess of free phosphine. The catalyst was a free-flowing powder that was made to react with the substrate and CO/H$_2$. The activity of this SILC system was slightly higher than conventional biphasic catalysis; however, leaching became significant unless free phosphine ligand was added in the ionic liquid layer.

SILC was also used without covalently anchoring the ionic liquid fragment to the silica support. In this case, [bmim][PF$_6$] was simply added to silica in acetone together with the catalyst.[26] [Rh(norbornadiene)(PPh$_3$)$_2$]PF$_6$ and the solvent evaporated to yield the supported catalyst-philic phase. Catalyst evaluation on the hydrogenation of model olefins showed enhanced activity in comparison to homogeneous and biphasic reaction systems, in analogy to Davis's observations.[9] Also

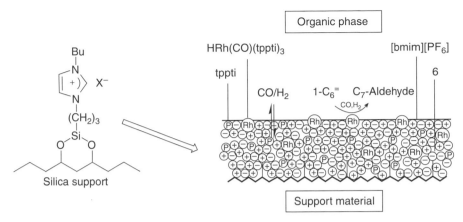

Figure 6.10 Immobilization of an ionic liquid used for SILC hydroformylation.

in this case, a concentration effect was invoked to explain the better performance, as most of the reaction occurs at the interphase (Figure 6.11). The SILC showed good stability. The same catalyst system could be used 18 times without loss of activity. As far as leaching of the metal is concerned, Rh remained below the 33-ppb (parts per billion) detection limit, and the residual organic phase did not show catalytic activity. In addition, scanning electron microscopy (SEM) and transmission electron microscope (TEM) images of used and fresh catalysts were identical, indicating no clustering of the metal.

An immediate extension of SAPC and SILC was the continuous-flow analogue performed by Fehrmann and co-workers, called supported ionic liquid phase (SILP).[27] The catalyst was prepared by impregnation of the rhodium precursor,

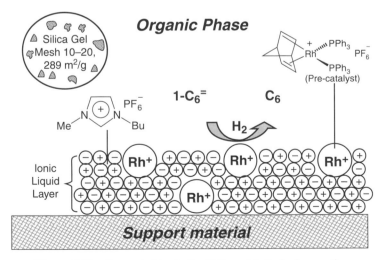

Figure 6.11 Supported ionic liquid for catalytic hydrogenation.

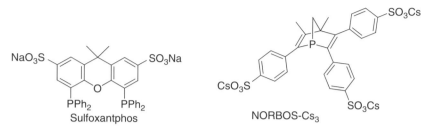

Figure 6.12   Structure of the ligands sulfoxantphos and NORBOS-Cs₃.

the sulfonated biphosphine ligand (sulfoxantphos, Figure 6.12) on silica, in the presence of the ionic liquids [bmim][PF₆], and halogen-free [bmim][n-C₈H₁₇-OSO₃]. The continuous-flow gas-phase hydroformylation of propene was demonstrated at 10 bar and 100°C. It was shown that the reaction proceeds with a ligand/Rh ratio between 10 and 20, indicating that the active ligand-containing species are formed in situ. What was also evident was that the catalytic performance was scarcely influenced by the anion of the ionic liquid. Finally, it was observed that deactivation of the catalyst could be prevented by tuning the ligand/Rh ratio.

The use of a catalyst-philic phase made by [bmim][n-C₈H₁₇OSO₃] also addresses environmental issues and should be noted. In fact, while the PF₆ anion is certainly useful for exploratory studies, it easily hydrolyzes and generates HF.

The scope of SILP was extended by investigating charged monophosphine ligands, as well as liquid-phase CF hydroformylation.[28] The latter was demonstrated on 1-octene using the SILP Rh-(NORBOS-Cs₃)/ [bmim][PF₆]/silica catalyst. The authors recognize that the supported catalyst-philic phase offers the significant advantage of very efficient ionic liquid use.

Both SILC and SILP offer the advantage over SAPC of using ionic liquids instead of water. The low vapor pressure ensures that the supported phase remains liquid under the reaction conditions, and that it is retained during continuous flow operation.

Onium salts, such as tetraethylammonium bromide (TEAB) and tetra-n-butylammonium bromide (TBAB), were also tested as PTCs immobilized on clay. In particular, Montmorillonite K10 modified with TBAB efficiently catalyzed the substitution reaction of α-tosyloxyketones with azide to α-azidoketones, in a biphasic CHCl₃/water system (Figure 6.13).[29,30] The transformation is a PTC reaction, where the reagents get transferred from the liquid to the solid phase. The authors dubbed the PTC-modified catalyst system "surfactant pillared clay" that formed a "thin membrane-like film at the interface of the chloroform in water emulsion," that is, a third liquid phase with a high affinity for the clay. The advantages over traditional nucleophilic substitution conditions were that the product obtained was very pure under these conditions and could be easily recovered without the need for dangerous distillation steps.

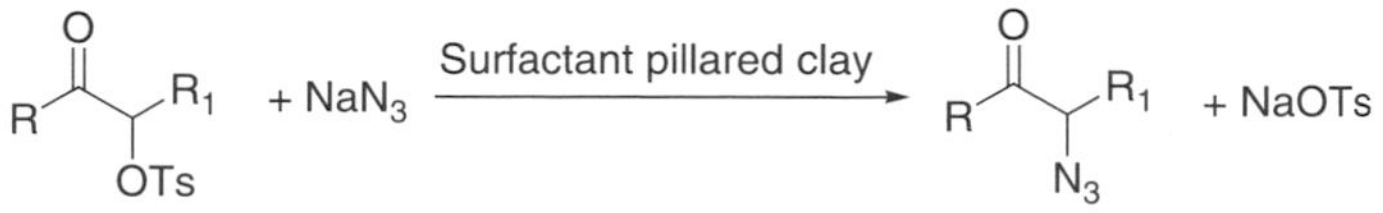

Figure 6.13 Substation of α-tosyloxyketones with azide to give α-azidoketones.

6.1.2 PEG-Stabilized Metal Nanoparticles

Metal nanoparticles, and in particular Pd-nanoparticles (also called Pd-colloids), are stabilized by the presence of, among others, ammonium salts,[31–35] surfactants,[36] PEGs,[37] polysiloxane,[38] and organic thiol monolayers.[39] The highly polar ethylene glycol was shown earlier to be a catalyst-philic phase for SAPC. Analogously, polyethylene glycols are an attractive polar and high-boiling class of compounds that can be considered catalyst-philic in a number of instances, and that can be coupled with a second immiscible phase.

An L–L–S system based on PEG, is, for example, the one described by Leitner and co-workers, where the ability of PEGs to stabilize dispersed Pd-nanoparticles was coupled with the use of scCO$_2$. The model reaction that was investigated in this system was the oxidation of alcohols with oxygen in the presence of Pd-clusters of structure [Pd$_{561}$-phen$_{60}$(OAc)$_{180}$] (phen = 1,10-phenanthroline) (Figure 6.14).[40]

The performance of the highly dispersed Pd-clusters—in the oxidation of the alcohols in scCO$_2$—when embedded in a PEG-1000 matrix was particularly good and showed high selectivity (>99%), with no overoxidation observed. PEG-1000 is a solid at room temperature, but became a liquid under the investigated conditions, wherein the catalyst was well dispersed. A 2-h induction period was observed, but once the reaction was terminated, the products could be extracted quantitatively with CO$_2$ (80°C, 14.5 MPa), and the catalyst-philic PEG phase remained in the reactor. Negligible amounts of PEG were extracted, and a very low contamination of the product by Pd was observed (<2.3 ppm). The PEG–Pd catalytic ensemble could be used repeatedly without loss of activity, and without the induction period of the first run. Control experiments showed that the activity and selectivity of the catalytic system were higher than those with conventional heterogeneous catalysts.

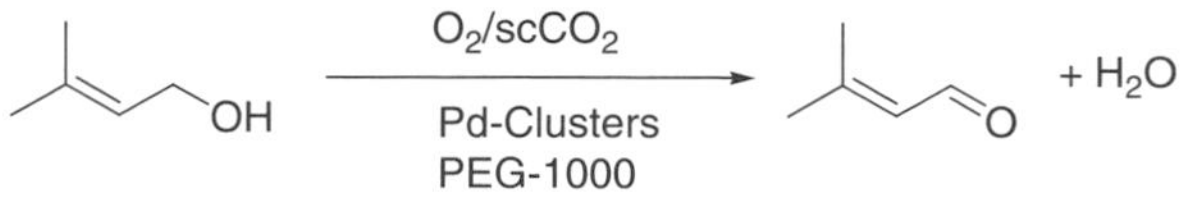

Figure 6.14 Selective oxidation using PEG-stabilized Pd-nanoparticles.

6.2 LIQUID–LIQUID–LIQUID–SOLID MULTIPHASIC SYSTEMS

The catalytic systems described here are liquid triphasic ones, with a heterogeneous catalyst such as supported noble metals (Pt, Pd), or high-surface-area metals (Raney–Ni). The liquid phases are constituted by an alkane, water, and an ammonium salt. This kind of system was developed over the past 12 years, initially as an efficient and mild catalytic methodology for the hydrodehalogenation reaction of haloaromatics, after which it was studied for other kinds of reactions, and careful observation has resulted in a theory of modes of action whereby reaction rates, and selectivity, could be intensified.

6.2.1 The Multiphasic System

The third liquid catalyst-philic phase was constituted in the majority of cases by Aliquat 336® [tricapryl-methylammonium chloride, $(C_8H_{17})_3N^+CH_3Cl^-$: A336], a well-known phase-transfer agent that is liquid at room temperature, and that dissolves in toluene and in iso-octane (Figure 6.15a). The peculiarity here is that, when water (even a drop) is added to the A336/isooctane solution, three liquid phases separate out (Figure 6.15b).

Figure 6.15 shows the triphasic system at rest: in order to visualize the different phases, macroscopic amounts of the three components were combined. In

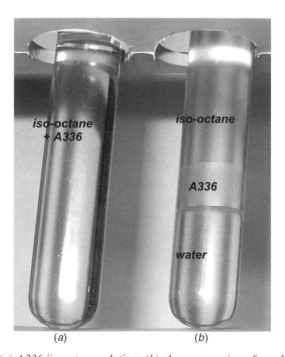

Figure 6.15 (a) A336/isooctane solution; (b) phase separation after addition of water.

practice, the experimental setup appeared different, in that the A336 phase was used in catalytic amounts (0.20 molar) with respect to the substrates. Under the operative conditions the system appeared biphasic, with a thin layer of A336 at the iso–octane–water interface. When the heterogeneous catalyst was added to the triphasic system it resided in the A336 catalyst-philic phase, as is shown in macroscopic quantities in Figure 6.16.

What should be highlighted is that the figures show the triphasic system at rest. When stirred, which was necessary in order to achieve high interfacial area and reduce mass transport limitations, the system was better represented by catalyst particles coated by a layer of A336, immersed in the isooctane–water biphasic system (Figure 6.17). This system—where the catalyst-philic phase was A336—also could be considered the other way around as an A336-philic catalyst that is dispersed in the bulk biphasic system. However, inverting the factors does not change the result. The thin film of A336 acted as an interfacial boundary layer and was in close proximity to the catalytically active sites. This vicinity, and the ability of the A336 membrane to mediate the transfer of the reagents and products to and from the catalyst, was used to explain the selectivity and kinetics enhancements described in the following sections.

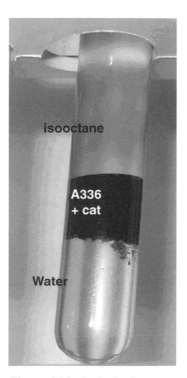

Figure 6.16 L–L–L–S system.

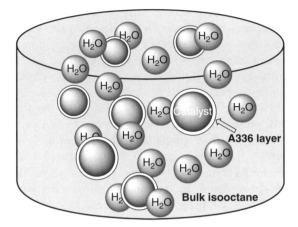

Figure 6.17   Pictorial view of the stirred L–L–L–S multiphasic system.

Unlike some of the multiphasic systems described earlier, this system uses only a catalytic amount of the third phase, thereby eliminating the need for large quantities of expensive phase-transfer catalysts or ionic liquids. Along with

TABLE 6.1   Third Phase Constituents

		Reference
1	Aliquat 336 (A336)	41–63
2	$C_{16}H_{33}(C_{18}H_{37})_3N^+Br^-$	43,45
3	$C_{16}H_{33}(n\text{-}Bu)_3P^+Br^-$	42–46
4	$C_{16}H_{33}(py)^+Br^-$	42
5	$PhCH_2(C_2H_5)_3N^+Br^-$	43
6	$(n\text{-}Bu)_4N^+HSO_4^-$	43
7	$MeO(CH_2CH_2O)_nH_{\ n\sim15}$	43,44
8	PEG 6000	43
9	PPG 2000	43
10	Brij 35, 52, 56, 58	49
11	Brij 52	49
12	Brij 58	49
13	Brij 56	49
14	$PhCH_2(CH_3CH_2)_3N^+Cl^-$	49,50
15	$PhCH_2(n\text{-}Bu)_3N^+Cl^-$	49
16	Et_2NH	50
17	Et_3N	50
18	$n\text{-}Bu_3N$	50
19	$(PhCH_2)_3N$	50
20	$n\text{-}C_8H_{17}NH_2$	50
21	Cinchonidine	53
22	Cinchonine	53

ammonium salts—in particular, Aliquat 336—other third phases were used as well: phosphonium salts, polyethylene glycols, amines, as shown in Table 6.1.

Not all of these salts formed a true separate phase, but all adhered to the heterogeneous catalysts that were used, and had an effect in modifying the catalytic activity and the reaction parameters.

The following discussion is organized by reaction type, while other parameters (metal catalyst, third phase, solvents, base concentration, etc.) are addressed as they arise.

6.2.2 Hydrodehalogenation

The hydrodehalogenation reaction of haloaromatics involved the substitution of halide atoms bound to the ring, with hydrogen. For example, tetrachlorobenzene could be reduced to benzene in 30 minutes, at 50°C, by bubbling H_2 at atmospheric pressure in the multiphasic system constituted by isooctane, 50% aqueous KOH, 0.2 molar A336, in the presence of Pd/C (0.02 molar) (Figure 6.18).[43]

The A336 phase was shown to play a decisive role on kinetics by coating the Pd/C catalyst. In fact, by plotting the rates of the competitive hydrodehalogenation of o-, m-, and p-chloroethylbenzenes as a function of the concentration of A336, a Langmuir isotherm was obtained: the rates increased by increasing A336, until a plateau was reached (Figure 6.19). Such behavior may indicate the formation of a film of A336 on the catalyst surface.[44]

Different halides were also shown to behave differently under the multiphasic conditions. While bromide was removed faster than chloride, hydrodeiodination became inhibited, likely due to the formation of the ammonium–iodide ion pair (lipophilic and therefore present in the organic phase), which in contact with the Pd/C catalyst poisoned the metal.

The multiphasic system was able to conduct the hydrodehalogenation reaction selectively in the presence of other functional groups prone to reduction. It was therefore possible to transform halogenated aryl ketones into the corresponding aryl ketones, without consecutive carbonyl reduction. This was possible in the multiphasic system—only when A336 was present—and not in the traditional alcoholic solvent systems (Figure 6.20).[45] This kind of selectivity could be used for synthetic purposes in the preparation of aryl ketones otherwise not accessible through direct Friedel–Crafts acylation (e.g., Figure 6.21).[48]

As far as the metal catalyst was concerned, Raney–Ni and Pt/C were also investigated. Raney–Ni proved effective in the hydrodehalogenation reaction of dichloro- and dibromobenzenes with hydrogen in the multiphasic system

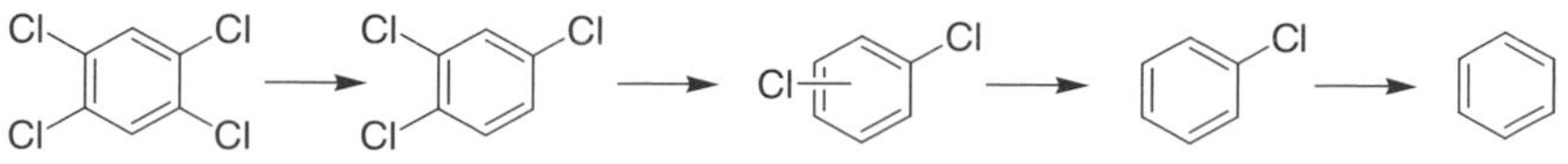

Figure 6.18 Hydrodehalogenation of tetrachlorobenzene.

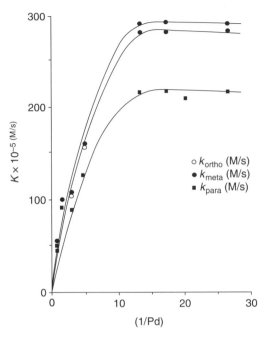

Figure 6.19 Rate constants for the competitive hydrodehalogenation of *o*-, *m*-, and *p*-chloroethylbenzenes as a function of the concentration of A336, in the multiphasic system.

(isooctane, 50% aqueous KOH, A336). And—contrary to Pd/C, which showed activity in the absence of A336 as well—Raney–Ni functioned only when A336 was present.[46,47]

Platinum on charcoal provided a further means for directing the selectivity of the multiphasic hydrodehalogenation reactions. For example, the same reaction in Figure 6.20 conducted using Pt/C instead of Pd/C, yielded selectively the dehalogenated benzylic alcohol (Figure 6.22).[49] The same reaction was conducted using

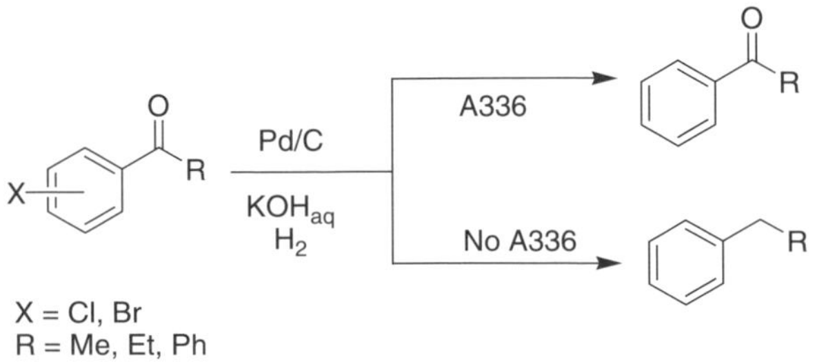

Figure 6.20 Selective reduction of haloaromatic ketones in the multiphasic system using Pd/C.

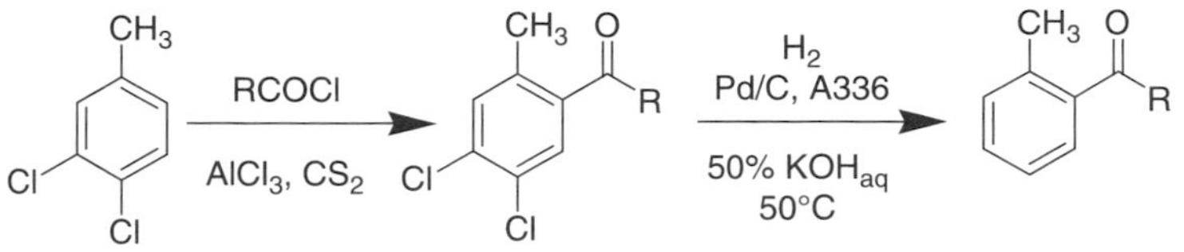

Figure 6.21 Chloride as directing–protecting group in Friedel–Crafts acylations.

different catalyst-philic phases, in order to understand their influence. In particular, different ammonium salts—triethylbenzyl- and (tributylbenzyl)ammonium; and polyethylene glycols such as Brij 35, 52, 56, 58—were used. A correlation between increased hydrophilicity and chemoselectivity toward the benzylic alcohol was proposed.

The study using Pt/C was further extended by modifying the amount of KOH. The base also influenced the selectivity by reducing the amount from 50% to less than 5% in the aqueous phase, the final pH of the reaction mixture remaining between 8 and 13. In this range, the preferred product returned to be the ketone (Figure 6.23).[50]

Kinetic studies performed on model compounds were aimed at understanding the effect of different parameters on the selectivity. They showed that selectivity was achieved only when A336 was present. In fact, in the absence of A336 and of the base the hydrodehalogenation of *p*-chloroacetophenone proceeded all the way to ethylcyclohexane in the biphasic aqueous–organic system. When A336 was added, selectivity was reversed—chloride was removed first—and the selective dehalogenated benzyl alcohol was obtained.[51]

The multiphasic hydrodehalogenation reaction proved attractive from an environmental perspective. It is well known that halogenated organic compounds are dangerous for the environment and for human health; according to the Stockholm convention, many of the banned compounds are classified as persistent organic pollutants (POPs) that belong in fact to this category. Examples of such toxics are polychlorobiphenyls (PCBs), chlorophenols, chlorobenzenes, poly-chlorodibenzodioxins and -dibenzofurans (PCDDs and PCDFs, respectively, also

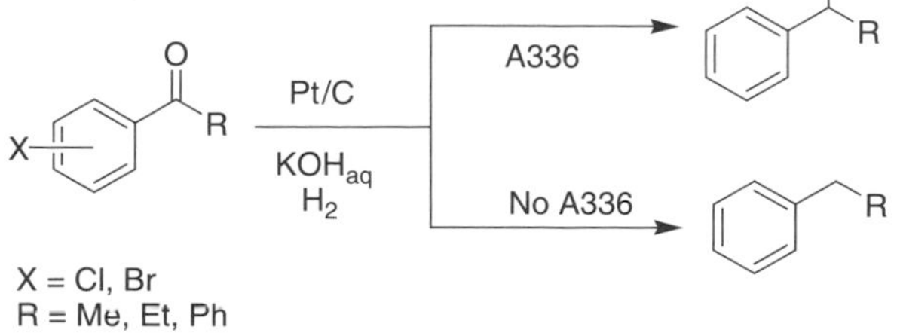

Figure 6.22 Selective reduction of haloaromatic ketones in the multiphasic system using Pt/C.

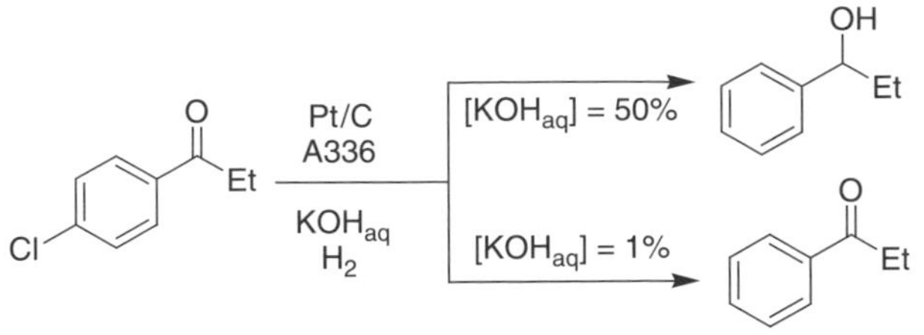

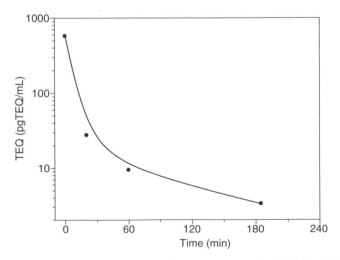

Figure 6.23 Selective reduction of haloaromatic ketones in the multiphasic system using Pt/C.

known as dioxins). They can be generated by incineration processes, high-temperature transformation of organic compounds, and in some cases they are residues of the agrochemical sector (e.g., insecticides such as DDT, or defoliants such as agent orange), or from other applications (PCB as dielectric medium in electric transformers). These molecules accumulate in the environment and in the fatty tissues of animals, and tend to concentrate along the food chain, posing a serious hazard for man. Elimination of this class of compounds by incineration is the cheapest method, and widely employed; it is, however, undesirable, as there is the risk of generating even more toxic PCDDs and PCDFs in the process. The alternative is reductive dehalogenation. The multiphasic catalytic hydrodehalogenation method described in the previous paragraph was immediately recognized as an efficient method for the detoxification of PCBs.[44] It was then applied to environmental samples of PCDDs and PCDFs extracted from fly ashes of a municipal solid-waste incinerator (MSWI).

The mild (50°C, 1 atm) catalytic multiphasic methodology can reduce the toxicity equivalent (TEQ) of a sample from a total of 572 pgTEQ/mL to less than

Figure 6.24 Hydrodechlorination of an environmental sample of dioxins: TEQ vs. time.

3 pgTEQ/mL (the detectability limit as determined by HRMS) in 3 hours reaction time (Figure 6.24).[52] Since the large excess of charcoal present as Pd/C may have simply adsorbed the dioxins without reaction, a series of experiments were conducted in order to ensure that the PCDD/Fs were being completely dechlorinated. Later the multiphasic methodology was employed to hydrodechlorinate other hazardous chlorinated organic compounds (Figure 6.25).

Lindane, a widespread insecticide constituted of technical mixtures of hexachlorocyclohexane (HCH), was dechlorinated using Pd/C, Pt/C, or Raney-Ni. The product using the multiphasic system was benzene, obtained within 1 hour.[58] The base initially promoted HCl elimination from HCH to yield trichlorobenzene, which then underwent the usual hydrodechlorination reaction.

Dieldrin—which belongs to the 12 POPs banned by the Stockholm convention and is in the same class of other pesticides named "drins," such as aldrin and endrin—possesses six aliphatic chlorine atoms on a polycyclic skeleton. The multiphasic dechlorination, in the presence of A336, isooctane, aqueous KOH, Pd/C, and hydrogen, proceeded with a different selectivity and degree of dechlorination, depending on the choice of catalyst system, and base concentration. It always required the base and was favored by the presence of A336. It produced a mixture of products derived from the subsequent removal of chlorines, up to a small percentage of monochlorinated derivative.[62]

DDT—[1,1'-bis-(4-chlorophenyl)-2,2,2-trichloroethane]—could be hydrodechlorintated completely, both with Pd/C and Raney-Ni, provided A336 was present. The aliphatic chlorines reacted faster than the aromatic ones, the first via base promoted elimination of HCl.[62]

Finally, 2,4,5-trichlorophenoxyacetic acid (2,4,5-T: agent orange) and pentachlorophenol (PCP)—two acidic polychlorinated pesticides—were also investigated. In these cases, the presence of A336 contributed to the solubility of

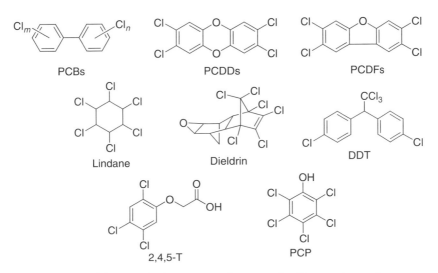

Figure 6.25 Chemical structures of some toxic chlorinated organics.

the reagent in the organic phase, by forming the ion pair between the anion of the reagent and the ammonium cation.[63]

6.2.3 Reduction

As already shown in paragraph Section 6.2.2, the multiphasic conditions for hydrodechlorination, are also active for hydrogenation reactions, such as was the case of haloaromatic ketones, which could selectively be reduced to the alcohol.[44–47,49,50] This reaction was investigated from the kinetic standpoint, using kinetic models that allowed an accurate analysis of the reaction selectivity.[51] In particular, it was shown that halogen removal was accelerated with respect to carbonyl reduction in the presence of A336.

The multiphasic system was also used for the enantioselective hydrogenation of acetophenone—a model carbonyl compound—using chiral modifiers in place of A336.[53] The idea was to employ chiral phase-transfer agents derived from chincona alkaloids in place of A336, coupled with Pt/C as catalyst. A rapid screening, however, demonstrated that the pure alkaloids chincona and chinconidine were more active than the corresponding ammonium salt in promoting a degree of enantioselectivity in the reduction of acetophenone (Figure 6.26). The amount of chincona modifier was correlated with the degree of conversion and ee, demonstrating that it covered the heterogeneous catalyst and formed chiral pockets able to stereo-recognize the substrate (Figure 6.27). This reaction was the first example of heterogeneous chiral catalysis on nonactivated ketones.

The use of Raney-Ni in place of more expensive Pd and Pt catalysts was considered from the beginning, and it showed potential for the hydrodechlorination reaction, as already discussed. It proved to be a less active catalyst—not surprisingly—but it also showed high potential in the multiphasic system, since its activity was always strongly enhanced by the presence of A336.[55] A striking example of this difference was shown in the case of the hydrogenolysis of benzylmethyl ether, where only Raney-Ni proved active enough to promote the C–O bond-breaking reaction under the multiphasic conditions (Table 6.2). The potential of this reaction was investigated over a series of different ethers, including BOC-protected O-benzyl-serine (Figure 6.28), where only the benzyl group was selectively removed under the multiphasic conditions.

>90% selectivity toward 1-phenylethanol
up to 20% ee

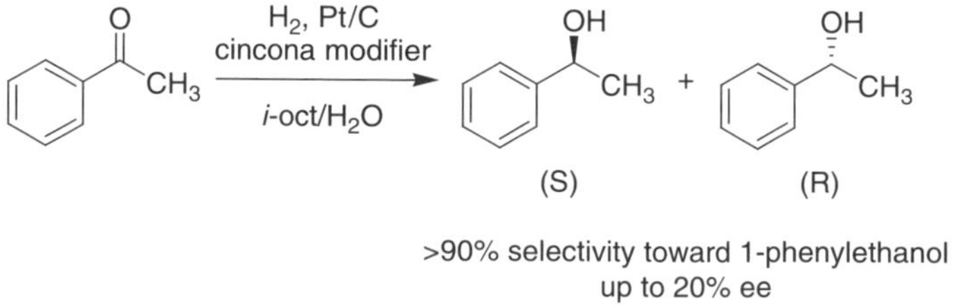

Figure 6.26 Acetophenone enantioselective reduction.

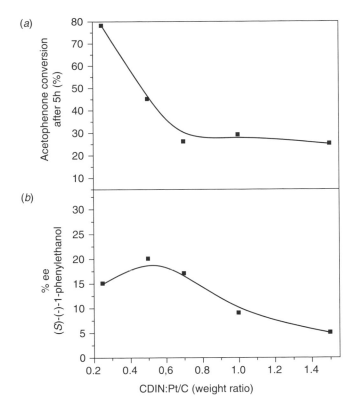

Figure 6.27   Effect of varying amounts of CD on conversion and ee for the enantioselective multiphasic reduction of acetophenone.

TABLE 6.2   Hydrogenolysis of Benzyl Methyl Ether

Catalyst	Solvent System	PT Agent	Time (min)	Conversion (%)
Pd/C	Ethanol	None	30	100
	Multiphasic	A336	310	3
		None	330	85
Pt/C	Ethanol	None	240	33[a]
	Multiphasic	A336	360	0
		None	300	5[a]
Ra–Ni	Ethanol	None	230	37
	Multiphasic	A336	150	100
		None	420	85
		SDS	435	98

[a]Toluene was further reduced to methylcyclohexane.

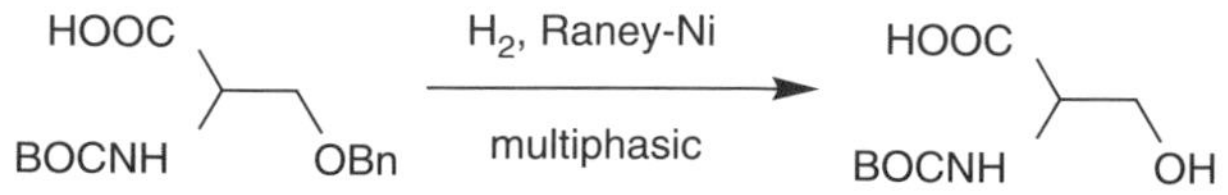

Figure 6.28   Selective debenzylation of BOC-protected serine.

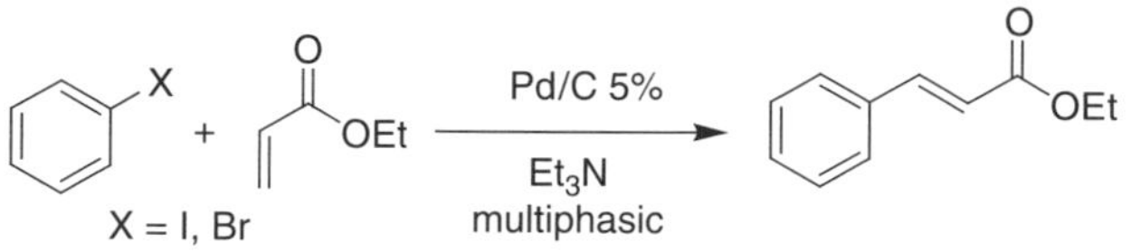

Figure 6.29   Heck reaction.

6.2.4   Carbon–Carbon Bond Formation

The scope of the multiphasic system was extended to coupling reactions—like the Heck reaction—using a heterogeneous supported catalyst, such as Pd/C.[61] The rationale here lay in the observation that aryl halides were activated in the multiphasic system (as seen for hydrodehalogenation), and that therefore they should also be activated toward C–C coupling reactions.

The multiphasic system, where A336 was coupled with the use of Pd/C in a mixed aqueous–organic solvent system, promoted the Heck reaction of aryl iodides and -bromides. In particular, the reaction of aryl iodides with electron-poor olefins became up to 10 times faster in the presence of A336 than without it (Figure 6.29). Under the multiphasic conditions, A336 formed a third catalyst-philic liquid phase that allowed the catalyst, products and reagents, and base to be kept separate. This was regarded as an advantage over solventless systems (where separation and recovery of the catalyst can be complicated), and over reactions carried out in pure ionic liquids that are expensive and that require a product recovery step). In addition, the multiphasic system used a heterogeneous catalyst rather than a homogeneous one, and Pd activating phosphines could be done without.

A similar "micromembrane" (Figure 6.30) formed by TBAB on Pd/C was able to selectively synthesize biphenyls from halobenzenes using a reducing

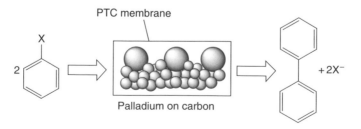

Figure 6.30   PTC membrane hypothesis.

agent (formate) and a weak base (Na_2CO_3). In this case, the concentration of water was crucial, since an excess caused the TBAB to leach from the supported catalyst.[64] Figurer 6.30 represents the same concept expressed earlier in Figure 6.17.

6.3 CONCLUSIONS AND OUTLOOK

Catalyst-philic liquid phases can be used to promote the catalytic activity of heterogeneous catalysts, and to facilitate product–catalyst separation. A variety of different constituents of such catalyst-philic phases can be used, the most attractive being quaternary ammonium and phosphonium salts, PEGs, as well as water and other kinds of low-temperature molten salts. In each system, the catalyst-philic phase is characterized as being separate from the remainder of the reaction mixture, and the catalyst should reside within this phase.

The types of reactions that can be run under multiphasic conditions range from substitution reactions, to metal catalyzed transformations (such as reductions, oxidations, hydrodehalogenations, hydrogenolyses, isomerizations, hydroformylations, C–C bond forming), to Friedel–Crafts and Wittig reactions. These multiphasic systems represent a tool—available to process and organic chemists—that can tune a catalytic reaction and provide a built-in method for product–catalyst separation. In principle, the latter can be done either in batch reactions by decanting and physically separating phases, or by operating in CF. An added advantage is that the volume of the catalyst-philic phase can be reduced until only catalytic amounts are used, a benefit in cases where cost and availability are an issue (e.g., many kinds of ionic liquids).

We hope we have managed to highlight the common underlying idea on how multiphasic systems can be formed and on how they can be used. We are nonetheless aware that it is difficult to determine how one system can be better than another, and how to choose the right one for a particular application. This negative impression may be increased by the many different possible multiphasic systems, and by the overwhelming number of different names attributed even to similar systems.

The point we would like to make is that, while it is of crucial importance to advertise new chemistry and new applications, also through catchy words and acronyms, one should be careful to maintain a clear record of what has already been done in the field. This is sometimes overlooked due to the legitimate aspiration of authors to name their discoveries, and may lead them to deem old things new and to mislead less experienced and young readers. For example, the thin line that divides PTC from the use of ammonium-based molten salts should be kept in mind when describing rate enhancements or phase separations in their presence. The origins of PTC and molten salts are different, since PTC started as a new method for catalytic anion activation, while molten salts attracted attention as new reaction media. Nonetheless, the latter field seems to parallel the development of PTC in some instances, and this should be remembered when putting one's results in context. The future of multiphasic systems is in its infancy, and many other kinds of multiphasic systems that have not been addressed here (supercritical fluids, fluorous fluids, etc.) are gathering a great deal of attention, and also should be considered when designing new chemical reactions and/or processes. In addition, combinations of the possibilities outlined herein should be considered; as was the case of a recently published article where imidazolium salts were supported on PEGs ([PEGmim][Cl]) and used as a reaction medium for the Heck reaction.[65]

REFERENCES

1. C. Starks, C. Liotta, M. Halpern, *Phase-Transfer Catalysis: Fundamentals, Applications and Industrial Perspectives*, Chapman & Hall, New York, 1994.

2. E. V. Dehmlow, S. S. Dehmlow, *Phase-Transfer Catalysis*, 3rd ed., Verlag Chemie, Weinheim, 1993.

3. B. Cornils, W. A. Herrmann, I. T. Horvath, et al., (Eds.) *Multiphase Homogeneous Catalysis* Wiley-VCH, Weinheim, 2005.

4. A. K. Bose, S. Pednekar, S. N. Ganguly, et al., *Tetrahedron Lett.*, 2004, **45**, 8351–8353.

5. K. Tanaka, *Solvent Free Organic Synthesis*, Wiley-VCH, Weinheim, 2005.

6. P. Tundo, *Continuous Flow Methods in Organic Synthesis*, Ellis Horwood Ltd., Chichester, UK, 1991.

7. P. Tundo, P. Venturello, *Synthesis*, 1979, 952–954.

8. J. P. Arhancet, M. E. Davis, J. S. Merola, B. E. Hanson, *Nature*, 1989, **339**, 454–455.

9. M. E. Davis, *Chemtech*, 1992, 498–502.

10. J. P. Arhancet, M. E. Davis, B. E. Hanson, *J. Catal.*, 1991, **129**, 94–99.

11. J. P. Arhancet, M. E. Davis, B. E. Hanson, *J. Catal.*, 1991, **129**, 100–105.

12. I. Guo, B. E. Hanson, I. Toth, M. E. Davis, *J. Mol. Catal.*, 1991, **70**, 363–368.

13. I. Guo, B. E. Hanson, I. Toth, M. E. Davis, *J. Organometal. Chem.*, 1991, **403**, 221–227.

14. E. Fache, C. Mercier, N. Pagnier, B. Desperyoux, et al., *J. Mol. Catal.*, 1993, **79**, 117–131.

15. K. T. Wan, M. E. Davis, *Nature*, 1994, **370**, 449–450.

16. K. T. Wan, M. E. Davis, *J. Catal.*, 1994, **148**, 1–8.

17. K. T. Wan, M. E. Davis, *J. Catal.*, 1995, **152**, 25–30.

18. M. J. Naughton, R. S. Drago, *J. Catal.*, 1995, **155**, 383–389.

19. B. M. Bhanage, F. Zhao, M. Shirai, M. Arai, *J. Mol. Catal. A: Chem.*, 1999, **145**, 69–74.

20. B. M. Bhanage, F. Zhao, M. Shirai, M. Arai, *Catal. Lett.*, 1998, **54**, 195–198.

21. I. T. Horvath, *Catal. Lett.*, 1990, **6**, 43–48.

22. M. S. Anson, M. P. Leese, L. Tonks, J. M. J. Williams, *J. Chem. Soc., Dalton Trans.*, 1998, 3529–3538.

23. L. Tonks, M. S. Anson, K. Hellgardt, et al., *Tetrahedron Lett.*, 1997, **38**, 4319–4322.

24. B. M. Bhanage, Y. Ikushima, M. Shirai, M. Arai, *J. Chem. Soc., Chem. Commun.*, 1999, 1277–1278.

25. P. Mehnert, R. A. Cook, N. C. Dispenziere, M. Afeworki, *J. Am. Chem. Soc.*, 2002, **124**, 12932–12933.

26. P. Mehnert, E. J. Mozeleski, R. A. Cook, *J. Chem. Soc., Chem. Commun.*, 2002, 3010–3011.

27. A. Riisager, P. Wasserscheid, R. van Hal, R. Fehrmann, *J. Catal.*, 2003, **219**, 452–455.

28. A. Riisager, K. M. Eriksen, P. Wasserscheid, R. Fehrmann, *Catal. Lett.*, 2003, **90**, 149–150.

29. R. S. Varma, D. Kumar, *Catal. Lett.*, 1998, **53**, 225–227.

30. R. S. Varma, K. P. Naicker, D. Kumar, *J. Mol. Catal. A: Chem.*, 1999, **149**, 153–160.

31. M. T. Reetz, W. Helbig, S. A. Quaiser, et al., *Science*, 1995, **267**, 367–369.

32. M. T. Reetz, M. Dugal, *Catal. Lett.*, 1999, **58**, 207–212.

33. M. T. Reetz, E. Westermann, *Angew. Chem. Int. Ed.*, 2000, **39**, 165–168.

34. M. T. Reetz, J. G. de Vries, *J. Chem. Soc., Chem. Commun.*, 2004, 1559–1563.

35. P.-F. Ho, K.-M. Chi, *Nanotechnology*, 2004, **15**, 1059–1064.

36. S.-W. Kim, J. Park, Y. Jang, et al., *Nano Lett.*, 2003, **3**, 1289–1291.

37. C. Luo, Y. Zhang, Y. Wang, *J. Mol. Catal. A: Chem.*, 2005, **229**, 7–12.

38. B. P. S. Chauhan, J. S. Rathore, M. Cauhan, A. Krawicz, *J. Am. Chem. Soc.*, 2003, **125**, 2876–2877.

39. M. Yu. Berezin, K.-T. Wan, R. M. Friedman, R. G. Orth, et al., *J. Mol. Catal. A: Chem.*, 2000, **158**, 567–576.

40. Z. Hou, N. Theyssen, A. Brinkmann, W. Leitner, *Angew. Chem. Int. Ed.*, 2005, **44**, 1346–1349.

41. C. A. Marques, M. Selva, P. Tundo, *Rend. Fis. Acc. Lincei*, 1992, Ser. 9, **3**, 283–294.

42. C. A. Marques, M. Selva, P. Tundo, *J. Chem. Soc. Perkin Trans. I*, 1993, 529–533.

43. C. A. Marques, M. Selva, P. Tundo, *J. Org. Chem.*, 1993, **58**, 5256–5260.

44. C. A. Marques, M. Selva, P. Tundo, *J. Org. Chem.*, 1994, **59**, 3830–3837.

45. C. A. Marques, M. Selva, P. Tundo, *J. Org. Chem.*, 1995, **60**, 2430–2435.

46. C. A. Marques, O. Rogozhnikova, M. Selva, P. Tundo, *J. Mol. Catal. A: Chem.*, 1995, **96**, 301–309.

47. C. A. Marques, M. Selva, P. Tundo, *Gazz. Chim. Ital.*, 1996, **126**, 317–327.

48. A. Bomben, C. A. Marques, M. Selva, P. Tundo, *Synthesis*, 1996, **9**, 1109–1114.

49. M. Selva, P. Tundo, A. Perosa, *J. Org. Chem.*, 1998, **63**, 3266–3271.

50. A. Perosa, M. Selva, P. Tundo, *J. Org. Chem.*, 1999, **64**, 3934–3939.

51. P. Tundo, S. Zinovyev, A. Perosa, *J. Catal.*, 2000, **196**, 330–338.

52. A. Perosa, M. Selva, P. Tundo, S. S. Zinovyev, *Appl. Catal. B: Environ.*, 2001, **32**, L1–L7.

53. A. Perosa, P. Tundo, M. Selva, *J. Mol. Catal. A: Chem.*, 2002, **180**, 169–175.

54. S. S. Zinovyev, A. Perosa, S. Yufit, P. Tundo, *J. Catal.*, 2002, **211**, 347–354.

55. A. Perosa, P. Tundo, S. S. Zinovyev, *Green Chem*, 2002, **4**, 492–494.

56. P. Tundo, A. Perosa, *React. Funct. Polym.*, 2003, **54**, 95–101.

57. P. Tundo, A. Perosa, S. S. Zinovyev, *J. Mol. Catal. A: Chem.*, 2003, **204–205**, 747–754.

58. S. S. Zinovyev, N. A. Shinkova, A. Perosa, P. Tundo, *Appl. Catal. B: Environ.*, 2004, **47**, 27–36.

59. S. S. Zinovyev, A. Perosa, P. Tundo, *J. Catal.*, 2004, **226**, 9–15.

60. G. Evdokimova, S. S. Zinovyev, A. Perosa, P. Tundo, *Appl. Catal. A: Gen.*, 2004, **271**, 129–136.

61. A. Perosa, P. Tundo, M. Selva, et al., *Org. Biomol. Chem.*, 2004, **2**, 2249–2252.

62. S. S. Zinovyev, N. A. Shinkova, A. Perosa, P. Tundo, *Appl. Catal. B: Environ.*, 2005, **55**, 39–48.

63. S. S. Zinovyev, N. A. Shinkova, A. Perosa, P. Tundo, *Appl. Catal. B: Environ.*, 2005, **55**, 49–56.

64. S. Mukhopadhyay, G. Rothenberg, N. Qafisheh, Y. Sasson, *Tetrahedron Lett.*, 2001, **42**, 6117–6119.

65. L. Wang, Y. Zhang, C. Xie, Y. Wang, *Synlett*, 2005, 1861–1864.

7

ORGANIC CHEMISTRY IN WATER: GREEN AND FAST

JAN B. F. N. ENGBERTS

Stratingh Institute, University of Groningen, Groningen, The Netherlands

INTRODUCTION

Traditionally, water is not a popular solvent for organic reactions. The limited solubility of many organic substrates and reagents as well as the fact that a variety of functional groups is reactive toward water have traditionally contributed to this lack of popularity of water as a reaction medium. Contrarily, the chemistry of all life processes occurs in aqueous media, and few people will doubt the high quality and efficiency of these transformations!

Recently there has been a revival of interest in water as the reaction medium in organic chemistry. Our increasing concern for the environment and for safe chemical procedures are reasons for this change in attitude. Interestingly, many organic reactions (and particularly carbon–carbon bond formation reactions) are accelerated in water relative to organic solvents. Water may also have a favorable effect on the stereochemistry of a variety of organic transformations.[1,2] And finally, very recent studies by Shapless et al. (*vide infra*) have shown that limited aqueous solubility can be turned into an advantage!

7.1 DIELS–ALDER REACTIONS IN WATER

In this chapter we will focus our attention first on Diels–Alder reactions in aqueous solutions: much to the surprise of many chemists, it has been found that

Methods and Reagents for Green Chemistry: An Introduction, Edited by Pietro Tundo, Alvise Perosa, and Fulvio Zecchini

Diels–Alder reactions and other types of cycloadditions are often greatly acceler-ated in water compared to organic solvents. Also the preferred stereochemistry (i.e., endo/exo ratio) of these reactions is greatly affected in aqueous media. In synthetic organic chemistry, advantage has been taken of this finding. Recent studies have been aimed at identifying the reasons for the rate enhancements and improved stereochemistry. The problem has been settled only recently. Ongoing research is focused on obtaining further understanding of organic reactivity in water.

Then we briefly examine Lewis-acid catalysis of Diels–Alder reactions in water: Lewis-acid catalysis is a valuable alternative for acid catalysis in organic chemistry and biochemistry. The Lewis-acid catalysts are usually polyvalent cations that bind to polar moieties in organic substances, thereby avoiding high Gibbs energy intermediates during the activation process for the particular trans-formation. In water these binding processes are in competition with hydration of the cations, usually resulting in weak (or no) binding and inefficient catalysis. However, in case of specially designed substrates, extremely efficient catalysis can be achieved.

In subsequent studies it has been found that a combination of Lewis-acid and micellar catalysis can lead to huge (in fact, enzyme like) rate acceleration in water. In the absence of Lewis-acid catalysts, micelles tend to inhibit Diels–Alder reactions, largely because of the particular nature of the substrate binding sites at the micelle. This problem can be solved by adding Lewis-acid catalysts that bind effectively at the micellar surface.

The Diels–Alder reaction (Figure 7.1) is of great value in synthetic organic chemistry. It is a $[4 + 2]$ cycloaddition in which a diene (4-π component) reacts with a dienophile (2-π component) to provide a six-membered ring. In the reac-tion six new stereocenters are formed in a single step. The reaction is stereospeci-fic and the absolute configuration of the newly formed asymmetric centers can be controlled efficiently.

Traditionally, the Diels–Alder reaction is performed in organic solvents.

In the one-step symmetry-allowed mechanism, with little charge separation in the activated complex, the Hammett ρ-values for p-XC$_6$H$_4$- substituted substrates are very small. As anticipated, the reaction has a negative volume of activation.

Quite generally, kinetic solvent effects on the Diels–Alder reaction are small, and, in fact, the small solvent effects have been taken as evidence for minor charge separation during the activation process, consistent with a concerted mechanism.

The first kinetic study of acceleration of some Diels-Alder[3-7] reactions in water by Breslow et al. has set the stage for worldwide interest in organic

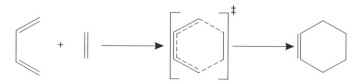

Figure 7.1   Classic mechanism of a Diels–Alder reaction.

TABLE 7.1 Acceleration of Some Diels–Alder Reactions in Water

	$k_{water}/k_{org.solvent}$	
	12,800	(Hexane)
	153	(Hexane)
	290	(CH₃CN)
	71.0	(CH₃CN)
	138	(Hexane)
(a) (b)	8.95 102	(CH₃CN) (Ethanol)
	44.3	(Toluene)

Source: Refs. 3, 5, and 7.

reactions in water. Some of the results obtained in our laboratory are shown in Table 7.1. It is clear that the aqueous rate accelerations strongly depend on the nature of the diene and the dienophile.

Apart from the aqueous rate accelerations, the aqueous medium has also a favorable effect on the endo-exo selectivity. Substantially higher preferences for the endo isomer were found. This effect can be explained taking the more efficient secondary orbital interactions into considerations as well as additional stabilization of the endo transition state, because it is more polar and has a smaller contact area with water.[3]

7.2 MECHANISM OF THE AQUEOUS RATE ACCELERATION

Several factors have been invoked to explain the aqueous rate acceleration: aggregation of the reactants leading to micellar catalysis, effects connected with the internal pressure of the solvent, polarity of the solvent, H-bonding interactions with the solvent, and hydrophobic interactions ($\Delta^{\#}V < 0$). The initial literature was rather controversial, and there was a strong need for a systematic study using physical–organic techniques.

Several experiments gave evidence against homotactic or heterotactic association of diene and dienophile.[8,9] Vapor-pressure measurement for cyclopentadiene (CPD) (Figure 7.2), in pure water and in 10% (w/w) n-PrOH/water, show that Henry's law is obeyed (vapor pressure varies linearly with solute concentration) until [CPD] is 0.03 M in pure water and 0.06 M in n-PrOH/water. In kinetic measurement the concentration of CPD was always below 0.002 M, indicating that association is highly unlikely. Similar results have been obtained with methyl vinyl ketone, ethyl vinyl ketone, and naphthoquinone.

It was also observed that the second-order rate constants are independent of the concentration of CPD, even in excess, supporting the notion that the

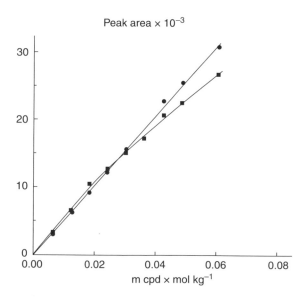

Figure 7.2 Plots of the peak area of cyclopentadiene, obtained after injection of a standard volume of vapor, withdrawn from the vapor above an aqueous solution of cyclopentadiene as a function of the molality of cyclopentadiene, at 25°C; solution of cyclopentadiene in pure water, ■, and solution of cyclopentadiene in aqueous solution, containing 10% (w/w) of ethanol, ●.

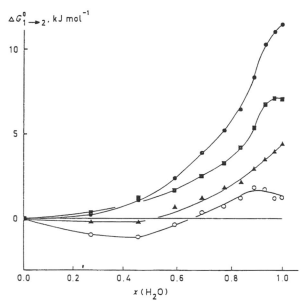

Figure 7.3 Standard Gibbs energies of transfer for reactants and activated complex for the Diels–Alder reaction of cyclopentadiene (1,■) with ethyl vinyl ketone (2,▲) from 1-PrOH to 1-PrOH–water as a function of the mole fraction of water; initial state (1 + 2,●); activated complex (○).

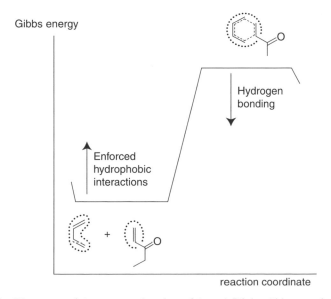

Figure 7.4 The cause of the rate acceleration of (most) Diels–Alder reactions in water.

acceleration of the Diels–Alder in water is not dependent on homotactic/hetero-tactic association. Intramolecular Diels–Alder reactions are also greatly acceler-ated in water, again indicating that association of diene and dienophile is not responsible for the aqueous rate acceleration.

The pseudothermodynamic analysis of solvent effects in 1-PrOH–water mixtures over the whole composition range (shown in Figure 7.3) depicts a combination of thermodynamic transfer parameters for diene and dienophile with isobaric activation parameters that allows for a distinction between solvent effects on reactants (initial state) and on the activated complex. The results clearly indicate that the aqueous rate accelerations are heavily dominated by initial-state solvation effects. It can be concluded that for Diels–Alder reactions in water the causes of the acceleration involve stabilization of the activated complex by enforced hydrophobic interactions and by hydrogen bonding to water (Table 7.1, Figure 7.4).[10]

7.3 LEWIS-ACID CATALYSIS IN WATER

Lewis-acid catalysis of Diels–Alder reactions (Figure 7.5) in organic solvents leads to an enhancement of the reaction rate, because of the lowering in energy for the lowest unoccupied molecular orbital (LUMO) of the dienophile, and an improvement in the selectivity with specific ligands.

Experiments aimed at examining the possibility of catalyzing Diels–Alder reaction in water gave rise to interesting results. For the reaction of bidentate die-nophiles (1) with cyclopentadiene,[11] shown in Figure 7.6, the reaction in water was less accelerated by the catalyst than the one in acetonitrile, probably because

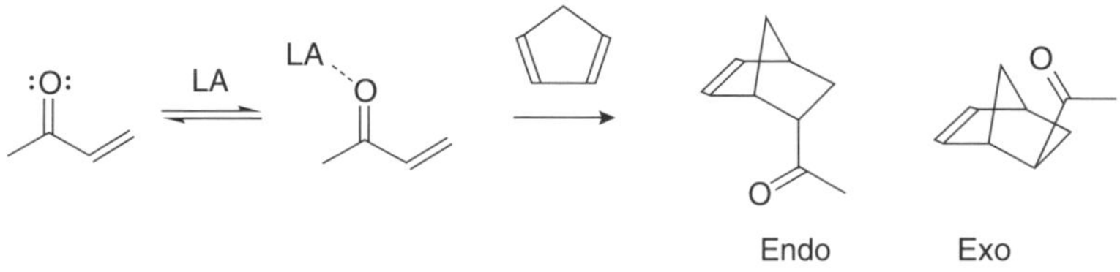

Figure 7.5 Diels–Alder reaction catalyzed by a Lewis acid (LA).

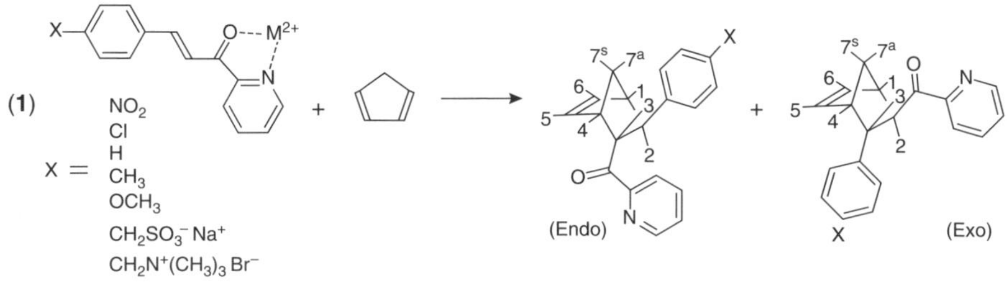

Figure 7.6 First example of Lewis-acid catalyzed Diels–Alder reaction in water. $M^{2+} = Cu^{2+}$, Ni^{2+}, Co^{2+}, Zn^{2+}.

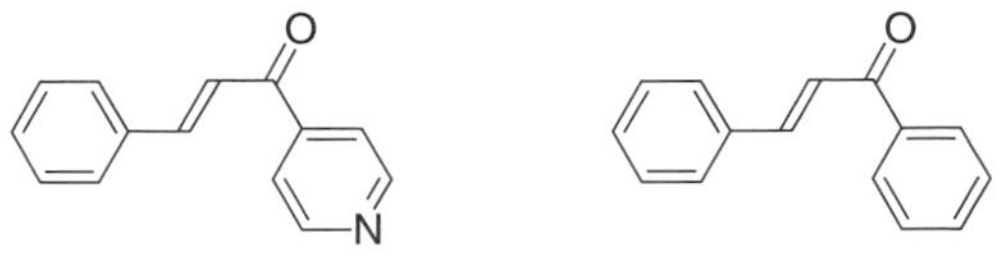

Figure 7.7 Nonbidentate substrates do not show Lewis-acid catalysis of the Diels–Alder reaction in water.

TABLE 7.2 Effects of the Lewis Acid in Different Solvents for Reaction (1)

Solvent	Catalyst	Half-Life	Endo/Exo	k(rel)
Acetonitrile	None	1.5 years	67/33	1
Ethanol		7 months	77/23	2.7
Water		48 hours	84/16	288
Acetonitrile	Cu^{2+}	5 minutes	94/6	158.000
Ethanol	(10 mM)	15 minutes	96/4	54.900
Water		4 minutes	93/7	691.000

TABLE 7.3 Solvent Effect on Enantioselectivity; Catalysis by Copper (L-abrine) as the Lewis Acid

Solvent	Time (days)	Enantiomeric Excess (ee) (%)
Water	3	74
Ethanol	10	39
Acetonitrile	7	17
THF	8	24
Chloroform	11	44

the hydrophobic effects are present but reduced, and the hydrogen bonding effect is diminished.

The reaction with the compounds shown in Figure 7.7 showed that there was no influence of Cu^{2+} on the rate of the reaction with cyclopentadiene, confirming that bidentate complexation is essential for efficient catalysis.

Lewis-acid catalysis (Table 7.2) was observed for Cu^{2+}, Ni^{2+}, Co^{2+}, and Zn^{2+}, with Cu^{2+} being the best catalyst (both strongest binding and most efficient in accelerating the reaction with CPD). Varying the solvent (Table 7.3) in the catalysis by Cu^{2+} ions with L-abrine as a ligand illustrates a large increase in enantioselectivity in water. So water promotes enantioselectivity.[12]

7.4 α-AMINO ACIDS AS CHIRAL LIGANDS

Various experiments conducted using α-amino acids as a ligand gave marked enantioselective effects for Cu(II) and Ni(II). Table 7.4 illustrates the results for the reaction (**1**) in Figure 7.8.[11,12]

TABLE 7.4 Influence of α-Amino Acid Ligands on the Equilibrium Constant for Binding of 3.8c to the Ligand–Cu²⁺ Complex (K_a) and the Second-Order rate Constant (k_{cat}) for Reaction of this Ternary Complex with Diene 2[a] and the Enantioselectivity of this Reaction in Water[b]

| Ligand | | | | |
Structure	Name	K_a (M^{-1})	k_{cat} (M^{-1} s^{-1})	ee[c] (%)
	H₂O	1.16×10^3	2.56	—
(structure)	Glycine	6.29×10^2	1.89	—
(structure)	L-Valine	5.71×10^2	1.90	3[d]
(structure)	L-Leucine	5.14×10^2	2.01	3[d]
(structure)	L-Phenylalanine	8.66×10^2	2.01	17
(structure)	L-Tyrosine	1.40×10^3	1.68	36
(structure)	N-Methyl-L-tyrosine	2.45×10^3	2.07	74
(structure)	N-Methyl-p-methoxy-L-phenylalanine	2.04×10^3	2.83	67

Structure	Compound			
	N,N-Dimethyl-L-tyrosine	1.66×10^3	2.92	73
	L-Tryptophan	3.02×10^3	1.44	33
	5-Hydroxy-L-tryptophan	4.89×10^3	1.15	29
	L-Abrine[e]	5.05×10^3	1.47	74

[a] All measurements were performed at constant ionic strength (2.00 M using KNO_3 as background electrolyte) and at pH 4.6–5.2.
[b] 10 mol % of $Cu(NO_3)_2$; 17.5 mol % of ligand; conditions as outlined in Refs. 11 and 12.
[c] Only the results for the major (>90%) endo isomer of the Diels–Alder adduct are shown.
[d] 50 mol % of catalyst was used.
[e] N^α-Methyl-L-tryptophan.

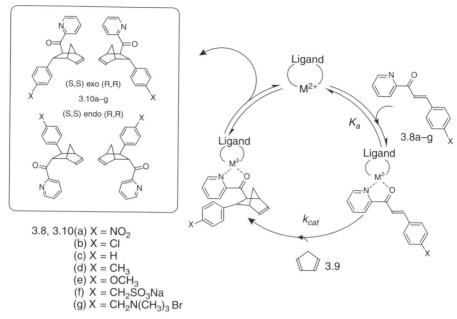

3.8, 3.10(a) X = NO₂
         (b) X = Cl
         (c) X = H
         (d) X = CH₃
         (e) X = OCH₃
         (f) X = CH₂SO₃Na
         (g) X = CH₂N(CH₃)₃ Br

Figure 7.8 Reaction (1) catalyzed by Lewis acid Cu(II) and Ni(II) with ligands in Table 7.4.

The mechanistic interpretation[13] involves the interaction of a hydrophobic substituent in the ligand bound to Cu(II) with an aromatic ring in the dienophile exerting a favorable effect on the dienophile–Cu(II) interaction. Both the catalytic efficiency of Lewis-acid catalysis and the stereochemistry of the cycloaddition benefit from this water- induced interaction (Figure 7.9).

7.5 MICELLAR CATALYSIS

Micellar catalysis, conducted in the absence of Lewis acid tends to inhibit the Diels–Alder reaction, relative to the reaction in water. The reason is that the local reaction medium in the Stern region is less favorable than bulk water. However, by combining Lewis-acid and micellar catalysis, enzyme-like rate accelerations can be obtained (Table 7.5) in case the Lewis acid acts as the counterion for the micelle.[14]

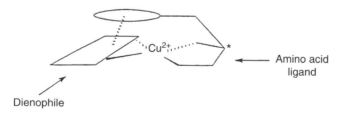

Figure 7.9 Mechanistic interpretation for catalysis by Lewis acids containing aromatic α-amino acids as ligands.

TABLE 7.5 Solvent Effects and Combined Lewis-Acid/Micellar Catalysis Increasing the Rate Constant of Reaction (1)

Solvent	$[Cu^{2+}]$ (mM)	Relative k_2
CH_3CN	0	1
EtOH	0	2.7
Water	0	287
CF_3CH_2OH	0	485
CH_3CN	10	158.000
EtOH	10	54.900
Water	10	691.000
CF_3CH_2OH	0.10	1.110.000
Cu (DS)$_2$ micelles	10	1.790.000

Note: DS is n = dodecyl sulfate.

7.6 CONCLUSIONS

Diels–Alder reactions (and other cycloadditions) are accelerated in water due to a combination of enforced hydrophobic interactions and hydrogen bonding, their relative contributions depending on the nature of the diene and dienophile. Subsequent work has shown that a large variety of other organic reactions show comparable favorable characteristics in aqueous media.

Lewis-acid catalysis of Diels–Alder reactions involving bidentate dienophiles in water is possible; also if the beneficial effect of water on the catalyzed reaction is reduced relative to pure water. There are no additional effects on endo–exo selectivity. As expected, catalysis by Cu^{2+} ions is much more efficient than specific-acid catalysis.[15] Using α-amino acids as chiral ligands, Lewis-acid enantioselectivity is enhanced in water compared to organic solvents. Micelles, in the absence of Lewis acids, are poor catalysts, but combining Lewis-acid catalysis and micellar catalysis leads to a rate accelaration that is enzyme-like.[14]

In many cases, synthetic organic chemistry in aqueous media[1,2] offers important advantages for clean, green chemistry. Industrial applications have been realized and further developments are envisaged. For a long time, solubility constraints were seen as a major disadvantage, but now that recent work by Sharpless et al.[16] has shown that several types of bimolecular organic reactions involving reactants insoluble in the aqueous phase are remarkably accelerated when performed in aqueous suspensions with efficient stirring ("on water reactions"), the situation has changed dramatically. The exact mechanism of these heterogeneous aqueous reactions is not yet known, but this work might well be a major breakthrough in aqueous synthetic chemistry.

The advantages of aqueous reaction media include:

- No pollution of the environment
- Low cost

- Safety
- Synthetic efficiency and often easy work-up
- Simple chemical processes, including easy heat control
- Good solvent for mechanistic studies
- Unique molecular association via hydrophobic interactions
- Good solvent for fast (catalytic), (stereo)selective transformations

In physical organic chemistry, future detailed kinetic studies and sophisticated molecular dynamics computer simulations[17] will lead to a still more thorough understanding of the exact role of the aqueous reaction medium in organic transformations. Particularly the effects of hydrophobic interactions are of great interest.[18] Since water is essential for life processes, these studies will bring organic chemistry and biochemistry closer together.

REFERENCES

1. Grieco, P. A. *Organic Synthesis in Water*, Blackie, London, 1998.

2. Li, C.-J.; Chan, T.-H. *Organic Reactions in Aqueuos Media*, John Wiley & Sons, New York, 1997.

3. Blokzijl, W.; Blandamer, M. J.; Engberts, J. B. F. N. *J. Am. Chem. Soc.* 1991, **113**, 4241–4246.

4. Otto, S.; Blokzijl, W.; Engberts, J. B. F. N. *J. Org. Chem.*, 1994, **59**, 5372–5376.

5. Wijnen, J. W.; Engberts, J. B. F. N. *J. Org. Chem.*, 1997, **62**, 2039–2044.

6. Van der Wel, G. K.; Wijnen J. W.; Engberts, J. B. F. N. *J. Org. Chem.*, 1996, **61**, 9001–9005.

7. Breslow, R.; Rideout, D. C. *J. Am. Chem. Soc.*, 1980, **102**, 7816–7817.

8. Blokzijl, W.; Ph.D. Dissertation, University of Groningen, The Netherlands, 1991.

9. Blokzijl, W.; Engberts, J. B. F. N. *J. Am. Chem. Soc.*, 1992, **114**, 5440–5442.

10. Engberts, J. B. F. N. *Pure Appl. Chem.*, 1995, **67**, 823–828.

11. Otto, S.; Bertoncin, F.; Engberts, J. B. F. N. *J. Am. Chem. Soc.*, 1996, **118**, 7702–7707.

12. Otto, S.; Engberts, J. B. F. N. *J. Am. Chem. Soc.*, 1999, **121**, 6798–6806.

13. Otto, S.; Bocaletti, G.; Engberts, J. B. F. N. *J. Am. Chem. Soc.*, 1998, **120**, 4238–4239.

14. Otto, S.; Engberts, J. B. F. N.; Kwak, J. C. T. *J. Am. Chem. Soc.*, 1998, **120**, 9517–952.

15. Mubofu, E. B.; Engberts, J. B. F. N. *J. Phys. Org. Chem.*, 2004, **17**, 180–186.

16. Narayan, S.; Muldoon, J.; Finn, M. G.; et al. *Angew. Chem. Int. Ed.*, 2005, **44**, 3275–3279.

17. For two examples, see, (a) Chandrasekhar, J.; Shariffskul, S.; Jorgensen, W. L. *J. Phys. Chem. B*, 2002, **106**, 8078–8085; (b) Rispens, T.; Lensink, M. F.; Berendsen, H. J. C.; Engberts, J. B. F. N. *J. Phys. Chem. B*, 2004, **108**, 5483–5488.

18. Otto, S.; Engberts, J. B. F. N. *Org. Biomol. Chem.*, 2003, **1**, 2809–2820.

8

FORMATION, MECHANISMS, AND MINIMIZATION OF CHLORINATED MICROPOLLUTANTS (DIOXINS) FORMED IN TECHNICAL INCINERATION PROCESSES

DIETER LENOIR, ERNST ANTON FEICHT, MARCHELA PANDELOVA, AND KARL-WERNER SCHRAMM,

Institut für Ökologische Chemie, GSF-Forschungszentrum für Umwelt und Gesundheit, München, Germany

INTRODUCTION

Chlorinated micropollutants are harmful for man and environment due to their toxicity, persistence, and bioaccumulation.[1] Persistent compounds are very stable and difficult to get metabolized and mineralized by biological and chemical processes in the environment, and as a result, they have become ubiquitous in water, sediments, and the atmosphere; bioaccumulation is the result of the lipophilicity of these compounds.[1] Polychlorinated dibenzodioxins and -furans (PCDD/F) are not produced purposely like many of other chlorinated technical products, such as chlorinated biocides DDT, lindane, and toxaphene.[2] The production and use of persistent organic pollutants (POPs), the "dirty dozen" has now been banned worldwide by the Stockholm protocol.[3] It should be mentioned that about 3000 halogenated products have now been isolated as natural products in plants, microorganisms, and animals,[4] but the total amount of these products is much smaller compared to xenobiotics.

Methods and Reagents for Green Chemistry: An Introduction, Edited by Pietro Tundo, Alvise Perosa, and Fulvio Zecchini

171

TABLE 8.1 Estimates for Emissions of PCDD/F for Six Different Countries

Emission Source	Austria (g I-TEQ/yr)	Germany (g I-TEQ/yr)	Japan (g I-TEQ/yr)	Netherlands (g I-TEQ/yr)	United Kingdom (g I-TEQ/yr)	United States (g I-TEQ/yr)
Pulp and paper mills	4		40			
Sewage sludge incineration	<1	0.01–1.1	7 (a)	0.3		10–52
Chemical production processes				0.5		
Municipal waste incineration	3 (1–14)	5.4–432	3100–7400	382	1150	1800–9000
Hazardous waste incineration	6 (2–10)	0.5–72	460	16	11	11–110
Hospital waste incineration	4 (1–11)	5.4	80–240	2.1	32	1600–16000
Cement kilns						110–1100
Metal smelting/ refining	19 (<1–38)	39–399	250	30		75–745 (c)
Traffic (vehicle fuel combustion)	<1	12.6	0.07	7.0	613	27–270 (d)
Domestic combustion	70 (40–230)					
Combustion of wood			0.2	12	16	113–1063 (c)
Combustion of coals and lignite	<1	1.1		3.7	1489	
Combustion of oils		1.2		1.0	2	
Outdoor burning of straw	10 (4–16)					

Burning of cables and electromotors				1.5		
Drum and barrel reclamation						0.5–5.0
Forest fires						27–270
Kraft black liquor boilers			3			0.9–4.3
Cigarette smoking			16			
Charcol briquette combustion		1.8				
Various high-temp. processess			20 (b)	2.7	55	
Crematoria				0.2		
Accidental fires				25	16	
(Former) use of wood preservatives						
Total	112 (50–320)	67–926	4000–8400	484	3870	3300–26000
Basis year	1987–1988	1990	1990	1991	1989	1994

Source: A. K. Djien Diem and J. A. van Zorge, *ESPR-Environ. Sci. Pollut. Res.*, 1995.

[a]Value represents total of incineration of sewage (5) and paper sludge (2).

[b]Value represents lubrication oil.

[c]Range represents total of secondary copper smelting (74–740) and secondary lead smelting (0.7–3.5).

[d]Value represents diesel only.

[e]Range represents total of industrial (100–1000) and residential wood burning (13–63).

PCDD/F are formed and emitted from various thermal processes, such as municipal and hazardous waste incinerators and metallurgy. They are transported globally through the atmosphere and precipitated to the surfaces of plants, soils, and water. In Table 8.1 the most important sources and amounts (inventories) for PCDD/F are summarized for six countries.[5] PCDD/F is a mixture of 210 compounds (see Figure 8.2). The 17 toxic isomers are expressed as a special sum parameter value, I-TE value (see the following definition). Besides the formation of PCDD/F by thermal processes, these isomers have been found in the past as by-products in technical products like chlorinated biphenyls (PCBs) and in technical grade pentachlorophenol (PCP). It should be mentioned that the amounts of I-TE emitted from technical incinerators have decreased during the last decade in many industrial countries due to strong legislative measures (ordinances such as clean air acts). For example, most European countries have defined limit values of 0.1 ng I-TE/m^3 for the emitted flue gas of waste incinerators. As a result, the estimated value of 400 g I-TE for German municipal waste incinerators for the year 1990 decreased to a value of 4 g I-TE in 1998. The United Nations Environmental Program (UNEP) publishes up-to-date inventories of PCDD/F for the most important countries.[6] It can be seen from Table 8.1, that pulp and paper mills today play only a minor part in overall dioxin emissions, while PCDD/Fs are emitted by the wastewater from these plants into the water of the rivers and seas.

8.1 FORMATION OF PCDD/F BY ACCIDENTS

All accidents concerned with PCDD/F are related to the production of chlorophenols. The most famous accident happened in Seveso close to Milan, Italy, on July 10, 1976. ICMESA Corp. manufactured 2,4,5-trichlorophenol for production of phenoxy-herbizides by alkaline hydrolysis of 1,2,4,5-tetrachlorobenzene (see Figure 8.1). This

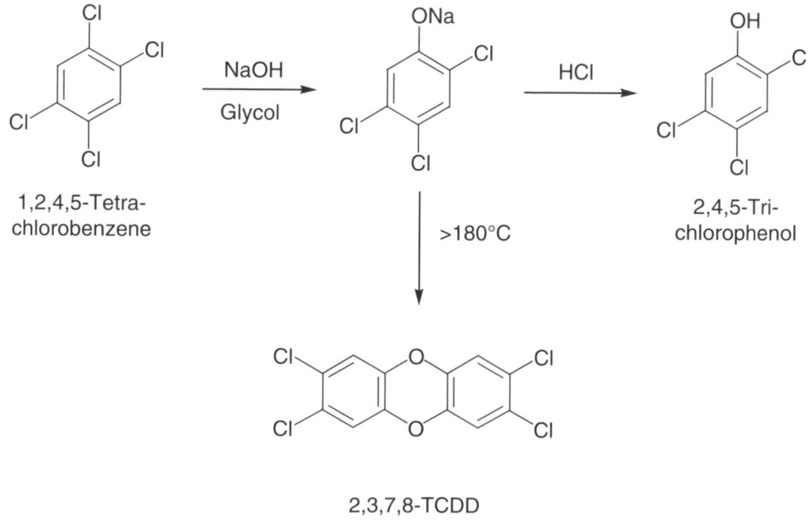

Figure 8.1 Chemistry of the Seveso accident in 1976.

process is a nucleophilic aromatic substitution of one chlorine atom by a hydroxi group. Due to overheating of the vessel, exothermic condensation did occur instead of substitution with the subsequent bursting of the valve of the apparatus. About 2.6 kg of 2,3,7,8-tetrachlorodibenzo-*p*-dioxin (TCDD) were released into the close vicinity of the factory.

Dioxins are mainly by-products of industrial processes, but can also result from natural processes, such as volcanic eruptions and forest fires. Besides the anthropogenic (man-made) sources of PCDD/F discussed earlier, biogenic and geogenic sources for dioxins also have been discovered recently. In natural clays of the kaolinite-type found in German mines in Westerwald, considerable levels of PCDD/F have been detected;[7] the same findings were obtained in special ball clays in the Mississippi area of the United States.[8] The pattern (isomeric ratios) of this natural type of dioxins is different from the pattern obtained from incineration plants.

8.2 STRUCTURES, PROPERTIES, AND BEHAVIOR OF PCDD/F

The PCDD/F class consists of 210 compounds, 75 isomers of PCDD, and 135 isomers of PCDF. The number of regioisomers are the following according to the number of chlorine atoms in either skeleton (see Table 8.2).

All PCDD/F isomers are solids with high melting points, but low vapor pressure and low solubility in water. The high octanol–water coefficients are an indication of the observed bioaccumulative behavior in plants and animals for these compounds. Detailed environmentally important physicochemical properties can be found in the literature.[9] All higher chlorinated compounds are very persistent in the environment with half-lives of 5–10 years; photolysis with sunlight is the only degradation process in the environment.

Identification and quantification is obtained by combined high-resolution gas chromatography/mass spectrometry (GC/MS) methods after special cleanup procedures of the matrix, as shown later for sediments (see Figure 8.2). The cleanup methods for other matrices are similar. Quantification is obtained by addition of 13-C labeled standards before the cleanup procedure. In general, only the toxic isomers are identified and quantified.

TABLE 8.2   Number of Regioisomers for PCDD and PCDF

Chlorine Substitution	PCDD	PCDF
Mono	2	4
Di	10	16
Tri	14	28
Tetra	22	38
Penta	14	28
Hexa	10	16
Hepta	2	4
Octa	1	1
Total	75	135

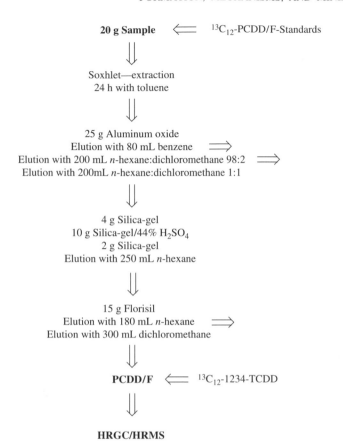

Figure 8.2 Scheme for the cleanup method of PCDD/F in sediments.

All 210 isomers of PCDD/F have been prepared by standard synthetic routes (see recent review.[10]). But none of the dioxins or furans are used for any practical purpose. OCDD had been prepared in 1872 by Merz and Weith, but without knowledge of the structure. Unsubstituted dibenzodioxin was prepared in 1906 by Ullmann and Stein. 2,3,7,8-TCDD as well as OCDD were synthesized in 1957 by W. Sandermann by electrophilic chlorination of unsubstituted dibenzodioxin. His group prepared about 15 g of 2,3,7,8-TCDD unintentionally and discovered its toxic behavior on themselves. Dr. Sorge, a medical doctor working for Boehringer Corporation in Hamburg showed the toxicity of 2,3,7,8-TCDD prepared and identified by W. Sandermann. At the same time about 30 workers of Boehringer were engaged in commercial production of trichlorophenol for further production of phenoxy herbicides (see Figure 8.1) and suffered from a severe illness that resembled chloracne and related symptoms. Later it was shown that these techni-cal products were contaminated with traces of 2,3,7,8-TCDD. Trace analysis for PCDD/F did not exist at this time. It should be mentioned that the "Vietnam syn-drome" can be traced back to the same cause: technical grade Agent Orange, a defoliant used during the war, was contaminated with traces of 2,3,7,8-TCDD, resulting in the severe illness of a large number of veterans.

8.3 TOXICOLOGY

PCDD and PCDF short-term exposure to humans in high levels may result in skin lesions, such as chloracne and patchy darkening of the skin, and altered liver function. Long-term exposure is linked to impairment of the immune system, the developing nervous system, the endocrine system, and reproductive functions. Chronic exposure of animals to dioxins has resulted in several types of cancer. TCDD was evaluated by International Agency for Research on Cancer (IARC) in 1997. Based on human epidemiology data, dioxin was categorized by IARC as a "known human carcinogen." However, TCDD does not affect genetic material and there is a level of exposure below which cancer risk would be negligible.

Toxic behavior of PCDD/F is a complex matter. Contrary to other poisons, LC-50 (lethal concentration) values that were studied for acute toxicity for a variety of mammals depend largely on the species being investigated. The value (in μg/kg) varies from 0.6 for guinea pigs to 300 for hamsters. For man a LC-50 value larger than 2000 has been estimated. In addition, 2,3,7,8-TCDD shows strong cancerogenic effects when administered to mice and rats. The toxic mechanism is a special binding to the Ah receptor of DNA.[11] 2,3,7,8-TCDD is the most toxic isomer among the 17 isomers with the 2,3,7,8 pattern (see Table 8.3). These values are obtained by enzyme-induction test studies. Properties of endocrine disruption are most likely.

The dioxin toxic equivalency factor (TEF) approach is currently used worldwide for assessing and managing the risks posed by exposure to mixtures of certain dioxin-like compounds (DLCs).[12b–12e] World Health Organization-TEF (WHO-TEF) values have been established for humans and mammals, birds, and fish.[12b,12f] (For new, refined values, see Ref. 12g.) It should be mentioned that 16 PCBs, the coplanar isomers with nonortho, monoortho, and diortho substitution by chlorine (overall, there are 209 isomers for this class of compounds) show dioxin-like toxic behavior. I-TE values are smaller, in the range of 0.0001–0.1. The most toxic isomers is 3,3′,4,4′,5-pentachlorodiphenyl with I-TE of 0.1.[13] Polybrominated dibenzodioxins and furans with the 2,3,7,8 pattern also show dioxin-like toxicity, but their I-TE values are lower compared to PCDD/F.

8.4 POLYCHLORINATED DIBENZODIOXINS AND FURANS AS POLLUTANTS FORMED IN INCINERATIONS

8.4.1 Primary and Secondary Measures for Minimization of PCDD/F in Incineration Plants

PCDD/F are emitted by the flue gas of the incineration plants. Primary measures have become very important in the production and technology of chemistry as the

TABLE 8.3   Toxic Equivalency Factors (TEFs) for Toxic PCDD/F Isomers According to NATO/CCMS (1988) and WHO I-TE

Structure	I-TE-value NATO/ CCMS 1988	WHO-TEF	Structure	I-TE-value NATO/ CCMS 1988e	WHO-TEF
2,3,7,8-Tetra-CDD	1	1	1,2,3,7,8,9-Hexa-CDD	0.1	0.1
1,2,3,7,8,9-Hexa-CDD	**0.5**	1	1,2,3,4,6,7,8-Hepta-CDD	0.01	0.01
1,2,3,4,7,8-Hexa-CDD	0.1	0.1	Octa-CDD	0.001	0.001
1,2,3,7,8,9-Hexa-CDD	0.1	0.1			
2,3,7,8-Tetra-CDF	0.1	0.1	2,3,4,6,7,8-Hexa-CDF	0.1	0.1
1,2,3,7,8-Penta-CDF	0.05	0.05	1,2,3,7,8,9-Hexa-CDF	0.1	0.1
2,3,4,7,8-Penta-CDF	0.5	0.5	1,2,3,4,6,7,8-Hepta-CDF	0.01	0.01

(*Continued*)

TABLE 8.3 *Continued*

1,2,3,4,7,8-Hexa-CDF   0.1   0.1   1,2,3,4,7,8,9-Hepta-CDF   0.01   0.01

1,2,3,6,7,8-Hexa-CDF   0.1   0.1   Octa-CDF   **0.001**   **0.0001**

Source: Landers J. P. and Bunce, N. J. Biochem. J., 1991,[12a] and van den Berg M. et al., *Environ. Health Perspect.*, 1998.[12b]

principal tool for the protection of the environment. They are related to the principles of green chemistry applied in industrial chemistry, called process-integrated protection of the environment.[14] The process in itself is designed to run without or with a minimum formation of pollutants. For incineration plants, this goal can be maintained by the following parameters, called good burning praxis (gbp):[15a,15b]

    Optimal burning temperature
    Optimal lambda value (air/fuel ratio)
    Optimal residence time of fuel in the flame, in general, regulated by turbulence

For either plant type, incineration, or fuel type, these factors must be empirically determined and controlled. Because dioxins as effluents are concerned, it is possible to reduce I-TE values from about 50 ng/m^3 to about 1 ng/m^3. Additional secondary measures (filter techniques) are therefore necessary for obtaining the lower limit value of 0.1 ng/m^3. Secondary measures are special filter techniques for pollutants formed in nongreen processes, also called end-of-pipe technology.[16] The main part of technical incineration plants consists of filter devices, mostly coke as adsorbent is used, which must be decontaminated later by itself by burning in hazardous-waste incinerators. The inhibition technology, discussed later, is related on principles of primary (green) measures for a clean incineration method.

8.4.2 Thermal Formation Mechanisms of PCDD/F

The specific mechanisms of PCDD/F formation in incineration processes are very complex.[17a,17b] Knowledge of the formation mechanisms of micropollutants allows the development of special minimization techniques and improvement of the whole process, therefore the study of formation mechanisms of toxic side products formed in chemical production is also a contribution to green chemistry.

PCDD/F and other chlorinated hydrocarbons observed as micropollutants in incineration plants are products of incomplete combustion like other products such as carbon monoxide, polycyclic aromatic hydrocarbons (PAH), and soot. The thermodynamically stable oxidation products of any organic material formed by more than 99% are carbon dioxide, water, and HCl. Traces of PCDD/F are formed in the combustion of any organic material in the presence of small amounts of inorganic and organic chlorine present in the fuel; municipal waste contains about 0.8% of chlorine. PCDD/F formation has been called "the inherent property of fire." Many investigations have shown that PCDD/Fs are not formed in the hot zones of flames of incinerators at about 1000°C, but in the postcombustion zone in a temperature range between 300 and 400°C.[17a] Fly ash particles play an important role in that they act as catalysts for the heterogeneous formation of PCDD/Fs on the surface of this matrix. Two different theories have been deduced from laboratory experiments for the formation pathways of PCCD/F:

1. *De novo Theory*: PCDD/Fs are formed from particulate (elementary) carbon species found in fly ash in the presence of inorganic chlorine of this matrix,
2. *Precursor Theory*: PCDD/Fs are formed from chemically related compounds as precursors. Chemically related products of PCDD/Fs are chlorophenols and chlorobenzenes. Both classes of compounds are present in the effluents of incinerators and can adsorb from the stack gas to the fly ash.[17b]

Both pathways have been shown to be relevant for PCDD/F formation in municipal-waste incinerations. Chlorophenols can be converted to PCDD by copper species known in synthetic chemistry as the Ullmann type II coupling reaction. By use of isotope labeling techniques in competitive concurrent reactions with both reactions performed in laboratory experiments it was shown that precursor theory pathways from chlorophenols may be more important compared to the de novo pathway, but either competing pathway strongly depends on such conditions as temperature, air flow rate, and residence time.[17] It may be difficult to model the complex reality of large incinerators using relevant laboratory experiments.

Recently, a general mechanistic scheme for most chlorinated compounds, including PCDD/F, observed in the effluents of incinerators was proposed using a special flow reactor (turbular furnace reactor) with acetylene as the starting material, and $CuCl_2$ and CuO as the most active catalytic components of fly ash (see Figure 8.3). The mechanism is based on ligand transfer chlorination of acetylene by copper chloride, leading to dichloroacetylene as the starting steps. Dichloroacetylene then condenses to a number of condensation products, such as various perchlorinated aliphatic and aromatic compounds,[18a–18b] (see Figure 8.3).

Hexachlorobenzene, shown in Figure 8.3, reacts further to chlorophenols and PCDD/F, which stay adsorbed on the copper species but can be further extracted[19] in the turbular furnace reactor. All low volatile chlorinated compounds shown in Figure 8.3 are eluted with the gas flow. The lower

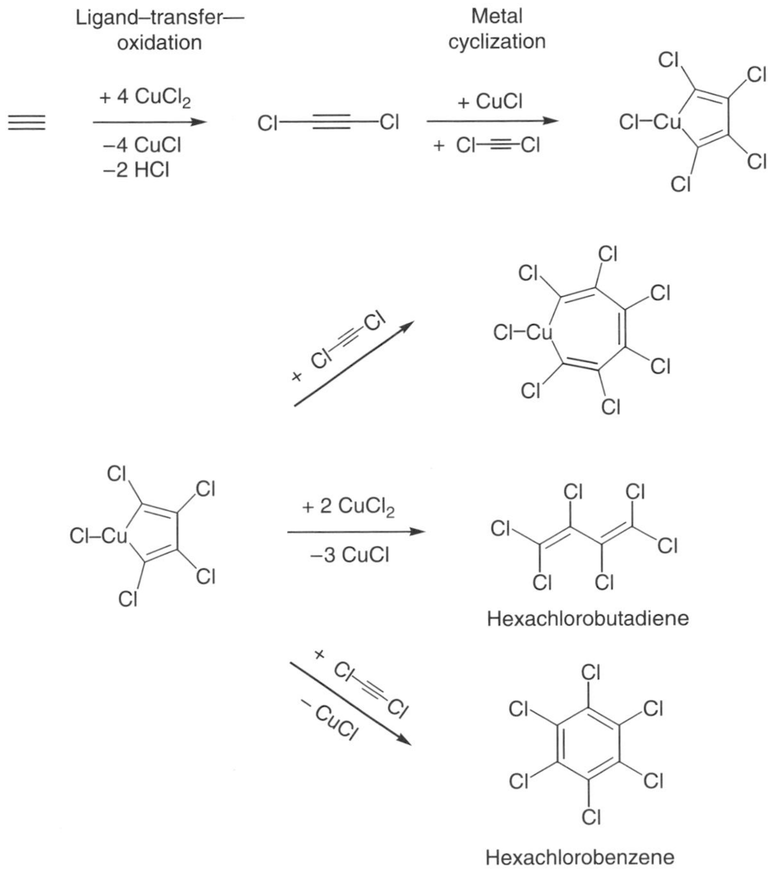

Figure 8.3 Scheme for global acetylene chlorination/condensation mechanism leading to hexachlorobenzene. (From A. Wehrmeier et al., *Environ. Sci. Technol.*, 1998.)

chlorinated isomers observed in the effluents are the result of subsequent dechlorination processes. Both classes of chlorine compounds have also been detected in the effluents of incinerators, Chlorobenzenes (CBs) and chlorophenols (CPs) are found in the stack gas of incinerators, but in much higher concentrations, showing a linear relationship with concentrations of PCDD/F. CBs and CPs have been used as indicator parameters for PCDD/Fs.[20a,20b] Chlorinated benzenes have been measured on-line by resonance enhanced multi-photon ionization (REMPI) spectroscopy in stack and flue gases of incinerators. This technique allows a direct and easy-to-do indirect estimation of PCDD/F concentrations in the effluents of incinerators.[20a] PCDD/F values are generally the result of a measurement during a sampling period of 6 hours, yielding an average value for PCDD/F for this time interval. Since a direct time control for PCCD/F is possible by measurement of indicator compounds an affected plant can be cleansed, for example, by the addition of more air (increase of the lambda value).

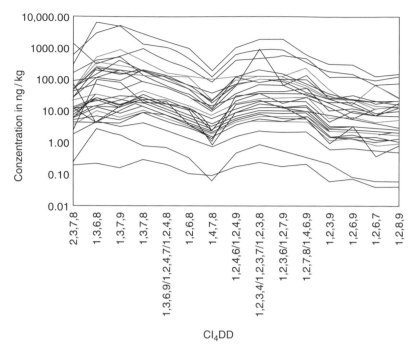

Figure 8.4 Ratio of tetrachlorinated dioxin isomers from a large variety and number of incineration samples. (From A. Wehrmeier et al., *Chemosphere*, 1998.)

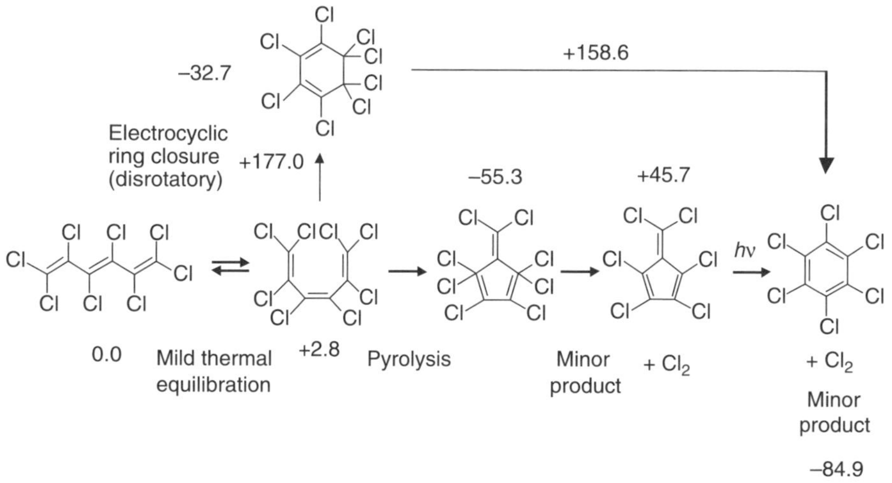

Figure 8.5 Observed and calculated reaction pathways from aliphatic C_6Cl_8-species to hexachlorobenzene. Numbers are values (KJ/mol) calculated by B3LYP/6-31G(2d). (From H. Detert et al., to be published.)

The relative ratio of regioisomers of PCDD/F and other chlorinated compounds formed in incinerators is called the incineration pattern. The pattern can be derived from statistical analysis of a large number of measurements of the same plants, and can be used for elucidation of thermal formation mechanisms in plants. In principle regioisomers can be formed either by stereospecific chlorination or dechlorination processes. The pattern has also been used as a part for explaining of the formation mechanism of PCDD/F and other chlorinated compounds formed in incinerations[21] (see Figure 8.4).

A more detailed mechanistic study was performed recently for the thermal conversion of perchlorinated aliphatic C-6 polyenes like C_6Cl_8 into hexachlorobenzene[22] (see Figure 8.5).

8.4.3 Inhibition Technology as Primary Measure for PCDD/F Minimization

As a consequence of the detection of catalytic pathways for formation of PCDD/F, special inhibition methods have been developed for PCDD/F. By this approach the catalytic reactions are blocked by adding special inhibitors as "poisoning" compounds for copper and other metal species in the fly ash. Special aliphatic amines (triethylamine) and alkanolamines (triethanolamine) have been found to be very efficient as inhibitors for PCDD/F, and have been used in pilot plants. The effect can be seen in Figure 8.6. The inhibitors have been introduced into the incinerator by spraying them into the postcombustion zone of the incinerator at about 400°C.[23a−c]

These amines used as inhibitors show negative side effects (disturbances) when used for larger plants. They can be regarded as pollutants by themselves, and can disturb special devices in the plants, especially, when used on a larger scale, filters like electrostatic precipitators. Therefore, we have improved the inhibition method by the use of much safer inorganic compounds as inhibitors, such as,

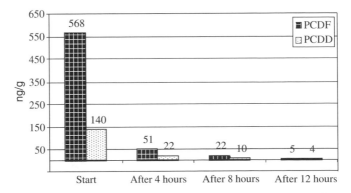

Figure 8.6 Effect of aliphatic amines addition as inhibitors on PCDD/F concentration (measured on fly ash). (From D. Lenoir et al., *Umweltchem. Ökotox.*, 1989.)

sulfamide and amido sulfonic acid, which can be added directly to the fuel and survive the hot area of the flames before entering the postcombusting zone.[24,25] In a recent study, incineration of lignite coal/solid waste/polyvinyl chloride (PVC) was used in a laboratory-scale furnace in order to study the prevention of PCDD/ F formation by inhibitors.[25] Nineteen inhibitors divided into four different types of groups (metal oxides, N-containing compounds, S-containing compounds, and N- and S-containing compounds) according to their chemical nature were tested. The total amounts of PCDD/F generated during the experiments with lignite coal, solid waste, and PVC are high enough to investigate a greater inhibition. The average I-TEQ value of the sum of PCDD/F is about 15 pg/g fuel (see Figure 8.7). A relatively low inhibitory effect is observed for the substances that contain only nitrogen. However, higher reduction effects of PCDD/F can be derived for the S-containing substances present in 10% of the fuel. Sulfur itself shows a very strong inhibition effect for PCDD/F. It is already known that sulfur is converted into SO_2 and that it reduces Cl_2 to HCl, and therefore dioxin and PCB formation can be reduced.[26] Also because of this mechanism, the rest of the S-containing compounds probably, inhibit PCDD/F flue gases. Although the single N- and S-containing compounds are not very effective as inhibitors, all other N- and S-containing substances seem to be able to greatly reduce PCDD/F flue gas emission if used as a 10% additive to lignite coal, solid waste, and PVC as fuel. A mixture of $(NH_2)_2CO+S$ (1:1) can successfully inhibit PCDD/F toxic gases. However, the most effective inhibitors for PCDD/F are $(NH_4)_2SO_4$ and $(NH_4)_2S_2O_3$. Both compounds can reduce the PCDD/F emission up to 98–99%. In addition, $(NH_4)_2SO_4$ and $(NH_4)_2S_2O_3$ were used at 5, 3, and 1% of the fuel. The results show that both substances are still effective inhibitors of PCDD/F formation at 5% and 3% of the fuel (see Figure 8.8).

If the percentage of these substances is decreased further, the suppressing effect of dioxin formation will also decrease. $(NH_4)_2SO_4$ might also reduce the PCDD/F

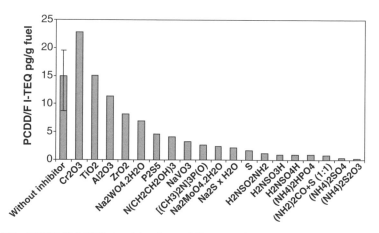

Figure 8.7 PCDD/F I-TEQ (pg/g) values of flue gas after the combustion of lignite coal, solid waste, and PVC in the samples without inhibitor and 19 different compounds used with a 10% inhibitor of the fuel. (From M. Pandelova et al., *Environ. Sci. Technol.*, 2005.)

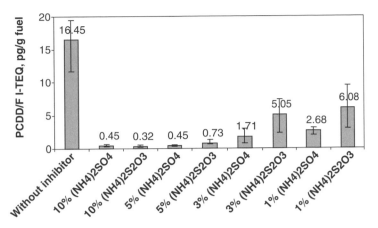

Figure 8.8 PCDD/F I-TEQ pg/g fuel values in flue gas for the samples without an inhibitor and with $(NH_4)_2SO_4$ and $(NH_4)_2S_2O_3$ as inhibitors of the fuel at 10%, 5%, 3%, and 1%.

flue-gas emission up to 90%, even at 3% of the fuel. $(NH_4)_2SO_4$ is a low-cost and nontoxic material. That makes it suitable for use in full-scale combustion units.

Inhibition technology also has been used recently by two other groups.[27,28] Urea as an aqueous solution added to the fuel has been found to be very effective as an inhibitor of PCDD/F in a pilot and technical plant. Furthermore, other *N*-compounds and *S*-compounds, such as sulfur dioxide, ammonia, dimethylamine, and methyl mercaptan sprayed as gaseous inhibitors in the flue gas, seem to be a promising technique for preventing the formation of PCDD/F in waste incineration.

8.5 CONCLUSION

An important principle in green chemistry is the avoidance of pollutants formed in chemical processes by the use of primary measures. This approach is shown in this chapter for dioxins formed in incinerations. Concentration of PCDD/F in various parts of the environment has increased during the last few decades as result of an increase in use of different technical thermal processes. Therefore, relevant formation mechanisms for PCDD/F have been studied, showing the importance of copper species in inducing catalytic pathways from aliphatic precursors like acetylene in the postcombustion zone at about 300°C. Now, indicator parameters for PCDD/F like chlorobenzenes can be measured on-line, allowing for the cleansing of the plants. The inhibition technology uses the addition of special compounds to block the active sites of copper in the fly ash of incinerators. PCDD/F concentrations are slowly decreasing in the environment due to primary measures discussed in this chapter in combination with advanced filter devices as secondary measures at incineration plants.

REFERENCES

1. Alloway, B. J.; Ayres, D. C. *Schadstoffe in der Umwelt, Chemische Grundlagen zur Beurteilung von Luft-, Wasser-, und Bodenverschmutzungen (Chemical Principles of Environmental Pollutants)*, Spektrum Verlag, Heidelberg, 1996.

2. Ramamoorthy, S.; Ramamoorty, S. *Chlorinated Organic Compounds in the Environment, Regulary and Monitoring Assessment*, Lewis Publishers, Boca Raton, Fla., 1997.

3. Schlottmann, U.; Kreibich, M. *Nachrichtenbl. Chem.*, 2001, **49**, 608; see also, www.chem.unep.ch/pops/

4. Gribble, G. W. *Acc. Chem. Res.*, 1998, **31**, 141.

5. Djien Diem, A. K.; van Zorge, J. A. *ESPR-Environ. Sci. Pollut. Res.*, 1995, **2**, 46.

6. Dioxin and Furan Inventories, National and Regional Emissions of PCDD/PCDF, United Nations Environmental Program, Prepared by UNEP Chemicals, Geneva, Switzerland, most recent issue, May 1999.

7. Jobst, H.; Aldag, R. *UWSF-Z. Umweltchem. Ökotox.*, 2000, **12**, 2.

8. Ferrario, J. B.; Byrne, C. J.; Cleverly, D. H. *Environ. Sci. Technol.*, 2000, **34**, 4524.

9. Mackay, D.; Shiu, W. Y.; Ma, K. C. *Illustrated Handbook of Physical-Chemical Properties and Environmental Fate for Organic Chemicals*, Vol. II, *Polynuclear Aromatic Hydrocarbons, Polychlorinated Dioxins, and Dibenzofurans*, Lewis Publishers, Boca Raton, Fla., 1992.

10. Parlar, H.; Angerhöfer, D. Dioxins and annulated derivatives, in *Houben-Weyl*, Vol. E, *Hetarenes IV*, Thieme, Stuttgart, 1997.

11. Safe, S.; Hutzinger, O.; Hill, T. A. (Eds.) *Polchlorinated Dibenzo-p-Dioxins and -Furans (PCDDs/PCDFs). Sources and Environmental Impact, Epidemiology, Mechanisms of Action, Health Risks, Environmental Toxin Series 3*, Safe, S.; Hutzinger, O. (Eds.) Springer, Berlin, 1990.

12a. Landers, J. P.; Bunce, N. J. The Ah receptor and the mechanism of dioxin toxicity (Review), *Biochem. J.*, 1991, **276**, 273.

12b. Van den Berg, M.; Birnbaum, L.; Bosveld, A. T. C. et al., *Environ Health Perspect.*, 1998, **106**, 775.

12c. Ahlborg, U. G.; Brouwer, A.; Fingerhut, M. A. et al., *Eur. J. Pharmacol.*, 1992, **228**, 179.

12d. DeVito, M. J.; Diliberto, J. J.; Ross, D. G. et al., *Toxicol., Appl. Pharmacol.*, 1997, **147**, 267.

12e. Safe, S. H. *Crit. Rev. Toxicol.*, 1990, **21**, 519.

12f. Ahlborg, U. G.; Becking, G. C.; Birnbaum, L. S. et al., *Chemosphere*, 1994, **28**, 1049.

12g. van den Berg, M.; Birnbaum, L. S.; Denison, M. et al. The 2005 World Health Organization re-evaluation of human and mammalian toxic equivalency factors for dioxins and dioxin-like compounds, *Toxicol. Sci.*, 2006, **93**, 223–241.

13. Ahlborg, U. G.; Becking, G. C.; Birnbaum, L. S. et al., *Chemosphere*, 1994, **28**, 1049.

14. Christ, C. (Ed.) *Production-Integrated Environmental Protection and Waste Management in the Chemical Industry*, Wiley-VCH, Weinheim, Germany, 1999.

15a. Lenoir, D.; Kaune, A.; Hutzinger, O. et al., *Chemosphere*, 1991, **23**, 1491.

15b. McKay, G. *Chem. Eng. J.*, 2002, **86**(3), 343.

16. Kaune, A.; Lenoir, D.; Nikolai, U. et al., *Organohalogen Compounds*, 1995, **23**, 477.

17a. Stanmore, B. R. *Combust. Flame*, 2004, **136**, 398.

17b. Dickson, L. C.; Lenoir, D.; Hutzinger, O. *Environ. Sci. Technol.*, 1992, **26**, 1822.

18a. Wehrmeier, A.; Lenoir, D.; Sidhu, S. et al., *Environ. Sci. Technol.*, 1998, **32**, 2741.

18b. Taylor, H. P.; Lenoir, D. *Sci. Total Environ.*, 2001, **269**, 1.

19. Taylor, P.; Wehrmeier, A.; Sidhu, S. S. et al., *Chemosphere*, 2000, **40**, 1297.

20a. Kaune, A.; Lenoir, D.; Schramm, K.-W. et al., *Environ. Engi. Sci.*, 1998, **15**, 85.

20b. Blumenstock, M.; Zimmermann R.; Schramm K.-W.; Kettrup A. *Chemosphere*, 2001, **42**, 507.

21. Wehrmeier, A.; Lenoir, D.; Schramm, K.-W. et al., *Chemosphere*, 1998, **36**, 2775.

22. Detert, H.; Zipse, H.; Lenoir, D., to be published.

23a. Lippert, T.; Wokaun, A.; Lenoir, D. *Environ. Sci. Technol.*, 1991, **25**, 1485.

23b. Dickson, L. C.; Lenoir, D.; Hutzinger, O. et al., *Chemosphere*, 1989, **19**, 1435.

23c. Lenoir, D.; Hutzinger, O.; Mützenich, G.; Horch, K. *Umweltchem. Ökotox.*, 1989, **4**, 3.

24. Samaras, P.; Blumenstock, M.; Lenoir, D. et al., *Environ. Sci. Technol.*, 2000, **34**, 5092.

25. Pandelova, M.; Lenoir, D.; Kettrup, A.; Schramm, K.-W. *Environ. Sci. Technol.*, 2005, **39**, 3345.

26. Gullett, B.; Bruce, K. R.; Beach, L. O. *Environ. Sci. Technol.*, 1992, **26**, 1938.

27. Ruokojarvi, P. H.; Asikainen, A. H.; Tuppurainen, K. A.; Ruuskanen, J. *Sci. Total Environ.*, 2004, **325**, 83.

28. Kuzuhara, S.; Sato, H.; Tsubouchi, N. et al., *Environ. Sci. Technol.*, 2005, **39**, 795.

PART 3

GREEN CATALYSIS AND BIOCATALYSIS

9

GREEN CHEMISTRY: CATALYSIS AND WASTE MINIMIZATION

ROGER A. SHELDON

Delft University of Technology, Delft, The Netherlands

INTRODUCTION

A growing environmental awareness and increasingly stringent environmental legislation have focused the attention of chemical manufacturers on what has become known as sustainable development or green chemistry. A working definition of green chemistry is: Technologies that efficiently utilize energy and (preferably renewable) raw materials and reduce, or preferably, eliminate, the generation of waste and avoid the use of toxic and/or hazardous reagents and solvents. This is actually very close to and has the same meaning as the definition given by the International Union of Pure and Applied Chemistry's (IUPAC's) working party on green chemistry. The emphasis is placed clearly on the reduction (elimination) of waste at the source, that is, primary pollution prevention rather than incremental, end-of-pipe solutions (waste remediation). A direct consequence of this trend toward green chemistry is that traditional concepts of process efficiency, based exclusively on chemical yield, are being replaced by a new paradigm that assigns economic value to eliminating waste and avoiding the use of toxic and/or hazardous chemicals.[1-4]

9.1 E FACTORS AND ATOM EFFICIENCY

Two useful measures of the potential environmental impact of chemical processes are the *E*-factor,[5-7] defined as the mass ratio of waste to desired product, and the

Methods and Reagents for Green Chemistry: An Introduction, Edited by Pietro Tundo, Alvise Perosa, and Fulvio Zecchini

TABLE 9.1   *E*-factors for Different Segments of the Chemical Industry

Industry Segment	Production (tons)	E-Factor kg Waste/kg Product
Oil refining	$10^6 - 10^8$	ca. 0.1
Bulk chemicals	$10^4 - 10^6$	$<1-5$
Fine chemicals	$10^2 - 10^4$	5–50
Pharmaceuticals	$10 - 10^3$	25 –> 100

atom efficiency,[1,2] calculated by dividing the molecular weight of the product by the sum of the molecular weights of all substances produced in the stoichiometric equation.

A prime cause of high E-factors is the use of stoichiometric inorganic reagents. Fine chemicals and pharmaceuticals manufacture, for example, is rampant with classic stoichiometric technologies that generate copious amounts of inorganic salt as waste. Examples that readily come to mind are stoichiometric reductions with metals (Zn, Fe) and metal hydrides ($NaBH_4$, $LiAlH_4$, and derivatives thereof) and stoichiometric oxidations with permanganate, dichromate, periodate, and so forth. Similarly, processes employing mineral acids (H_2SO_4, HF), Lewis acids ($AlCl_3$, $ZnCl_2$, BF_3), or inorganic bases (NaOH, K_2CO_3), often in stoichiometric amounts, represent a major source of inorganic waste that cannot easily be recycled. Reactions of this type, widely employed in the fine chemical industry, include Friedel–Crafts acylation mediated by Lewis acids such as $AlCl_3$, sulfonations, and diazotizations, to name but a few.

The workup for such reactions involves neutralization and concomitant generation of salts such as NaCl, Na_2SO_4, and $(NH_4)_2SO_4$. The elimination of such waste streams and a reduction in the dependence on the use of hazardous chemicals, such as phosgene, dimethyl sulfate, peracids, sodium azide, halogens, and HF, are primary goals in green chemistry.

Table 9.1 contains the values of E-factors (mass ratio of waste to desired product) for different industry segments; most of the processes for fine chemicals and pharmaceuticals, with a very large E-factor, use reagents in stoichiometric quantities, often in combination with environmentally unfriendly solvents. The E-factor is the actual amount of waste formed in the process and includes everything except the desired product, not only the raw materials and reagents involved in the stoichiometric equation but also chemicals used in the workup, for example, acids and bases for neutralization, and solvent losses. Strictly speaking, it should also include the fuel used to generate the energy required to operate the process, but this is often difficult to quantify. Process water is not included, as this leads to *E*-factors that are not generally meaningful.

9.2   THE ROLE OF CATALYSIS

The increasing use of catalytic processes can substantially reduce waste at the source, resulting in primary pollution prevention. The theoretical process efficiency

Stoichiometric:

$$3 \, PhCH(OH)CH_3 + 2 \, CrO_3 + 3 \, H_2SO_4 \longrightarrow 3 \, PhCOCH_3 + Cr_2(SO_4)_3 + 6 \, H_2O$$

Atom efficiency = 360/860 = 42% E_{theor} = ca. 1.5

Catalytic:

$$PhCH(OH)CH_3 + 1/2 \, O_2 \xrightarrow{\text{Catalyst}} PhCOCH_3 + H_2O$$

Atom efficiency = 120/138 = 87%

E_{theor} = ca. 0.1(0)

By-product: H_2O

Figure 9.1   Acetophenone synthesis by stoichiometric and catalytic oxidation.

can be quantified by the atom efficiency, the ratio between the molecular weight of the product, and the sum of the molecular weights of all substances produced in the stoichiometric equation. It should be pointed out, however, that the atom efficiency only takes the chemicals appearing in the stoichiometric equation into account.

Figure 9.1 compares the synthesis of acetophenone by classic oxidation of 1-phenylethanol with stoichiometric amounts of chromium oxide and sulphuric acid, with an atom efficiency of 42%, with the heterogeneous catalytic oxidation with O_2, with an atom efficiency of 87%, and with water as the only by-product. This is especially important if we consider the environmental unfriendliness of chromium salts: the potential environmental impact of reactions can be expressed by the environmental quotient (EQ), where E is the E-factor (kg waste/kg product) and Q is the environmental unfriendliness quotient of the waste. If Q is

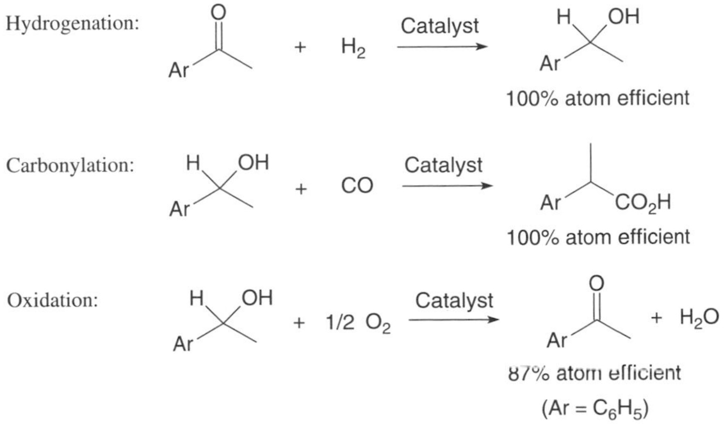

Figure 9.2   Atom-efficient catalytic processes.

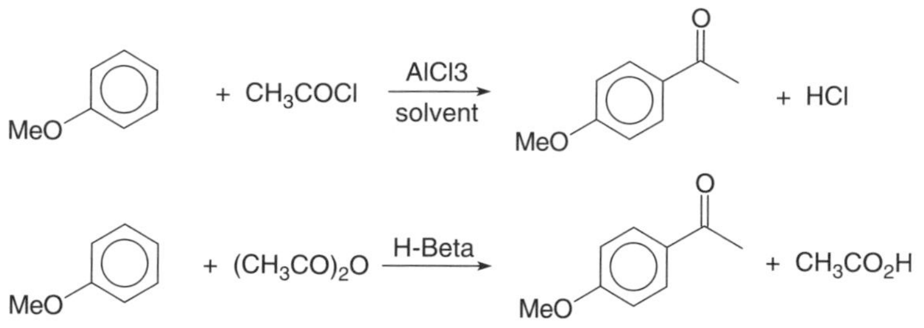

Homogeneous	Heterogeneous
AlCl$_3$ >1 equivalent	H-Beta, catalytic, and regenerable
Solvent (recycle)	No solvent
Hydrolysis of products	No water necessary
85–95% yield	>95% yield/higher purity
4.5 kg aqueous effluent per kg	0.035 kg aqueous effluent per kg

Figure 9.3 Friedel–Crafts acylation of anisole.

1 for NaCl, for example, then for chromium salts Q could be arbitrarily set at, say 100 or 1000. Similarly, clean catalytic technologies can be utilized for hydrogenation of acetophenone and carbonylation of 1-phenylethanol (Figure 9.2), with 100% atom efficiency in both cases.

One way to significantly reduce the amount of waste is to substitute traditional mineral acids and Lewis acids with recyclable solid acid catalysts. A good example of this is the Rhodia process for the synthesis of 4-methoxy acetophenone by Friedel–Crafts acetylation of anisole (Figure 9.3) with acetic anhydride, catalyzed by the acid form of zeolite beta.[8] This replaced a traditional Friedel–Crafts acylation using acetyl chloride in combination with more than one equivalent of aluminium chloride in a chlorinated hydrocarbon solvent. The new process requires no solvent and avoids the generation of HCl in both the acylation and the synthesis of the acetyl chloride. The original process generated 4.5 kg of aqueous effluent (containing AlCl$_3$, HCl, chlorinated hydrcarbon residues, and acetic acid) per kg of product. The catalytic alternative generates 0.035 kg of aqueous effluent (i.e., >100 times less), consisting of 99% water, 0.8% acetic acid, and <0.2% other organics per kg of product . Workup consists of catalyst filtration and distillation of the product. Because of the simpler process, a higher chemical yield is obtained (>95% vs. 85–95%) and higher product purity is obtained. Moreover, the catalyst is recyclable and the number of unit operations is reduced from 12 to 3. The conclusion is clear: The new technology is not only cleaner and greener, it also leads to lower production costs than the classic process. An important lesson indeed.

Other important successes have been achieved in developing clean, "green," methods to oxidize alcohols, for example, the Ru/TEMPO (tetramethylpiperidinyloxyl radical) catalysis, shown in Figure 9.4, for the aerobic oxidation of alcohols.[9]

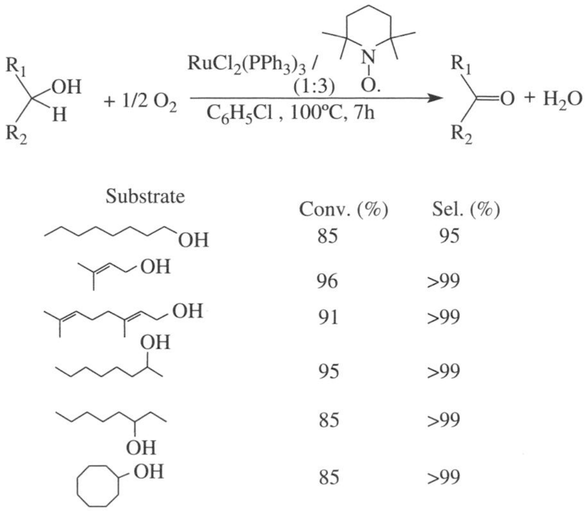

Figure 9.4 Aerobic oxidation of primary and secondary alcohols catalyzed by $RuCl_2$ $(Ph_3P)_3$/TEMPO in PhCl at 100°C.

9.3 CATALYSIS IN WATER

Another environmental issue is the use of organic solvents. The use of chlorinated hydrocarbons, for example, has been severely curtailed. In fact, so many of the solvents favored by organic chemists are now on the black list that the whole question of solvents requires rethinking. The best solvent is no solvent, and if a solvent (diluent) is needed, then water has a lot to recommend it. This provides a golden opportunity for biocatalysis, since the replacement of classic chemical methods in organic solvents by enzymatic procedures in water at ambient temperature and pressure can provide substantial environmental and economic benefits. Similarly, there is a marked trend toward the application of organometallic catalysis in aqueous biphasic systems and other nonconventional media, such as fluorous biphasic, supercritical carbon dioxide and ionic liquids.[10]

A prime advantage of such biphasic systems is that the catalyst resides in one phase and the starting materials and products are in the second phase, thus providing for easy recovery and recycling of the catalyst by simple phase separation. A pertinent example is the aerobic oxidation of alcohols catalyzed by a water-soluble Pd-bathophenanthroline complex (Figure 9.5).[11] The only solvent used is water, the oxidant is air, and the catalyst is recycled by phase separation.

The Boots Hoechst Celanese (BHC) ibuprofen process[12] involves palladium-catalyzed carbonylation of a benzylic alcohol (IBPE). More recently, we performed this reaction in an aqueous biphasic system using Pd/tppts as the catalyst (Figure 9.6; tppts = triphenylphosphinetrisulfonate). This process has the advantage of easy removal of the catalyst, resulting in less contamination of the product.

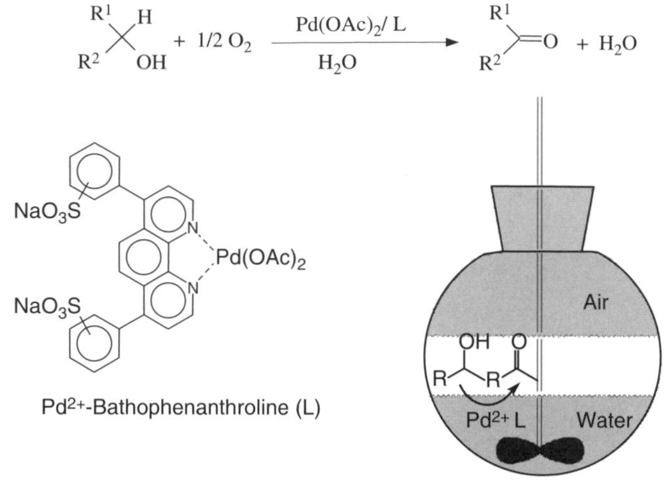

Figure 9.5   Aerobic oxidation of alcohols catalyzed by Pd(II)/bathophenanthroline in water.

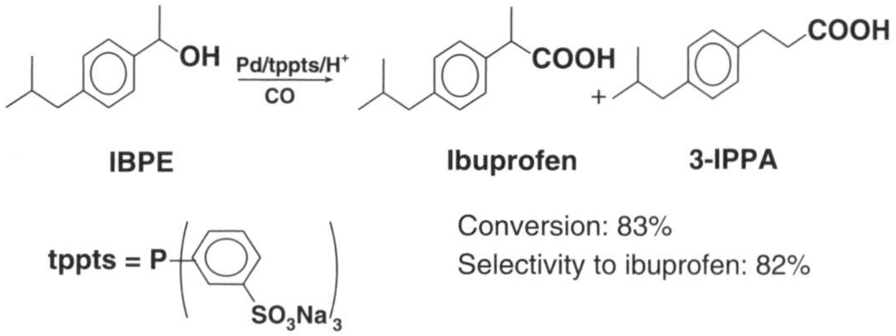

Figure 9.6   Ibuprofen synthesis by Pd/tppts-catalyzed biphasic carbonylation in water.

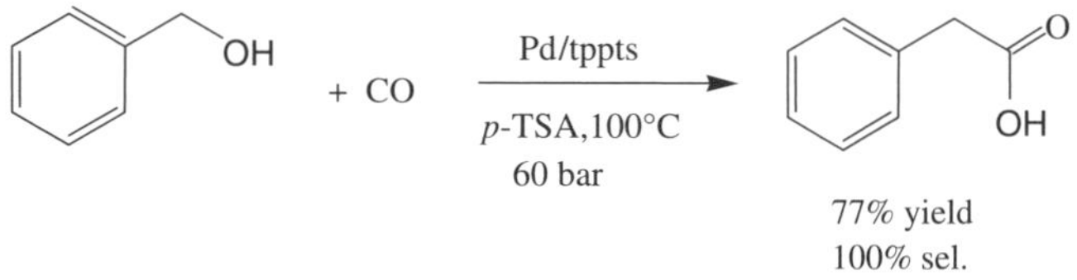

Figure 9.7   Carbonylation of benzyl alcohol-catalyzed by tppts/Pd.

In the same way, the biphasic carbonylation of benzyl alcohol (Figure 9.7) was achieved.[13] Phenylacetic acid was obtained in 77% yield, 100% selectivity, and 100% atom utilization.

Similarly, acylamino acids can be prepared with 100% atom utilization via palladium-catalyzed amidocarbonylation.[14] The method was used for the synthesis of a surfactant from sarcosine (Figure 9.8).

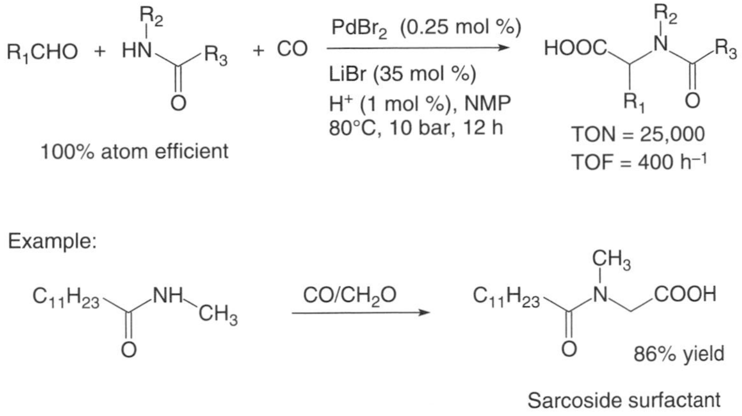

Figure 9.8 Acylamino acids via palladium-catalyzed amidocarbonylation.

9.4 PROCESS INTEGRATION

The ultimate "greening" of fine chemical synthesis is the replacement of multistep syntheses by the integration of several atom-efficient catalytic steps. For example, Figure 9.9 shows the new Rhodia, salt-free caprolactam process involving three catalytic steps. The last step involves cyclization in the vapor phase over an alumina catalyst in more than 99% conversion and more than 99.5% selectivity.

Another example of the substitution of classic routes for chemical synthesis by multistep catalytic processes is the Rhodia vanillin process (Figure 9.10),[8] which involves four steps, all employing a heterogeneous catalyst.

Finally, the Lonza nicotinamide process (Figure 9.11),[15] involves the integration of both heterogeneous catalysis with a final step employing enzymatic catalysis.

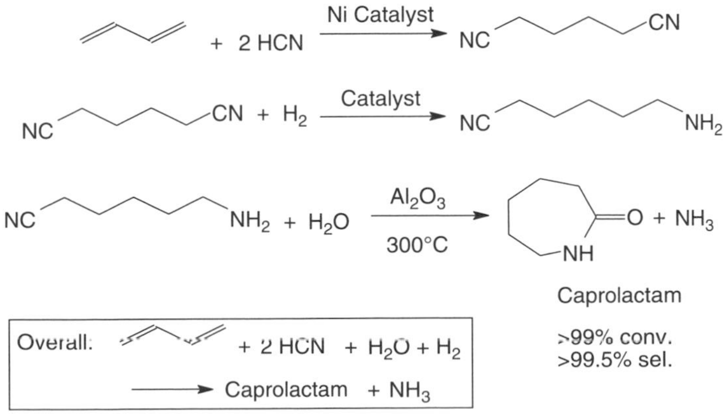

Figure 9.9 Salt-free caprolactam process.

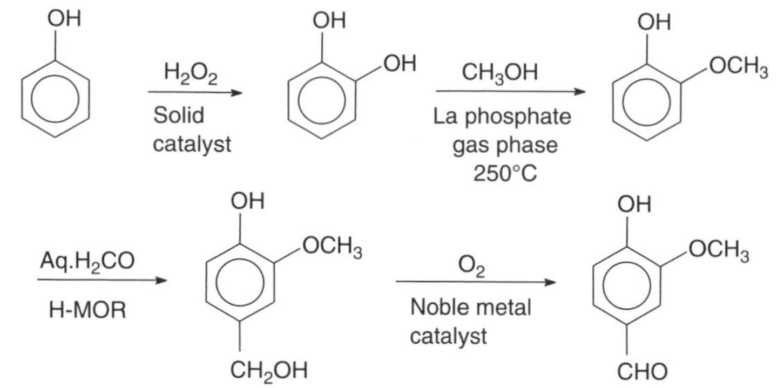

Overall: $C_6H_6O + H_2O_2 + CH_3OH + H_2CO + 1/2\,O_2 \rightarrow C_6H_8O_3 + 3\,H_2O$

Figure 9.10 Rhodia vanillin process.

9.5 CONCLUSIONS

The key to achieving the goal of reducing the generation of environmentally unfriendly waste and the use of toxic solvents and reagents is the widespread substitution of "stoichiometric" technologies by greener, catalytic alternatives. Examples include catalytic hydrogenation, carbonylation, and oxidation. The first two involve 100% atom efficiency, while the latter is slightly less than perfect owing to the coproduction of a molecule of water. The longer-term trend is toward the use of the simplest raw materials—H_2, O_2, H_2O, H_2O_2, NH_3, CO, and CO_2—in catalytic, low-salt processes. Similarly, the widespread substitution of classic mineral and Lewis acids by recyclable solid acids, such as zeolites and acidic clays, and the introduction of recyclable solid bases, such as hydrotalcites (anionic clays) will result in a dramatic reduction of inorganic waste.

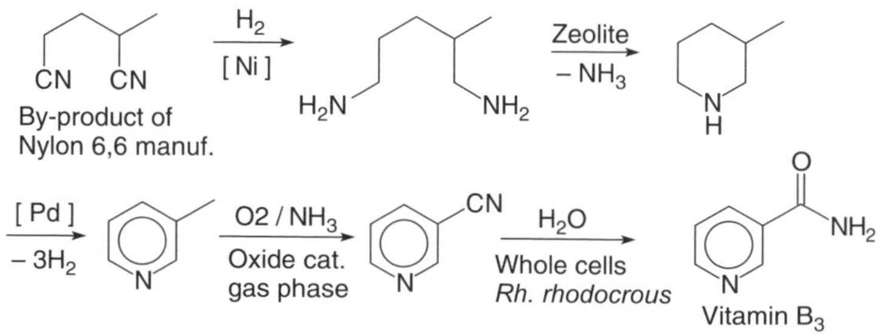

Overall: $C_6H_8N_2 + 3/2\,O_2 + H_2 \rightarrow C_6H_6N_2O + 2\,H_2O$

Figure 9.11 Lonza nicotinamide process.

A possible alternative for the use of organic solvents (many of which are on the black list), is the extensive utilization of water as a solvent. This provides a golden opportunity for biocatalysis, since the replacement of classic chemical methods in organic solvents by enzymatic procedures in water, at ambient temperature, can provide both environmental and economic benefits. Similarly, there is a marked trend toward organometallic catalysis in aqueous biphasic systems and other nonconventional media, such as fluorous biphasic, supercritical carbon dioxide, and ionic liquids.

In conclusion, the widespread application of chemo- and biocatalytic methodologies to the manufacture of fine chemicals has enormous potential for creating greener, environmentally benign processes.

REFERENCES

1. Trost, B. M. *Science*, 1991, **254**, 1471.

2. Trost, B. M. *Angew. Chem. Int. Ed.*, 1995, **34**, 259.

3. Sheldon, R. A., Downing, R. S. *Appl. Catal. A.: General*, 1999, **189**, 163.

4. Sheldon, R. A. *Pure Appl. Chem.*, 2000, **72**, 1233.

5. Sheldon, R. A. *Chem. Ind. (London)*, 1997, **12**, also, 1992, **903**.

6. Sheldon, R. A. *J. Chem. Technol. Biotechnol.*, 1997, **68**, 381.

7. Sheldon, R. A. *Chemtech*, 1994, **38**; also, *J. Mol. Cat.*, 1996, **75**, 107.

8. Ratton, S. *Chem. Today*, March/April, 1997, **33**.

9. Dijksman, A.; Arends, I. W. C. E.; Sheldon, R. A. *Chem. Commun.*, 1999, **16**, 1591–1592.

10. For a recent review see Sheldon, R. A. *Green Chem.*, 2005, **7**, 267–278.

11. Ten Brink, G. J.; Arends, I. W. C. E.; Sheldon, R. A. *Science*, 2000, **287**, 1636–1639.

12. Elango, V.; et al., US Patent 4, 981, 995, 1991 (to Hoechst Celanese).

13. Papadogianakis, G.; Maat, L.; Sheldon, R. A. *J. Mol. Cat. A: Chem.*, 1997, **179**, 116.

14. Beller, M.; et al. *Angew. Chem., Int. Ed. Eng.*, 1997, **36**, 1494; ibid., 1999, **38**, 1454.

15. Heveling, J. *Chimia*, 1996, **50**, 114.

10

SEAMLESS CHEMISTRY FOR SUSTAINABILITY

JOHAN THOEN

Dow Benelux B.V., Terneuzen, The Netherlands

JEAN LUC GUILLAUME

Dow Europe GmbH, Horgen, Switzerland

INTRODUCTION

The chemical industry manufactures and supplies the citizen of the world with thousands of chemicals essential to daily life. The industry product portfolio ranges, for example, from epoxy resins designed for paints protecting concrete or metal, to membranes for water purification or pharmaceuticals protecting human life.

Two main global trends provide the industry with novel opportunities to continue serving the citizens of the world:

- The first trend toward a growing societal awareness of the environment and citizen safety leads to a need for products to be increasingly designed for reuse or for more benign environmental impact, while requirements for product testing and knowledge about product use are increasing.
- The second trend is nurtured by a rapidly growing demand in emerging economies for products that are expected to have a dramatic impact on the state of the world, as shown in Table 10.1, if the current world production patterns and consumer behaviors continue.

Methods and Reagents for Green Chemistry: An Introduction, Edited by Pietro Tundo, Alvise Perosa, and Fulvio Zecchini
Copyright © 2007 John Wiley & Sons, Inc.

TABLE 10.1   State of the World: 1950–2050

	1950	1972	1997	2050
1. Population	2.5	3.8	5.8	10.7
2. Megacities	2	9	25	200
3. Food	1980	2450	2770	2200
4. Fisheries	19	58	91	35
5. Water use	1300	2600	4200	7500
6. Rain forest	100	85	70	45
7. CO_2 emissions	1.6	4.9	7.0	14.0
8. Ozone layer	—	1.4	3.0	7.0

Key: 1. Billion persons; 2. cities with population greater than 8 million; 3. average daily food production in caloriescapita; 4. annual fish catch in million tons; 5. annual water use in cubic kilometers; 6. index of forest cover, 1950 = 100; 7. annual CO_2 emissions in billion tons of carbon; 8. atmospheric concentration of CFCs in parts/billion.
Source: World Resource Institute.

To mitigate these two trends, the world should put in place a more "sustainable development," that is, a development that meets the needs of the present without compromising the ability of the future generation to meet its needs.[1]

The search for a more sustainable development translates into two immediate challenges for the chemical industry:

- *Feedstock availability*, or what alternatives exist to nonrenewable fossil feedstock.
- *Energy cost*, or what alternatives exist to increasingly expensive nonrenewable fossil fuels.

In the first section, this chapter describes the current feedstock and energy status within the chemical industry in Europe. This section also summarizes some of the foreseeable problems the industry faces. The second section reviews the various technology options available to mitigate these foreseeable problems. In conclusion, a most probable scenario for change is tentatively be laid out.

10.1   THE FEEDSTOCK AND ENERGY CHALLENGE

10.1.1   The Current Situation

As shown in Figure 10.1, the European chemical industry utilizes around 85 million tons equivalent crude oil as feedstock and around 82 million tons equivalent crude oil as energy, with around 40% of that energy being electricity. Around 70% of the feedstock consumed by the European chemical industry

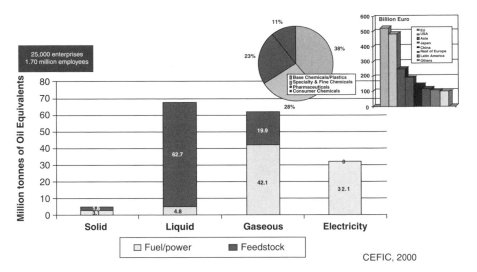

Figure 10.1 European chemical industry today.

originates from Naphtha (crude oil), while the remainder originates from gas liquids (ethane to butane). Thus, both the current feedstock and the energy consumed by the European chemical industry essentially rely on fossil resources, that is, crude oil (including the crude oil burned by power providers to generate electricity).

Most experts predict that crude oil reserves will last no more than 40–50 years of world consumption, although wide disagreement exists on when the world crude oil production peak will possibly occur.[2] The combination of a growing worldwide product and transport demand, as seen in China and India, for example, with a diminishing world supply of crude oil may lead to sharply escalating price levels for crude oil with a detrimental impact on the activity level of industry.

European industry has already improved its energy efficiency by more than 50% (see Figure 10.2) since the first oil crisis in 1975. As a consequence, a dominant part of the current energy consumption of petrochemical companies and its associated CO_2 emissions originates from one single primary operation, that is, the cracking of hydrocarbon feedstock to primary building blocks (ethylene, propylene, benzene, etc.). Transforming these primary building blocks into intermediates or polymers demands much less energy (see Figure 10.3), and therefore emits much less CO_2.

Exploiting alternative feedstock that would require a less energy-intensive primary operation than cracking, for example, by fermentation, may therefore simultaneously help address the chemical industry's energy challenge and the feedstock challenge, while making a positive contribution to the climate change issue through reduction of CO_2 emission.

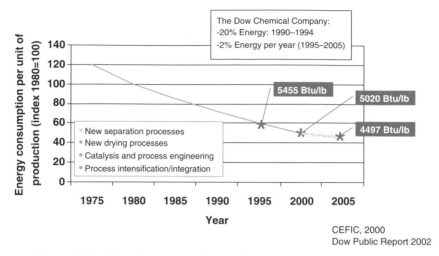

CEFIC, 2000
Dow Public Report 2002

Figure 10.2 Historical energy efficiency in the European chemical industry.

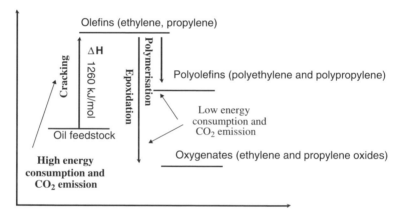

Figure 10.3 Example of material and energy efficiency in the current crude-oil-based chemistry.

10.1.2   Feedstock of the Future

Three generic types of alternative feedstock to crude oil can be exploited as a carbon source for the chemical industry:

- Fossil feedstock not exploited today
- Renewable feedstock
- Atmospheric CO_2

Methane is a fossil feedstock of potential interest to the chemical industry. Methane is to be found as an unexploited (so-called stranded gas) stream from

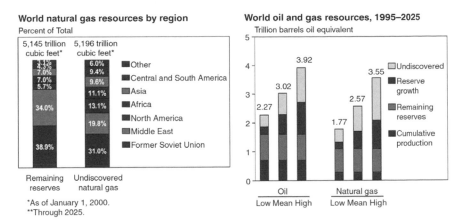

Figure 10.4 Feedstocks in the Future. (From U.S. Geological Survey. *World Petroleum Assessment 2000*. http://greenwood.cr.usgs.gov/energy/Work(Energy/DDS 60).

crude oil extraction or in large reserve quantities (see Figure 10.4) well distributed across the world.

Renewable resources, which are by definition homogeneously distributed across the world, represent another potential feedstock of choice for the chemical industry. Renewable resources are essentially made of carbohydrates (cellulose, hemicellulose, starch), of lignin, and of a small part of vegetable oils (see Figure 10.5). They present the advantage as simultaneously a potential source of carbon and a CO_2 sink through photosynthesis so that the impact of their exploitation on climate change (green house gas effect) can be regarded as neutral in the long run.

The total amount of renewable resources available around the world amounts to 170 GT/y (see Figure 10.6). Among them, 4.6 GT/y equivalent carbon are left in the fields as agricultural residues, while the world chemical industry currently consumes around 0.6 GT/y equivalent carbon (10% of the total current world

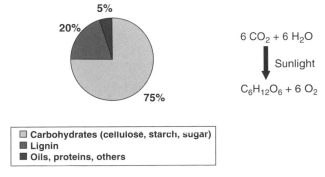

Figure 10.5 Renewable resources split.

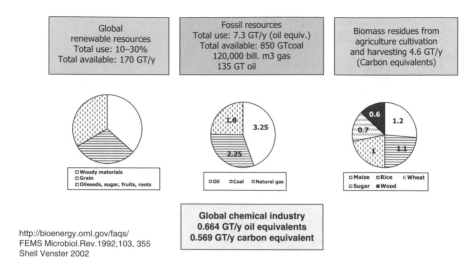

Figure 10.6   Renewable resources, an alternative?

crude oil consumption). These numbers show that stranded methane and/or biomass residues, as well as the freely available atmospheric CO_2, could theoretically supply enough carbon to replace crude oil and feed the growing world demand for chemicals while diminishing associated CO_2 emissions. An additional benefit would be that the exploitation of these potential alternative carbon sources for chemistry would not compete with the carbon needs and associated land area required to feed humans and cattle. The exploitation of these three potential carbon sources is, however, facing very significant technology barriers, preventing their large-scale industrial exploitation in the short term.

10.2   THE TECHNOLOGY CHALLENGES

This section reviews the challenges faced by the exploitation of CO_2, stranded methane, and biomass residues in an attempt to highlight areas where technology breakthroughs and research efforts are needed.

10.2.1   Exploitation of CO_2

Phosgene is commercially obtained by passing carbon monoxide and chlorine over activated carbon.[3] Around 7 million tons of phosgene are produced worldwide, 85% of which is utilized in the production of isocyanates for polyurethanes, 10% in the production of aromatic polycarbonate, and 5% in fine chemicals. Phosgene is a highly poisonous material that requires extreme handling precautions and that can generate corrosive by-products such as HCl. An advantageous alternative would be to replace phosgene with dimethylcarbonate (DMC), which

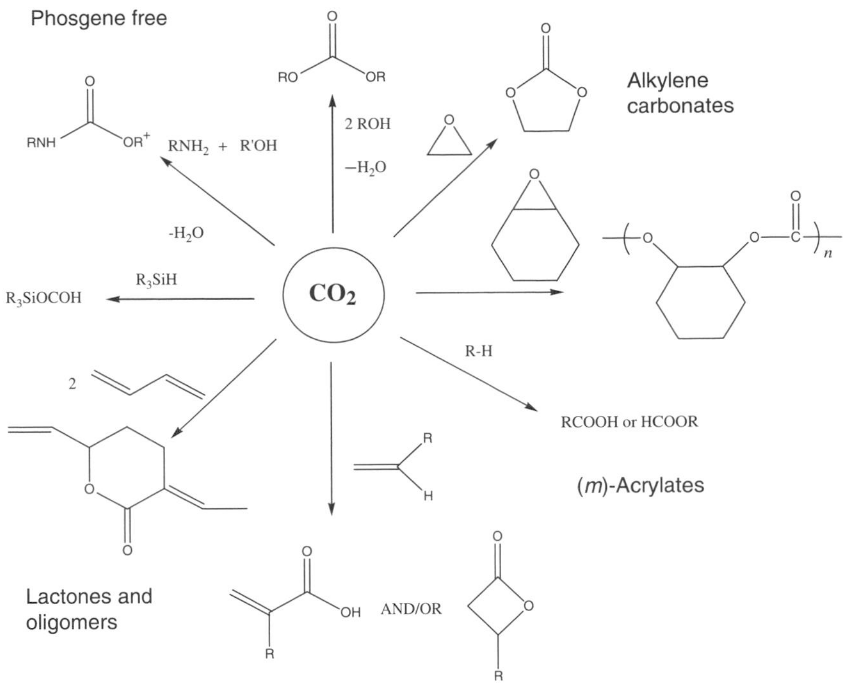

Figure 10.7 Carbon dioxide chemistry.

is a nontoxic, nonchlorine-containing molecule. Around 70,000 tons/year of DMC is currently produced by oxidative carbonylation of carbon monoxide[4] in a process not economically viable for large-scale production. To overcome this problem, the direct synthesis of DMC from methanol and CO_2 is being actively researched.[5] More specifically, catalyst development and ways to remove the water created during the reaction are being investigated to find a way to shift the reaction equilibrium toward better DMC yield.

Beyond DMC and its utilization as methylating agent or as polycondensation building block,[6] CO_2 can be used as feedstock in a number of reactions that are highlighted in Figure 10.7. Most of these reactions need further research before CO_2 can be utilized as a mass feedstock for the chemical industry.

10.2.2 Exploitation of Stranded Methane

Replacing one or several of the hydrogen atoms in methane by one or several other atoms than hydrogen automatically creates secondary or tertiary C–H bonds. Secondary and tertiary C–H bonds are more reactive than a primary C–H bond. During oxidation reactions, this leads to an easier oxidation of the reaction products than methane, and consequently to a low(er) reaction selectivity. Such reactions therefore produce complicated reactant mixtures that require costly purification operations to isolate the targeted end molecule. Chemists have designed two strategies to overcome this challenge. However, none of them have yielded economically viable processes so far.

- *Full Oxidation Followed by Recombination (Two-step Route)*: This strategy is exemplified by the production of syngas (CO/H_2) followed by Fisher– Tropsch conversion of syngas into higher alkane or into methanol, depending on reaction conditions.[7] Fisher–Tropsch chemistry works by radical polymerization, which leads to product mixtures. Such mixtures, however, necessitate cleaning steps to produce chemical intermediates with high purity, which add cost to the process. Additionally, the current catalysts for methane to syngas conversion suffer from coking, further limiting the economic viability of the process. Novel catalysts need to be developed while keeping economies of scale.
- *Selective Coupling in One Step*: Novel catalysts, for example, for methane oxidative coupling, combined with the usage of novel process engineering technologies (microreactors for highly exothermic reaction, membrane reactors for early purification, etc.) are being actively developed.

A list of some of these on-going efforts and their current status follows:

- *Methane to Methanol and/or Formaldehyde*: Recent research indicates that a catalyst system in the presence of H_2SO_4 can convert methane directly into methanol. Homogeneous catalyst systems show promise. Also, heterogeneous Fe-ZSM-5 catalysts are reported to be attractive for this chemistry. Novel plasma reactors to generate hydroxyl radicals are also being investigated.
- *Methane to Ethylene*: One target is to achieve an ethylene selectively of 90% at a methane conversion level of 50% in a single pass. Additionally, design of novel recycle reactors or membrane systems (to remove the ethylene produced) remain part of the active research.
- *Methane to Benzene*: Both oxidative and nonoxidative routes have been reported. Most attention has been directed at nonoxidative aromatization. In particular, Chinese workers are active in this field. Recently, attractive results have been reported for Mo-loaded HZSM-5 catalysts.
- *Methane and Toluene to Styrene*: Basic catalysts in the presence of oxygen and/or air are reported to be attractive catalysts for this reaction. Most research was performed in the late 1980s and early 1990s. The fundamentals resemble the oxidative coupling reaction of methane to ethylene.

These reactions require a lot more research to reach fruition, most specifically in the field of long-term catalyst stability and selectivity. Furthermore, in many instances the reaction mechanism and the active catalytic site are still poorly understood. Issues such as the importance of site isolation and phase cooperation (in mixed oxide catalysts) inducing synergistic effects are increasingly receiving attention.

10.2.3 Exploitation of the Biomass

The molecular structures of the main biomass constituents are given in Figures 10.8 and 10.9. These structures induce a physicochemical behavior that is markedly different from the behavior of the hydrocarbons contained in crude oil (see Table 10.2). Note that the relative thermal fragility of the biomass molecular structures encourages the chemist to prefer thermally mild and therefore low-energy-consuming conversion techniques such as fermentation or hydrothermolysis to exploit biomass (see Figure 10.10).

As a matter of fact, most of the processes currently developed to generate biochemicals out of biomass involve fermentation of starch originating from corn, wheat, or rice, for example. The various chemicals obtainable from theses processes and their end applications are listed in Table 10.3. A lot of these fermented biochemicals, however, are not yet economically competitive compared with their petrochemical equivalent, essentially due to the large capital investment in equipment and land needed to implement the fermentation process on an industrial scale. An additional disadvantage of this route is that it competes with feedstock needed by the food industry. More research to reduce the costs of fermentation technology is needed.

Agricultural residues (stem, leaves, etc.) currently left in the fields after harvesting are made of cellulose, hemicellulose, and lignin. They are not competing with the feedstock for the food industry.

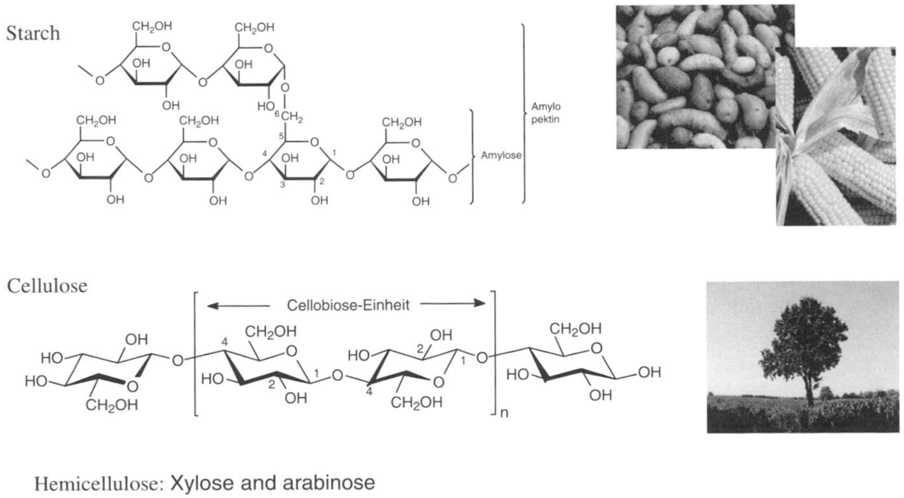

Hemicellulose: Xylose and arabinose

Figure 10.8   Bio-based feedstock constituents.

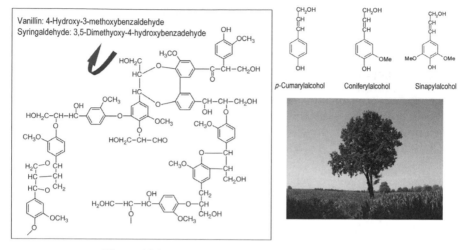

Figure 10.9   Bio-based feedstock constituent: lignin.

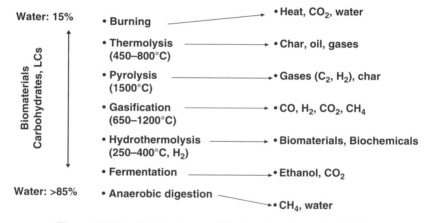

Figure 10.10   Carbohydrate and lignin processing technologies.

TABLE 10.2   Physicochemical Behavior of Feedstock

Crude-Oil-Based Feedstock	Biomass-Based Feedstock
Low degree of oxygenation	High degree of oxygenation
High volatility	Low volatility
Thermally stable	Thermal fragility

Unlike starch, which is amorphous, cellulose is fermented with difficulty because of its semicrystalline structure. As a consequence, ethanol fermented from cellulose using the latest generation enzymes would still be more expensive

TABLE 10.3 Industrial Strach-Based Product Opportunities

Technology Platform	Chemical	Application
Sugar fermentation	Lactic acid	Acidulant, electroplating additive, textile/leather auxiliary
	Polylactic acid Ethyllactate	Thermoplastic polymer Solvent, intermediate
	1, 3-Propanediol	Specially resins
	Succinic acid Succinic acid derivatives Tetrahydroturan 1, 4-butanediol γ-butyrolactone N-methylpytrolldone	Surfactant, food, pharma, antibiotics, amino acid and vitamin production Solvents, adhesives, paints, printing inks, tapes, plasticizer, emulsifier, deicing compound, herbicide
	3-Hydroxypropionic acid and Terrivales Acrylic acid Acrylonitrille Acrylamide	Acrylates, acrylic polymers, fibers, and resins
	N-Butanol Itaconic acid	Solvent, Plasticizer, polymer, reactive comonomer
Sugar fermentation and thermochemical processing	Propylene glycol Levullnic acid and derivatives Me-Tetrahydroturan δ-amino levullinic acid	Solvent, chain extender in PU, antifreeze, plasticizer oxygenates for fuels biodegradable herbicide, bisphenol A alternative

than ethanol fermented from starch, which is itself noncompetitive with petrochemical ethanol[10] (when government subsidies are not taken into account).

Cellulose and hemicellulose can be converted to biochemicals via acid treatment. These treatments, however, generate large aqueous waste streams (acid neutralization), which makes them economically uncompetitive.[11] Recently, a process involving cellulose and hemicellulose hydrolysis in supercritical water was patented.[12] This represents a very promising route to exploit renewable resources and agricultural residues. Figure 10.11 shows the cascade of biochemicals that could be obtained from cellulose, hemicellulose, and lignin when combining the hydrolysis process in super critical water with fermentation and acid treatment.

It is thought that this cascade of products could be manufactured in a biorefinery by integrating all the unit operations needed to convert biomass into chemicals. A lot more research is needed, however, to develop the hydrothermolysis process and high-efficiency/low-cost fermentation and acid treatments. An efficient exploitation of lignin (the only natural source of aromatic rings) has also to be developed.

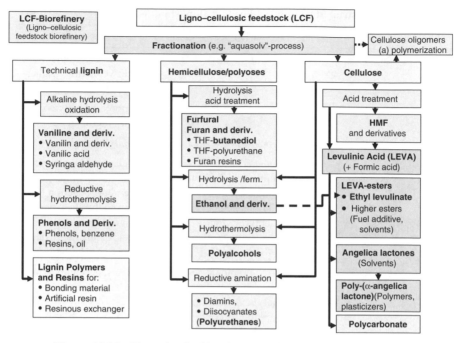

Figure 10.11   Example of a biorefinery. A ligno–cellulosic biorefinery.

10.2.4   Exploitation of Vegetable Oils

Vegetable oils represent only 5% of the renewable resources available. Today, vegetable oils currently provide a marginal carbon feedstock contribution to the chemical industry in such applications as solvents, surfactants, and lubricants. Vegetable oils may, however, play a much more important role in the future. They are mixtures of fatty acid trigclycerides whose typical molecular structures are given in Figure 10.12.

Unlike carbohydrates, which are highly functionalized with an average of one hydroxyl group per carbon, and therefore require cumbersome protection/deprotection protocols to ensure high selectivity during chemical modification, fatty acids are bifunctional. Fatty acids can therefore be directly converted in a small number of steps (i.e., less cost intensive than carbohydrates) into a number of building blocks suitable for a variety of specialty applications, such as engineering thermoplastics (nylon, polyester) and thermoset resins (epoxies and polyurethanes). Some of the chemistries available are shown in Figure 10.13. Among the chemistries, vegetable oil metathesis (see Figure 10.14) is of particular interest, as it allows a reduction in the fatty acid chain length. This chain-length reduction greatly enhances the chemical resistance and the mechanical performances of polymers based on such intermediates.[13] Metathesis catalysts that are not poisoned by the polar functions present in vegetable oils have been developed,[14] but are not yet advanced enough to permit an economically viable process.

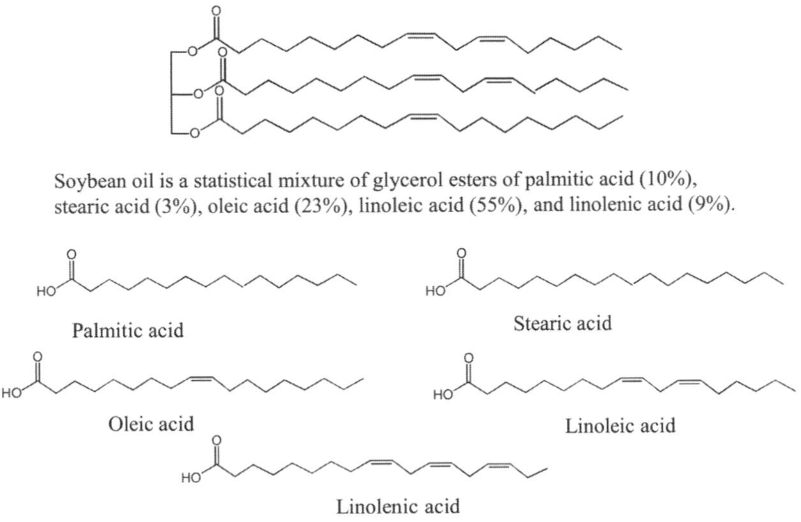

Soybean oil is a statistical mixture of glycerol esters of palmitic acid (10%), stearic acid (3%), oleic acid (23%), linoleic acid (55%), and linolenic acid (9%).
Palmitic acid
Stearic acid
Oleic acid
Linoleic acid
Linolenic acid

Figure 10.12 Example of seed oil chemical composition (soybean oil).

Further research is needed in that area. It is thought that vegetable oil biorefineries working independently, or integrated with lignocellulosic biorefineries, will provide a diversified portfolio of chemicals, as petrochemical refineries do today. Beyond developing an economically viable metathesis, one of the main challenge in the successful development of an oleo-refinery is to find valuable outlets for glycerin.

Vegetable oil
CO/H2, Rh, Co
Hydroformylation
Polyaldehyde
Reduction
Polyol (primary alcohol)
Oxidation
Polycarboxylic
Reductive
Amination
Polyamine

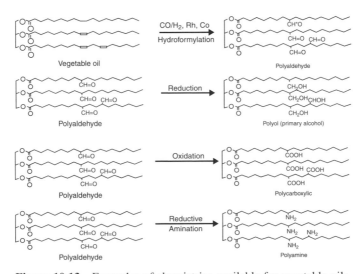

Figure 10.13 Examples of chemistries available for vegetable oils.

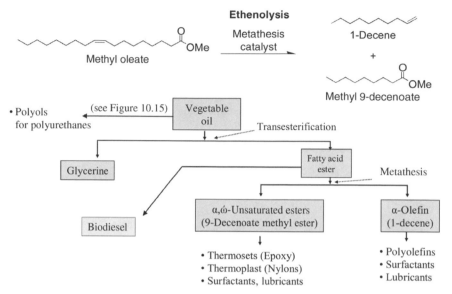

Figure 10.14 Cross-metathesis of ethylene and methyl oleate (ethenolysis) and associated biorefinery.

10.3 CONCLUSION

We can now complete the picture and easily visualize a possible long-term future for feedstock and processes, as shown in Figure 10.15. The occurrence of this vision, however, is strongly dependant on the ability of researchers to remove the technological barriers highlighted in the previous section.

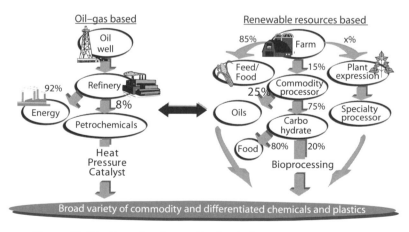

Figure 10.15 Emerging future: Duality in feedstocks and processes.

- On the one hand, the current petrochemical route would continue to provide the world with the chemicals consumers require. To satisfy the need for a more sustainable development, the petrochemical industry would continue its drive toward a continuous improvement in energy efficiency (see Figure 10.3). This drive will primarily include the continuous improvement of the current crude-oil-based processes while stranded methane or CO_2 would be utilized as complementary feedstock.
- In parallel, the existing commodity-grain and -oil processing infrastructure would increasingly produce the carbohydrates needed for bioprocesses to generate low-cost bulk commodity chemicals and biofuels (ethanol, 1-2-propanediol, succinic acid, etc.) as well as the fatty acid esters needed for biodiesel and specialty polymers. For the more distant future, technology and specialty processing is being put in place in order to also use plants as factories and express specific oils, biopharmaceuticals, or polymers in identity-preserved crops.

The realization of this vision will require a multidisciplinary approach where green chemistry will be a key enabler to meet broad technological challenges in a balanced approach. This approach will balance economic profitability, societal satisfaction, and environment protection. One of these challenges relates to bioprocessing. Nature operates without pressure or heat, and most of its feedstock is solids. Nature also tends to operate mostly in an aqueous medium. Solids processing and aqueous separations are new challenges the bioprocessing industry will have to face.

REFERENCES

1. World Commission on Environment and Development, *Our Common Future*, 1987, Oxford University Press, New York.
2. Hiller, K.; Kehrer, P. *Erdoel Erdgas Kohle*, 2000, **116**(9), 427.
3. Schneider, W. Phosgene, in *Ullmann's Encyclopedia of Industrial Chemistry*, 7th ed., CRC Press, Boca Raton, Fla., 2005.
4. Delledonne, D.; Rivetti, F.; Romagno, U. *Appl. Catal.: General*, 2001, **221**, 241–251.
5. Wu, X. L.; Xiao, M.; Meng, Y. Z.; Lu, Y. X. *J. M. catal. A: Chem.*, 2005, **238**, 158–162.
6. Ono, Y. *Appl. Catal. A: General*, 1997, **155**, 133–166.
7. Makino, E. Coal liquefaction, in *Ullmann's Encyclopedia of Industrial Chemistry*, 7th ed., CRC Press, Boca Raton, Fla, 2005.
8. Otsuka, K.; Wang, Y. *Appl. Catal., A: General*, 2001, **222**(1–2), 145–161.
9. Yide, X., Xinhe, B.; Liwu, L. *J. Catal.*, 2003, **216**, 386–395.
10. Cherry, J. Novozymes, in *Proc. Biorefinery 2004*, San Francisco, June 8, 2005.
11. eere.energy.gov/biomass/concentrated_acid.html#commercial

12. Transforming biomass to hydrocarbon mixtures in near-critical or supercritical water, U.S. Patent 6,180,845 B1.

13. Earls, J. D.; White, J. E.; Null, M.; Dettloff, M. *J. Chem. Technol.*, 2004, **1**(3), 243–245.

14. Grubbs, R. H. *Tetrahedron*, 2004, **60**, 7117–7140.

OTHER USEFUL READINGS

Affordable Resins and Adhesives from Optimized Soybean Varieties, U.S. Department of Energy, Industrial Technologies Program Fact Sheet, May 2002.

Glasser, W. G.; Northey, R. A.; Schultz, T. P. (Eds.) Lignin: Historical, Biological, and Materials Perspectives, ACS Symposium Series 742, 1999.

Barger, P. Methanol to Olefins (MTO) and Beyond. Catalytic Science Series, 2002, **3**(Zeolites for Cleaner Technologies), 239–260.

Bioprocessing, reaping the benefits of renewable resources, *Chem. Week*, February 11, 2004.

Current Situation and Future Prospects of EU Industry Using Renewable Raw Materials, European Renewable Resources & Materials Association (EERMA), Brussels, 2002.

Deamin, A. L., Small bugs, big business: the economic power of the microbe, *Biotechnol. Adv.*, 2000, **18**, 499–514.

Sasaki, M.; Fang, Z.; Fukushima, Y.; et al. *Ind. Eng. Chem. Res.*, 2000, **39**, 2883–2890.

Eggersdorfer, M.; Meyer, J.; Eckes, P. Use of renewable resources for non-food materials, *FEMS Microbiol. Rev.*, 1992, **103**, 355–364.

EU FP6 Expression of Interest; http://eoi.cordis.lu/dsp_details.cfm?ID = 25887

Functionalized Vegetable Oils for Utilization as Polymer Building Blocks, U.S. Department of Energy, Industrial Technologies Program Fact Sheet, May 2001; oit.doe.gov/agriculture

Dale, B. E. Greening the chemical industry: research and development priorities for bio-based industrial products, *J Chem. Technol. Biotechnol.*, 2003, **78**, 1093–1103.

Industrial biotech and sustainable chemistry, *Eur. Biotechnol. News*, Vol. 3, No 1–2, 2004, pp. 28–29.

Industrial Biotechnology and Sustainable Chemistry, Royal Belgian Academy Council of Applied Sciences, January 2004.

Johnston, S. Emissions and Reduction of Greenhouse Gases from Agriculture and Food Manufacturing, S.C. Johnson Associates, Inc., for the U.S. Department of Energy, December 1999.

Industrial Bioproducts; Today and Tomorrow, U.S. Department of Energy, July 2003.

Matsumura, Y.; Minowa, T. *Int. J. Hydrogen Energy*, 2004, **29**, 701–707.

Mangold, E. C.; Munradaz, M. A.; Quellette, R. P.; et al. *Coal Liquification and Gasification Technologies*, Ann Arbor Science Publishers, Inc., Ann Arboa Michi., 1982.

NACHRICHTEN—Forschungszentrum Karlsruhe Jahrg. 33 1/2001, pp. 59–70, A. Kruse, ITC.

NACHRICHTEN—Forschungszentrum Karlsruhe Jahrg. 35 3/2003.

NSF Workshop on Catalysis for Biorenewables Conversion, April 13–14; www.egr. msu.edu/apps/nsfworkshop

OECD2001. The Application of Biotechnology to Industrial Sustainability; www.oecd.org/ sti/biotechnology

Okkerse, H; Van Bekkum, H. From fossil to green, *Green Chem.*, 1999, 107–114.

Gunstone, F. D.; Hamilton, R. J. (Eds.) *Oleochemical Manufacture and Applications*, CRC Press, Boca Raton, Fla., 2001.

Steps Towards a Sustainable Development, A White Book for R&D of Energy-Efficient Technologies, Eberhard Jochem, March 2004, A Project of Novatlantis—Sustainability at the ETH-Domain (-CH).

The Refining Process, Corn Refiners Association, Washington, D.C., August 2002.

Trash to treasure, *DOE Pulse*, No 50, Feb. 28, 2000; ornl.gov/news/pulse

Vision for Bioenergy and Biobased Products in the United States, DOE, October 2002; bioproducts-bioenergy.gov/board.html

White Biotechnology: Gateway to a More Sustainable Future, 2003; www.europabio.org

Wiedenroth, H. *Zuckerrüben-Magazin*, No. 32, September 2002.

11

ENANTIOSELECTIVE METAL CATALYZED OXIDATION PROCESSES

DAVID STC. BLACK

The University of New South Wales, Sydney, Australia

INTRODUCTION

One appropriate definition of green chemistry is "the design of new synthetic pathways to perform chemical processes in a very controlled way, so as to reduce drastically, or to eliminate completely, all the environmental impact." In terms of synthetic industrial processes, this idea leads to the application of highly efficient catalysts, many of which are based on metal complex or organometallic structures. Some of the most important industrial processes involve mild oxidation reactions leading to intermediates in more extensive synthetic schemes. An important consideration is the potential establishment of new chiral centers, consequently, enantiospecific reactions are highly desirable. A range of the most useful oxidation processes are considered here. However, these are confined to the formation of carbon–oxygen bonds.

11.1 EPOXIDATION OF ALLYLIC ALCOHOLS

The epoxidation of alkenes is an especially important process because of its capacity to introduce useful functionalization that is also capable of establishing stereochemistry at two vicinal carbon atoms. Allylic alcohols provide an excellent

Methods and Reagents for Green Chemistry: An Introduction, Edited by Pietro Tundo, Alvise Perosa, and Fulvio Zecchini
Copyright © 2007 John Wiley & Sons, Inc.

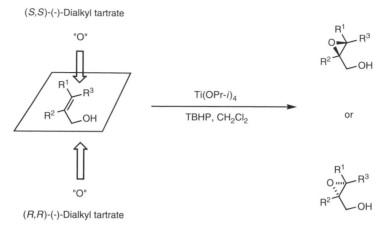

Figure 11.1   Epoxidation of allylic alcohols.

opportunity for reactivity and enantioselectivity through metal coordination to the alcohol oxygen atom.[1] This phenomenon is illustrated superbly by the Sharpless oxidation process (Figure 11.1), which utilizes titanium tetraisopropoxide, diethyl tartrate, and tertiarybutylhydroperoxide to deliver high enantioselectivity and high reliability.[2]

A more recent approach to the epoxidation of allylic alcohols makes use of a vanadium polyoxometallate, together with a sterically demanding chiral tartrate (or TADDOL) -derived hydroperoxide, to give a highly chemoselective, regioselective and enantioselective outcome (Figure 11.2).[3]

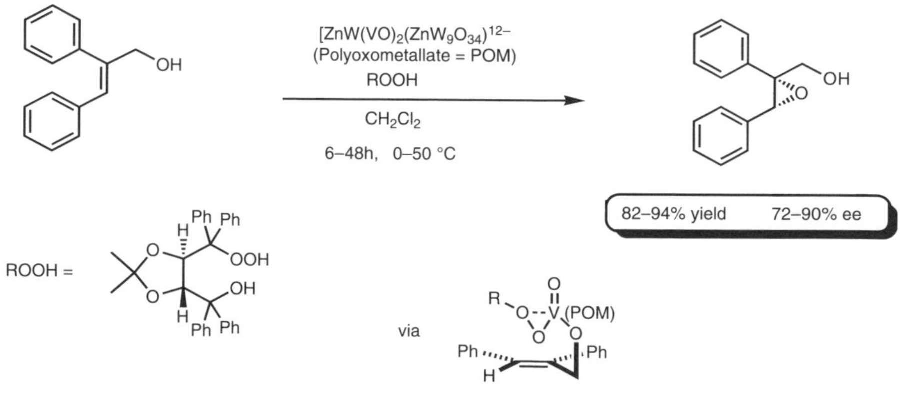

Figure 11.2   Epoxidation of allylic alcohols.

11.2 EPOXIDATION OF ALKENES OTHER THAN ALLYLIC ALCOHOLS

The asymmetric epoxidation of unfunctionalised alkenes has received major attention and delivered considerable success through the development of metal-catalyzed oxo-transfer processes and the use of chiral manganese(III) salen complexes.[4] The latter process requires the incorporation of a dissymmetric diimine bridge derived from a C_2-symmetric 1,2-diamine and bulky substituents adjacent to the phenolate group. This is illustrated in the oxidation of *cis*-β-methylstyrene (Figure 11.3).[5] The mechanistic hypothesis is that the bulky groups force approach of the alkene to an intermediate manganese oxo complex past the chiral bridge, thus inducing an enantioselective product outcome. While this hypothesis has served well in the development of successful catalysts, many problems remain, and there are indications that the hypothesis is rather simplistic. However, the delivery of very high enantioselectivity through relatively mild and inexpensive catalysts has encouraged further development and also industrial usage. There is also the possibility of using different oxidation conditions, such as sodium hypochlorite, isobutyraldehyde, and oxygen, peracids with an additional *N*-oxide catalyst, or iodosylbenzene.

A wide variety of salen complex analogs has been generated, and individual examples identified as being optimal for the asymmetric epoxidation of specific alkenes. Further examples are shown in Figure 11.4.[6]

There are now many examples of the industrial use of manganese(III) salen catalyzed asymmetric epoxidations. For example, the asymmetric epoxidation of a chromene derivative was central to the synthesis of the potassium channel activator BRL 55834 (Figure 11.5).[7,8]

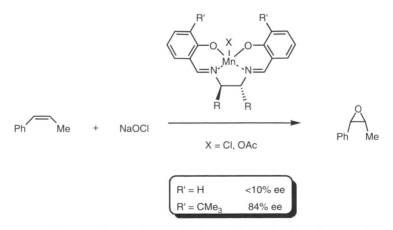

Figure 11.3 Enantioselective epoxidation of alkenes by Mn salen complexes.

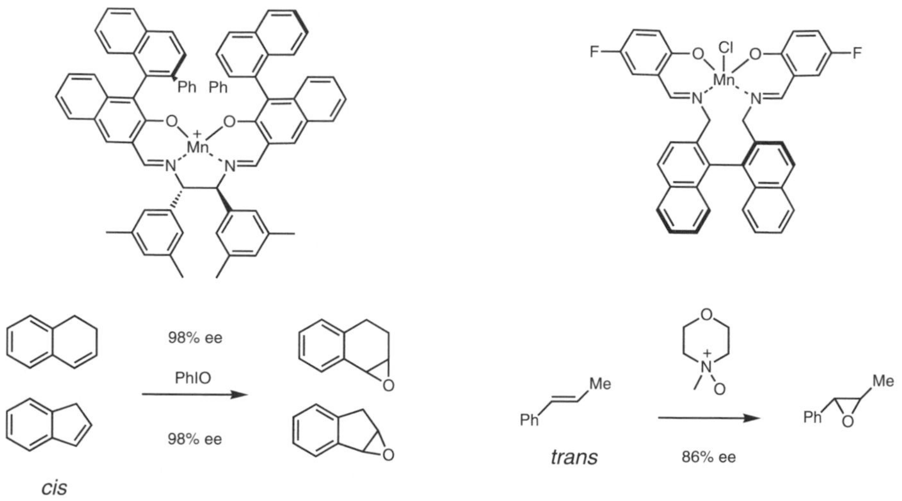

Figure 11.4   Enantioselective epoxidation of alkenes by Mn salen complexes.

The indole-7-carbaldehyde structural fragment is isosteric with the salicylalde-hyde structure, and consequently can replace it in the design and construction of new ligand systems. A wide range of substituted indole-7-carbaldehydes has been achieved by our group through the activation of C7 to formylation by the presence of methoxy groups at C4 and C6. This strategy provides an acidic NH instead of the acidic phenolic OH. Furthermore, a wide range of alkyl or aryl groups can be built in at C3, and C2 can also be substituted initially or be available for acid-catalyzed addition reactions later in the synthetic sequence (Figure 11.6).[9] Benzimidazoles can be used similarly. Although these catalysts are highly reactive and effective, so far only relatively low enantioselectivity has been achieved.

Chiral porphyrin metal complex catalysts have also received much attention. In this situation, the flat, symmetrical porphyrin structure must be modified dramatically in order to incorporate dissymmetry. This has been achieved through strapping techniques.[10,11] Some examples are shown in Figure 11.7.

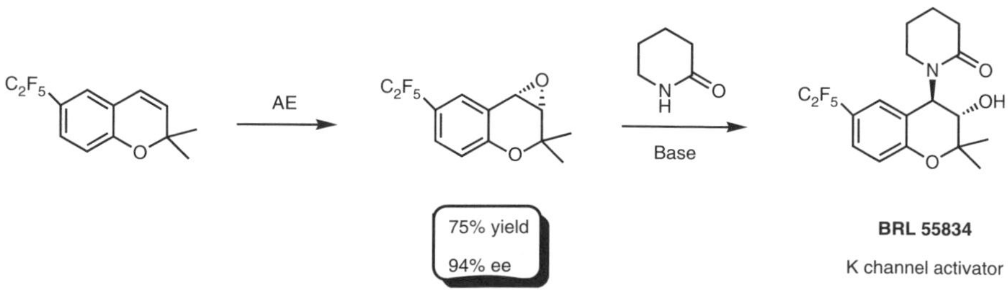

Figure 11.5   Synthetic application of enantioselective epoxidation.

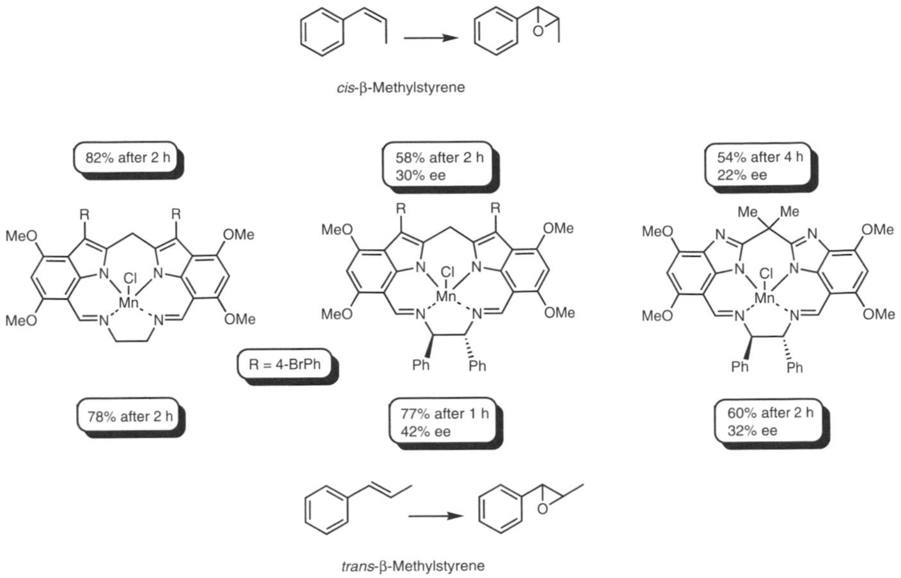

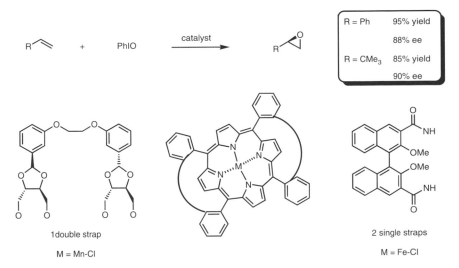

cis-β-Methylstyrene

82% after 2 h

58% after 2 h
30% ee

54% after 4 h
22% ee

MeO ... OMe

R = 4-BrPh

Ph Ph

Me Me

Ph Ph

78% after 2 h

77% after 1 h
42% ee

60% after 2 h
32% ee

trans-β-Methylstyrene

Figure 11.6 The indole and benzimidazole strategy for enantioselective epoxidation.

R = Ph   95% yield
          88% ee
R = CMe₃ 85% yield
          90% ee

1 double strap

M = Mn-Cl

2 single straps

M = Fe-Cl

Figure 11.7 Chiral porphyrin complex catalysts.

11.3 EPOXIDATION OF ENONES

Metal alkyl peroxides can be used for the epoxidation of electron-deficient alkenes such as enones.[12] The use of a combination of diethylzinc, oxygen, and *N*-methylephedrine gave epoxides in very high yield and generally high enantioselectivity (Figure 11.8).[13]

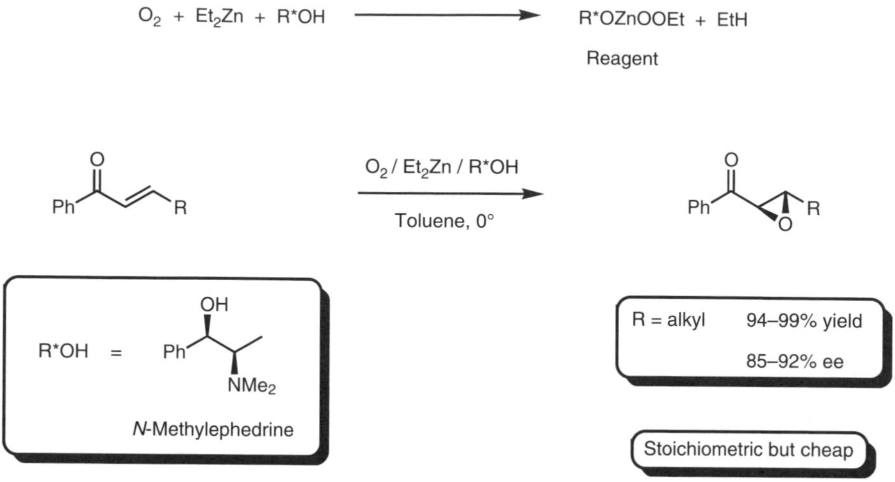

$$O_2 + Et_2Zn + R^*OH \longrightarrow R^*OZnOOEt + EtH$$

Reagent

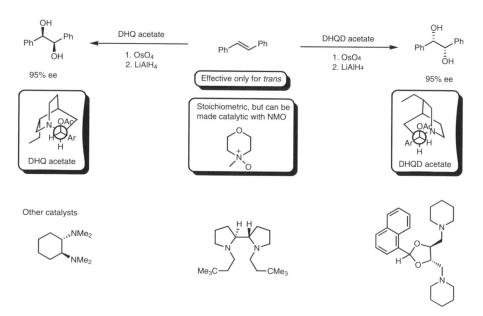

Figure 11.8   Epoxidation of enones by metal hydroperoxides.

11.4 DIHYDROXYLATION OF ALKENES

Asymmetric dihydroxylation can be achieved using osmium tetroxide in conjunction with a chiral nitrogen ligand.[14] The very successful Sharpless procedure uses the natural cinchona alkaloids dihydroquinine (DHQ) and its diastereomer dihydroquinidine (DHQD), as exemplified in the epoxidation of *trans*-stilbene

Figure 11.9   Dihydroxylation of alkenes.

(Figure 11.9).[15,16] Numerous synthetic chiral diamines have also been used, and some examples are shown.[17–19]

11.5 BENZYLIC HYDROXYLATION

Direct C–H oxidation is very difficult but extremely important, as it occurs in some important biological processes.[20] The presence of some kind of activating group such as an aryl or alkenyl group offers some encouragement for success. Benzylic hydroxylation can be effected using iodosylbenzene and an iron oxo species derived from a porphyrin (Figure 11.10).[21] Such reactions mirror the role of cytochrome P-450 in monooxygenase enzyme processes. The yields obtained for benzylic hydroxylation are only moderate, as is the enantioselectivity.

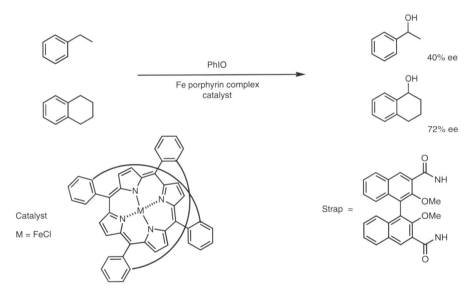

Figure 11.10 Enantioselective benzylic hydroxylation.

11.6 ALLYLIC OXIDATION

Allylic oxidation can be carried out using a peroxide together with a copper (I) complex.[20] Enantioselective processes have now been developed using tertiary-butylperbenzoate and either copper (I) or copper (II) complexes in acetonitrile (Figure 11.11).[22,23]

This work has been developed further through the use of a copper(I) bipyridyl complex catalyst, incorporating a chiral and sterically demanding ligand, and

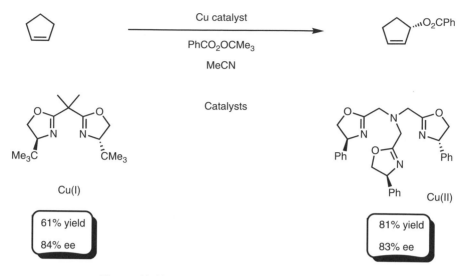

Figure 11.11   Enantioselective allylic hydroxylation.

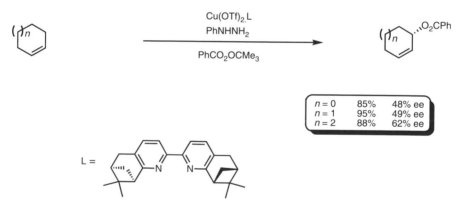

Figure 11.12   Enantioselective allylic hydroxylation.

reducing copper(II) to copper(I) in situ with phenylhydrazine (Figure 11.12). The yield is excellent, but the enantiomeric excess needs further improvement.[24,25]

11.7   HYDROXYLATION OF β-KETOESTERS

The tartrate (or TADDOL) derived approach to catalyst design has also been applied to the enantioselective α-hydroxylation of β-ketoesters. In this case, an enantiospecific titanium(IV) complex combines with a sulfonyloxaziridine as the

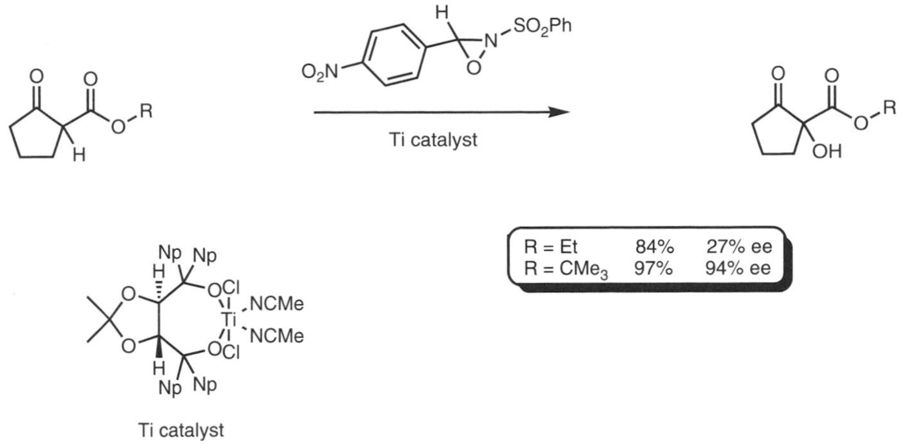

Figure 11.13   Enantioselective hydroxylation of β-ketoesters.

oxidant to give high yields, and also high enantiomeric excess for tertiary-butyl esters (Figure 11.13).[26]

11.8   BAEYER–VILLIGER OXIDATION

The oxidation of an acyclic ketone to an ester, or a cyclic ketone to a lactone (the Baeyer–Villiger reaction) can be achieved with the help of metal species of the Lewis acid type.[27] Chiral copper(II), palladium(II), and titanium(IV) catalysts

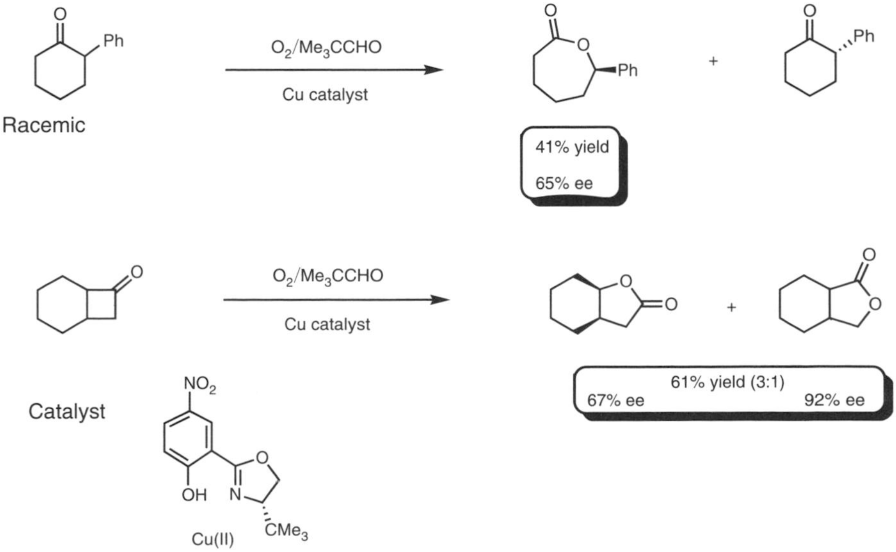

Figure 11.14   Enantioselective Baeyer–Villiger oxidation.

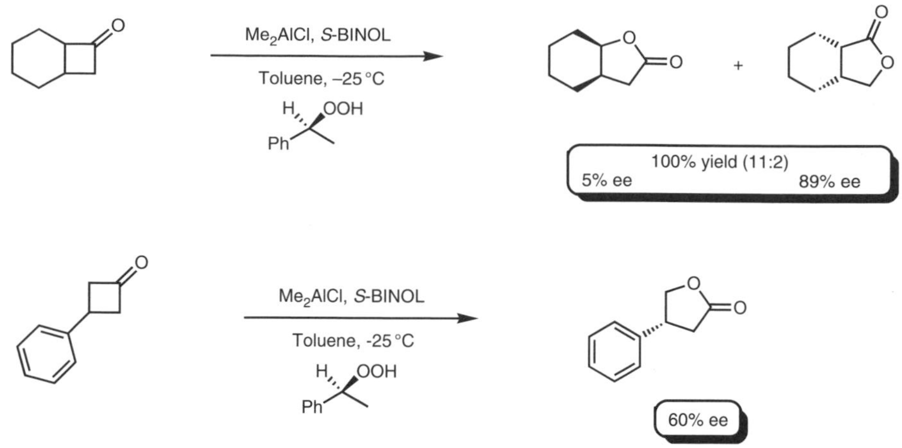

Figure 11.15 Enantioselective Baeyer–Villiger oxidation.

have been developed for enantioselective processes, but the reaction protocols are quite specific, and yields and enantioselectivity are only moderate. For example, copper(II) oxazoline catalysts in combination with pivaldehyde and oxygen selectively oxidize one enantiomer of a racemic compound to the desired lactone, and fail to react with the other, thereby generating enantioselectivity (Figure 11.14)[28]

A more recent variant of the enantioselective Baeyer–Villiger oxidation involves an aluminium complex of BINOL, together with a chiral hydroperoxide, and gives similar levels of enantioselectivity (Figure 11.15)[29]

11.9 CONCLUSIONS

Certainly the enormous amount of research being carried out in order to discover mild, catalytic oxidation processes is being rewarded with some excellent results. However, there is a major problem. Very few of the successful processes are understood mechanistically, and approaches are largely empirical. Attempts to find new, effective processes involve the screening of the large number of possible catalytic systems. When these also involve metal complexes, there are numerous possible ligands, so the task requires even greater resources. There is an urgent need to gain an understanding of the mechanisms of catalytic processes, so that rational design can replace the current empiricism. Thus this is a research area for the future, and the rewards will be worthy of the challenge.

REFERENCES

1. Katsuki, T., in *Comprehensive Asymmetric Catalysis II*, Jacobsen, E. N., Pfaltz, A., Yamamoto, H. (Eds.) Springer, Berlin, 1999, pp. 621–648.

2. Katsuki, T.; Sharpless, K. B. *J. Am. Chem. Soc.*, 1980, **102**, 5974–5976.

3. Adam, W.; Alsters, P. L.; Neumann, R.; *J. Org. Chem.*, 2003, **68**, 8222–8231.

4. Jacobsen, E. N.; Wu, M. H., in *Comprehensive Asymmetric Catalysis II*, Jacobsen, E. N.; Pfaltz, A.; Yamamoto, H. (Eds.) Springer, Berlin, 1999, pp. 649–677.

5. Brandes, B. D.; Jacobsen, E. N. *J. Org. Chem.*, 1994, **59**, 4378–4380.

6. Katsuki, T. *J. Synth. Org. Chem. Jpn.*, 1995, **53**, 940–951.

7. Lee, N. H.; Muci, A. R.; Jacobsen, E. N. *Tetrahedron Lett.*, 1991, **32**, 5055–5058.

8. Bell, D.; Davies, M. R.; Finney, F. J. L.; et al. *Tetrahedron Lett.*, 1996, **37**, 3895–3898.

9. Black, D. StC., in *Green Chemistry Series No. 1: Collection of Lectures of the Summer Schools on Green Chemistry*, 2nd ed., P. Tundo (Ed.) INCA, Mestre, Italy, 2002, pp. 187–197.

10. Collman, J. P.; Lee, V. J.; Zhang, X.; et al. *J. Am. Chem. Soc.*, 1993, **115**, 3834–3835.

11. Collman, J. P.; Wang, Z.; Straumanis, A.; et al. *J. Am. Chem. Soc.*, 1999, **121**, 460–461.

12. Aggarwal, V., in *Comprehensive Asymmetric Catalysis II*, Jacobsen, E. N.; Pfaltz, A.; Yamamoto, H. (Eds.) Springer, Berlin, 1999, pp. 679–693.

13. Enders, D.; Zhu, J.; Raabe, G. *Angew. Chem., Int. Ed. Engl.*, 1996, **35**, 1725–1728.

14. Marko, I. V., Svendsen, J. S., in *Comprehensive Asymmetric Catalysis II*, Jacobsen, E. N.; Pfaltz, A.; Yamamoto, H. (Eds.) Springer, Berlin, 1999, pp. 713–787.

15. Hentges, S. G.; Sharpless, K. B. *J. Am. Chem. Soc.*, 1980, **102**, 4263–4265.

16. VanRheenan, V.; Kelly, R. C.; Cha, D. Y. *Tetrahedron Lett.*, 1976, 1973–1976.

17. Tokles, M.; Snyder, J. K. *Tetrahedron Lett.*, 1986, **27**, 3951–3954.

18. Oishi, T.; Hirama, M. *J. Org. Chem.*, 1989, **54**, 5834–5835.

19. Yamada, T.; Narasaka, K. *Chem. Lett.*, 1986, 131–134.

20. Katsuki, T., in *Comprehensive Asymmetric Catalysis II*, Jacobsen, E. N.; Pfaltz, A.; Yamamoto, H. (Eds) Springer, Berlin, 1999, pp. 791–799.

21. Groves, J. T.; Viski, P. *J. Org. Chem.*, 1990, **55**, 3628–3634.

22. Gokhale, A. S.; Minidis, A. B. E.; Pfaltz, A. *Tetrahedron Lett.*, 1995, **36**, 1831–1834.

23. Kawasaki, K.; Katsuki, T. *Tetrahedron*, 1997, **53**, 6337–6350.

24. Malkov, A. V.; Baxendale, I. R.; Bella, M.; et al. *Organometallics*, 2001, **20**, 673–690.

25. Malkov, A. V.; Pernazza, D.; Bell, M.; et al. *J. Org. Chem.*, 2003, **68**, 4727–4742.

26. Toullec, P. Y.; Bonaccorsi, C.; Mezzetti, A.; Togni, A. *Proc. Natl. Acad. Sci. U.S.A.*, 2004, **101**, 5810–5814.

27. Bolm, C.; Beckmann, O., in *Comprehensive Asymmetric Catalysis II*, Jacobsen, E. N.; Pfaltz, A.; Yamamoto, H. (Eds.) Springer, Berlin, 1999, pp. 803–810.

28. Bolm, C.; Luong, T. K. K.; Schlingloff, G. *Synlett*, 1997, 1151–1152.

29. Bolm, C.; Beckmann, O.; Kühn, T.; et al. *Tetrahedron: Asymmetry*, 2001, **12**, 2441–2446.

12

ZEOLITE CATALYSTS FOR CLEANER TECHNOLOGIES

MICHEL GUISNET

University of Poitiers, Poitiers, France

INTRODUCTION

The name of zeolites, which originates from the Greek words *zeo* (to boil) and *lithos* (stone), was given some 250 years ago to a family of minerals (hydrated aluminosilicates) that exhibited intumescence when heated in a flame.[1] However, the history of zeolites really began 60 years ago with the development of synthesis methods. Commercial applications in three main fields—ion exchange, adsorption, and catalysis—were rapidly developed, the corresponding processes being more environmentally friendly than their predecessors.

Thus, the largest use of zeolites (∼70%) is as a water-softener substitute for sodium tripolyphosphate (STPP) in laundry detergents.[2] Contrary to STPP, which disrupts the biological equilibrium of rivers and lakes by eutrophication, the NaA zeolite, which is now generally employed, has no negative effect on the environment. Furthermore, the high efficiency of zeolites for the ultimate drying of gases and liquids led to a large variety of applications; separation of isomers (e.g., *n*-butane/isobutane) through adsorption over zeolites was substituted for the classic methods of distillation and crystallization, which consume energy.

The highest market value for zeolites, however, is in the field of catalysis: zeolites are now involved as basis components of most of the catalysts used in the production of fuels and petrochemicals; moreover they are playing an increasing

Methods and Reagents for Green Chemistry: An Introduction, Edited by Pietro Tundo, Alvise Perosa, and Fulvio Zecchini

role in the synthesis of intermediate and fine chemicals, as well as in pollution abatement.[3] This is due to three main factors:

1. The molecular-size pore system of zeolites, in which the catalytic reactions occur: zeolite catalysts can be considered as a succession of *nano or molecular reactors* (their channels, cages, or channel intersections). The result is that the rate, selectivity, and stability of all the zeolite catalyzed reactions are affected by the shape and size of the nano reactors and of their apertures. This effect has two main origins: (i) *spatial constraints* on the diffusion of reactant or product molecules or on the formation of intermediates or transition states that limit the formation of undesired products; (ii) *confinement* within the micropores of reactant molecules, with a positive effect on the rates of the reaction, especially the bimolecular ones, but also of product molecules with sometimes autoinhibition of the desired reaction. These characteristics of zeolite catalyzed reactions are generally grouped under the label *shape selectivity*, which was proposed 45 years ago by Weisz and Frilette.[4] This name, which considers only the shape of the micropores, but not their size and that of their apertures, as well as only the selectivity, although both reaction rate and stability are also significantly affected, could seem too restrictive. However, it has the great advantage of being simple and striking. The concept of shape selectivity, which was at the origin of the design and development of many processes, remains an important source for conceiving new processes and improving existing ones. This is clearly demonstrated by the major breakthrough in the field established by the discovery that the shape selective properties of zeolites were not limited to their inner micropores, but could also originate from the external surface of their crystallites. This discovery that was made possible by the development of methods for the synthesis of very small crystallites ($\sim$20–50 mm instead of 0.3–1 μm for conventional zeolites), hence with a large external surface, has already led to commercial applications.

2. *The rich variety of active sites* that can be present in zeolites: (i) protonic acidic sites, which catalyze acid reactions; (ii) Lewis-acid sites, which often act in association with basic sites (acid–base catalysis); (iii) basic sites; (iv) redox sites, incorporated either in the zeolite framework (e.g., Ti of titanosilicates) or in the channels or cages (e.g., Pt clusters, metal complexes). Moreover, redox and acidic or basic sites can act in a concerted way for catalyzing bifunctional processes.

3. *The large number of different zeolites* that have been synthesized (more than 130), as well as *the variety of well-mastered postsynthesis treatments* that have been developed for tailoring their composition, porosity and acidity. Other characteristics of zeolites—their easy handling (no corrosivity even for the most acidic ones, etc.) and setup in continuous processes, their high thermal stability, which allows their use under a wide range of operating conditions, as well as their regeneration by coke oxidation, etc., have also played a significant role in the development of zeolite-catalyzed processes.

After a short description of the main features of zeolites, the significant contribution of zeolite catalysts in green chemistry will be shown in examples of commercial or the potential processes of refining, petrochemicals, and fine chemicals involving acid or metal acid bifunctional catalysts.

12.1 MAIN ZEOLITES FEATURES

12.1.1 Description and Denomination of Zeolites

Zeolites are crystalline aluminosilicates based on a three-dimensional arrangement of TO_4 tetrahedra (SiO_4 or AlO_4^-) connected through their O atoms to form subunits and then large lattices by repeating identical building blocks (unit cells, u.c.). The structure formula of zeolites (i.e., the u.c. composition) is the following: $M_{x/n} (AlO_2)_x (SiO_2)_y$, where n is the valence of M (cation or proton), $x + y$ the number of tetrahedra per u.c., and y/x the atomic Si/Al ratio, which may vary from a minimal value of 1 (Lowenstein rule) to infinite.

Most of the zeolites can be classified into three categories according to the number of O (or T) atoms in the largest micropore apertures: 8-member-ring (8-MR) or small-pore zeolites with 8 O atoms apertures with free diameters of 0.3–0.45 nm; -10-MR or medium-pore zeolites, with free diameters 0.45–0.60 nm; -12-MR or large-pore zeolites, with free diameters of 0.6–0.8 nm. A comparison between these pore openings and the kinetic diameter of organic molecules clearly shows that zeolites can be used for *molecular sieving.*

Zeolite structures are designated by a three capital-letter code,[5] for example, FAU stands for the faujasite structure, to which the well-known X and Y zeolites belong. A very useful short notation is used for the description of the pore system(s): each pore network is characterized by the channel directions, the number of atoms (in bold type) in the apertures, the crystallographic free diameter of the aperture (in Å), asterisks (1, 2, or 3) indicating whether the systems is one-, two-, or three-dimensional. To completely specify the pore system, the eventual presence of cages (or channel intersections) should be indicated, along with their

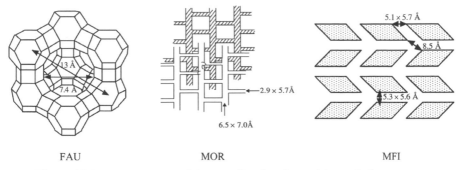

| FAU | MOR | MFI |

Figure 12.1 Pore structure of three zeolites largely used in catalytic processes.

TABLE 12.1   Commercially Used Zeolites

Structure Code (other names)	N_T/u.c.	Channels	Applications
		(a) *Large Pore Zeolites*	
BEA	64	<100> **12** 6, 6 × 6,7** ↔[001] **12** 5,6×5,6*	Catal: cumene synthesis, anisole acetylation
FAU (X,Y)	192	<111> **12** 7,4 × 7,4***	X-Drying, separation (*p*-xylene) Y-Separation, catalysis (FCC, Hydrocracking, etc.)
LTL	36	[001] **12** 7,1 × 7,1*	Catal: aromatization
MOR	48	[001] **12** 6,5 × 7,0* ↔ {[010] **8** 3,4 × 4,8 ↔ [001] **8** 2,6 × 5,7}*	Catal: isomerization of C_5–C_6 alkanes, of C_8 aromatics
		(b) *Medium Pore Zeolites*	
AEL (SAPO-11)	—	[001] **10** 4,0 × 6,5*	Catal: isodewaxing
FER	36	[001] **10** 4,2 × 5,4* ↔[010] **8** 3,5 × 4,8*	Catal: isomerization of *n*-butenes
MFI (ZSM-5)	96	{[100] **10** 5,1 × 5,5 ↔ [010] **10** 5,3 × 5,6}***	Catal: MTO, FCC, selective synthesis of alkylbenzene, (ethylbenzene, paraxylene, etc.)
MWW (MCM-22)	72	⊥ [001] **10** 4,0 × 5,5**| ⊥ [001] **10** 4,1 × 5,1**	Catal: synthesis of ethylbenzene, of cumene
TON (ZSM-22)	24	[001] **10** 4,6 × 5,7	Catal: isodewaxing
		(c) *Small Pore Zeolites*	
CHA (SAPO-34)		⊥ [001] **8** 3,8 × 3,8***|	Catal: MTO
LTA (3A, 4A, 5A)	24	<100>**8** 4,1 × 4,1***	Ion exchange: detergents separation (*n*, isoalkanes) Drying

Note: N_T/u.c: number of T(Si + Al) atoms per unit cell.

dimensions. The International Zeolite Association (IZA) coding of the main commercially used zeolites is given in Table 12.1. Figure 12.1 shows as an example the pore structure of the three zeolites that are the most used in catalysis.

12.1.2   Acid and Bifunctional Metal/Acid Catalysis: Active Sites

Most of the commercial zeolite catalyzed processes occur either through acid catalysis: fluid catalytic cracking (FCC), aromatic alkylation, methanol to olefins (MTO), acetylation, and so on; or through hydrogenating/acid bifunctional catalysis: hydroisomerization of alkanes, hydrocracking, dewaxing, hydroisomerization of the C_8 aromatic cut, and so on. Protonic zeolites are generally used, the active sites being the protons that are associated with the bridging hydroxyl groups Al (OH) Si. Therefore, their maximum concentration is equal to that of the framework Al atoms; however, the actual concentration is always smaller due to the remaining cations or to dehydroxylation and dealumination during the catalyst activation.

The rate of acidic reactions depends on the concentration of accessible protonic acid sites, on their strength, and for certain bimolecular reactions, such as hydrogen transfer, on their proximity. The concentration (and inversely the proximity) of protonic sites can be adjusted either during the synthesis (choice of the Si/Al ratio) or through postsynthesis treatment of the zeolite: ion exchange, dealumination by various methods, that is, by steaming, acid leaching, or isomorphic substitution by silicon compounds. These postsynthesis treatments may also modify the acid strength of the zeolites, thus dealumination can have a positive effect on the acid strength, the "isolated" protonic sites (no next nearest neighbours) having the maximum acid strength,[6] and extraframework Al species created by steaming (which are Lewis-acid sites) increasing the acid strength of neighboring protonic sites.[7] These post-synthesis treatments which are well- mastered allowed the adequate adjustment of the zeolite acidity to the catalysis of the desired reactions.

Noble metals (e.g., Pt) can be introduced within the micropores of zeolites by exchange with a complex cation (e.g., $Pt(NH_3)_4^{2+}$) followed by calcination and reduction. This mode of introduction generally leads to very small clusters of Pt (high Pt dispersion) located within the micropores. Pt supported on acid zeolites are used as bifunctional catalysts in many commercial processes. The desired transformations involve a series of catalytic and diffusion (D) steps,[8] as shown in n-hexane isomerization over Pt acidic zeolite (Equation 12.1).

$$nC_6 \underset{Pt}{\overset{-H_2}{\rightleftharpoons}} nC_6 = \underset{D}{\rightleftharpoons} nC_6 = \underset{H^+}{\rightleftharpoons} iC_6 \underset{D}{\rightleftharpoons} iC_6 = \underset{Pt}{\overset{+H_2}{\rightleftharpoons}} iC_6 \qquad (12.1)$$

Under the operating conditions, the reaction intermediates (n-hexenes and i-hexenes in n-hexane isomerization) are thermodynamically very adverse, hence appear only as traces in the products. These intermediates (which are generally olefinic) are highly reactive in acid catalysis, which explains that the rates of bifunctional catalysis transformations are relatively high. The activity, stability, and selectivity of bifunctional zeolite catalysts depend mainly on three parameters: the zeolite pore structure, the balance between hydrogenating and acid functions, and their intimacy.[9] In most of the commercial processes, the balance is in favor of the hydrogenation function, that is, the transformations are limited by the acid function.

Bifunctional catalysis is one of the most important routes to green (more economical and more environmentally friendly) processes.[10] Indeed, the deactivation of bifunctional catalysts by coking is much slower than that of monofunctional catalysts and their selectivity generally higher (e.g., hydrocracking compared to

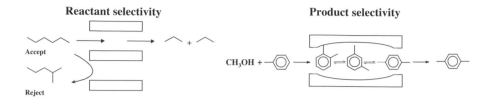

Restricted transition-state selectivity

Selective xylene isomerization over HZSM5 zeolites (no disproportionation)

Figure 12.2   Shape selectivity of zeolites.

cracking). Moreover, bifunctional catalysts can be used in one-pot multistep synthesis with therefore a reduction in the number of chemical and separation steps, hence, of waste production.

12.1.3   Shape Selectivity

There are four widely accepted theories of shape selectivity:[11-13] reactant shape selectivity (RSS), product shape selectivity (PSS), transition state selectivity (TSS) (Figure 12.2), and concentration effect; all of them are based on the hypothesis that the reactions occur within the zeolite micropores only. As indicated earlier, this hypothesis is often verified, the external surface area of the commonly used zeolites being much lower (one to two orders of magnitude) than their internal "surface area."

12.1.3.1   Shape Selectivity Due to Molecular Sieving. The simplest types of shape selectivity are related to the impossibility for certain molecules of a reactant mixture to enter the micropores (RSS) or for certain product molecules to exit from these pores (PSS). In practice, RSS and PSS are observed not only when the molecule size is larger than the pore openings (size exclusion) but also when their diffusion rate is significantly lower (by two orders of magnitude) than that of the other molecules.

12.1.3.2   Spatioselectivity or Transition State Selectivity. TSS occurs when the formation of reaction intermediates (and/or transition states) is sterically limited by the space available near the active sites. Contrary to molecular sieving, TSS does not depend on the size of pore openings, but depends on the size and shape of cages, channels, and channel intersections. TSS often plays a significant role when the reactant(s) may undergo both monomolecular and bimolecular reactions, the latter involving much bulkier intermediates and/or transition states than the former.

12.1.3.3   Shape Selectivity Related to Molecular Concentration in Zeolite Micropores. The interaction between organic molecules and the walls of molecular-size micropores is very strong and zeolites may be considered as *solid solvents.*[14] Therefore, the concentration of reactants in the zeolite micropores is considerably higher than in the gas phase with a significant positive effect on the reaction rates, this effect being more pronounced on bimolecular than on monomolecular reactions. This concentration of molecules in zeolite micropores is largely responsible for the observation that zeolite catalysts are much more active (10 to 10,000 times) than conventional catalysts. It is also responsible for the completely different distribution of hydrocarbons in the gasoline FCC over zeolites and amorphous silica alumina catalysts: more aromatics and alkanes with zeolites at the expense of naphthenes and alkenes owing to a higher ratio between the two major reactions in FCC, that is, hydrogen transfer (bimolecular) and cracking (monomolecular).

12.1.3.4   Other Types of Shape Selectivity. Various other types of shape selectivity have been proposed, some of them requiring additional demonstration. This is not the case for the shape selectivity of the external surface of zeolite crystallites: nest effect, pore mouth, and key lock catalysis, which is discussed in the examples in the next section.

12.2   RECENT DEVELOPMENTS

12.2.1   Isodewaxing

To meet the cold flow requirements, the high molecular-weight linear paraffins (waxes) have to be removed from distillates and base oils. This removal, called *dewaxing*, might be achieved either by solvent extraction or by selective catalytic conversion: hydrocracking or hydroisomerization over zeolite catalysts.[15] Hydroisomerization, which decreases the pour point without decreasing the product yield, is the most advantageous. However, whereas it is very easy to selectively isomerize C_5–C_6 n-alkanes over bifunctional zeolite catalysts (such as Pt HMOR), this is not generally the case for heavier hydrocarbons, because of a fast secondary cracking of branched alkanes. Indeed, with C_6 alkanes, only the very slow C- type acid cracking (Equation 12.2) can occur

$$\underset{\displaystyle \text{C-}\overset{\displaystyle \overset{\textstyle C}{|}}{\text{C}}\text{-C-}\overset{+}{\text{C}}\text{-C}}{} \longrightarrow \text{C-}\overset{+}{\text{C}}\text{-C} \; + \; \text{C-C=C} \tag{12.2}$$

whereas with C_8^+ alkanes, the very fast A cracking, which involves two tertiary carbenium ions intermediates, appears:

$$C-\underset{\underset{C}{|}}{C}-C-\overset{+}{\underset{\underset{C}{|}}{C}}-C \longrightarrow C-\overset{+}{\underset{\underset{C}{|}}{C}}-C + C=\underset{\underset{C}{|}}{C}-C \qquad (12.3)$$

This A cracking was shown to be 20 times faster than the rearrangement of type B involved in n-alkane hydroisomerization.

As a consequence, over ideal bifunctional Pt HFAU catalysts, that is, over catalysts on which acid isomerization and cracking of alkene intermediates are the limiting steps, the maximum yield of isodecanes that could be obtained is not very high ($\sim$50%).[16] The maximum yield is even lower ($\sim$20%) with HMFI as the acidic component of the bifunctional catalyst; this can be ascribed to limitations in the diffusion of isodecene intermediates, which undergo secondary cracking during their long residence time in the narrow pores of HMFI. However, with another medium-pore zeolite (TON) as acid component of the bifunctional catalyst, an exceptionally high yield in isodecane can be obtained,[17] and this practically without deactivation. This seems all the more surprising as (1) TON is a monodimensional zeolite and the pore blockage by carbonaceous deposits of these type of zeolites is known to be very fast, and (2) TON has pore openings smaller than the three-dimensional MFI zeolites (Table 12.1), which should be even more favorable for secondary cracking. Whereas the classic reaction scheme

$$n\text{-}C_{10} \rightleftharpoons m\text{-}C_9 \rightleftharpoons dm\text{-}C_8 \rightleftharpoons \text{Cracking products} \qquad (12.4)$$

found with bifunctional catalysts, can also be observed with Pt TON, the maximum in monobranched (n-C_9), bibranched (dm-C_8) are higher and much better separated . Therefore, long-chain n alkanes can be very selectively transformed into isomers, but also to monobranched or to multibranched isomers. Since it is only the first methyl group that causes a significant decrease in the pour point of alkanes (from 30°C to 60°C), this possibility to selectively form monobranched isomers seems in favor to the use of Pt TON as isodewaxing catalyst. However to minimize the pour point, the methyl branch has to be preferentially located toward the center of the chain,[15] although isomerization of n-alkanes was found to result in selective 2-methylbranching up to high conversion levels.[17] Subsequent isomerization of 2-methylalkanes by alkyl shift (which occurs very fast over bifunctional large-pore-based catalyst) seems desirable.

Because molecular graphics suggested that isomerization of decene intermediates could not occur entirely inside the TON channels,[18] it was theorized that this reaction occurred at the mouth of the micropores. This pore-mouth catalysis was confirmed[19] by characterization of Pt TON samples recovered after various time-on-stream (TOS). The quasi-immediate retention of 1.5–2.0 wt % of carbonaceous compounds on the catalysts could be observed, these compounds causing a quasi-total blockage of the access of nitrogen (hence, of reactants) to the TON channels. These carbonaceous compounds were demonstrated to be essentially constituted by $C_{12}-C_{20}$ linear and monobranched alkanes trapped within the zeolite channels.

In agreement with this pore blockage, it was suggested that isomerization would be catalyzed by protonic sites at the mouth of the channels; however, another possibility might be that the molecules of products trapped within the micropores near the external surface were the active species. Indeed, a simple mechanism, in which active sites are tertiary carbenium ions formed at the pore mouth by adsorption of methylalkenes trapped in the zeolite pores during the first minutes of reaction, can be theorized.[19] This mechanism involves only very simple steps: A isomerization (through alkyl shift), hydride transfer, A cracking, which is simple than the limiting step of isomerization (B-type) catalyzed by the protonic-acid sites located at the pore mouth. Moreover, contrary to this latter mechanism, it requires no diffusion of alkene intermediate molecules in the narrow zeolite channels (Figure 12.3).

High selectivity in hydroisomerization of long-chain alkanes can also be obtained with the other medium-pore zeolites or silicoaluminophosphates similar to TON, that is, having a monodimensional pore system with small apertures: MTT, FER, SAPO 11. Commercial processes were developed using these molecular sieves as catalysts, in particular Isodewaxing by Chevron, Mobil's Selective Dewaxing (MSDW), and Wax Isomerization (MWI) by Exxon Mobil. These

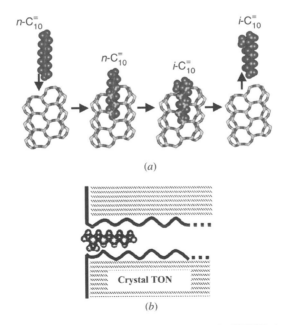

Figure 12.3 Mechanisms of *n*-decane isomerization over Pt HTON. (*a*) Protonic sites at the pore mouth. (*b*) Monobranched alkenes (formed from "coke" molecules) adsorbed on protonic sites at the pore mouth.

processes are more efficient than those based on solvent extraction or on selective hydrocracking. Thus, results obtained on pilot plants showed that for similar values of the pour point of the product ($-12°C$), the yield was much higher (90.5 wt %) with isodewaxing than with dewaxing by hydrocracking; moreover, the viscosity index VI (the temperature dependence of the viscosity, a high VI indicating a low dependence) was of 105 instead of 92.[15]

12.2.2 Ethylbenzene and Cumene Production

12.2.2.1 Generalities. Alkylation of aromatic compounds is practiced commercially on a large scale. The major products are ethylbenzene converted into styrene for polymer use and cumene (isopropylbenzene), a precursor to acetone and phenol.[20, 21] Alkylation of benzene by ethylene or propylene is an electrophilic substitution catalyzed by protonic acid sites. Acidic halides such as $AlCl_3$, which are typically Lewis acids, have little or no activity when used in a pure state; they become active by addition of co-catalysts such as hydrochloric acid, which interact with $AlCl_3$ to generate a strong protonic acidity ($AlCl_4^- H^+$).

The mechanism of catalytic alkylation is later in the example of benzene isopropylation:

$$(12.5)$$

Ethylation, which involves an unstable ethylcarbenium ion as intermediate, is much slower (1500 times over $AlCl_3$) than isopropylation. It is also the case in benzene alkylation with propene for the undesired formation of n-propylbenzene, which involves a primary n-propyl carbocation. Furthermore, as alkyl substituents activate the aromatic ring, consecutive alkylation of the primary product occurs with a greater rate than the first alkylation step ($k_2 > k_1$).

$$(12.6)$$

There are several means to maximize the yield in the desired monoalkylation product: high aromatic/alkylation agent ratio, association of a transalkylation unit to the alkylation unit, and use of a shape selective zeolite as catalyst.

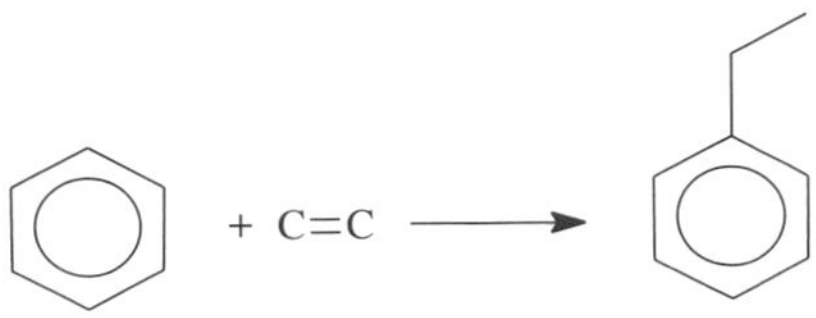

Figure 12.4 Ethylbenzene synthesis. Commercial processes. **AlCl₃ + HCl catalyst (1950)**: AlCl$_3$ corrosivity and problems associated with safe handling and disposal. For one tonne of EB, consumption of 2–4 kg catalyst, 1 kg of HCl, and 5 kg of caustic solution with production of salts. **Zeolite catalysts:** (1) 1980 Mobil Badger vapor-phase process MFI (ZSM5) 370–420°C, 7–27 bar, B/C$_2^=$ 5–20, WHSV 300-400 h^{-1}, recycling of DEB, yield > 99.5%, cycle length: 1 year; high-energy efficiency. (2) 1995 EB Max liquid-phase process MWW (MCM22) 200°C, B/C$_2^=$ 3.5, Yield > 99.9%, cycle length >3 years.

12.2.2.2 Toward Cleaner Processes.

AlCl$_3$ was the first catalyst used for ethylbenzene production, and nearly 40% of the ethylbenzene capacity still utilizes the AlCl$_3$ process. There are serious problems due to the corrosivity of AlCl$_3$ and to waste production (Figure.12.4). This has led most manufacturers to move toward zeolite catalyzed processes that do not present these inconveniencies. Figure 12.4 shows the significant improvements caused by the substitution of AlCl$_3$ by the MFI zeolite (Mobil Badger process, 1980), and then by the MWW (MCM-22) zeolite (EB Max process, 1995). The Mobil Badger process operates in the vapor phase at a high temperature. One of the initial problems was the relatively fast deactivation of the MFI zeolite by coking with the need for frequent regenerations. This problem was solved in the second and third (1986, 1991) generations of processes, the cycle lengths passing from 60 days to more than one year. A very high ethylbenzene yield (>99.5%) can be obtained. Other advantages over the AlCl$_3$ process were the recovery of practically all the heat of reaction ($\Delta H = -114$ kJ·mol^{-1}) as medium or low-pressure vapor, and the possibility to use dilute ethylene as feed. EB Max is a liquid-phase process. The MWW zeolite is very stable (cycle lengths >3 years) and very selective to monoalkylate, which allows the process to use low feed ratios of benzene to ethylene. The undesired diethylbenzene products undergo transalkylation with benzene in an additional reactor operating either in liquid or in gas phase.[21]

Up to 10 years ago, cumene was essentially produced by isopropylation of benzene using solid phosphoric acid (SPA), actually phosphoric acid supported on kieselgur, or AlCl$_3$ (for only a small number of plants). Small amounts of water have to be continuously fed to the reactor in order to maintain the desired level of activity; consequently, there is a continuous release of corrosive phosphoric acid. The other drawbacks of the SPA process deal with the disposal of the used catalyst as well as with the production of undesired bi- and triisopropyl benzenes (4–5 wt. %). Greener processes based on zeolite catalysts have been recently licensed by DOW (dealuminated mordenite), Enichem (BEA), and Mobil Badger (MWW). Higher cumene yields can be obtained owing to the attachment to the alkylation process of a transalkylation unit for converting the diisopropylbenzene by-product to cumene.

12.2.2.3 Alkylation over the MWW Zeolite. The MWW (or MCM-22) zeolite developed by Mobil as catalyst for ethylbenzene and cumene production deserves particular attention. Indeed, this zeolite presents unique structural features (Figure 12.5). Its structure is constituted of *three independent pore systems:*[21] large supercages (inner diameter of 7.1 Å defined by a 12-member-ring [12-MR], height 18.2 Å) each connected to six others through 10-MR apertures (4.0 Å × 5.5 Å), bidimensional sinusoidal channels (10-MR, 4.1 × 5.1 Å),[22] and *large hemisupercages* (7.1 Å Ø, 7.0 Å) on *the external surface.*[23]

Acid sites were shown to be located in the three-pore system of protonated samples (HMWW), and methods were recently proposed for determining the distribution of these sites as well as their respective role in *o-*, *m-*, and *p-*xylene transformations.[24] While xylene transformation was shown to occur in the three locations, benzene alkylation with ethylene was catalyzed by the acidic sites of the external hemicups only.[25] Indeed, the activity for this reaction is completely suppressed by adding a base molecule (collidine) to the feed that is too bulky to enter the inner micropores. Moreover, adsorption experiments show that collidine does not influence the rate of ethylbenzene adsorption, so that the suppression of alkylation activity was not caused by pore mouth blocking.[25]

However, by using a model reaction (gas-phase toluene alkylation with propene) we have recently demonstrated[26] that initially the protonic sites of the supercages catalyze the formation of heavy alkylaromatics, those of the sinusoidal channels, the transformation of the olefin through an oligomerization–cracking process, and the formation of a small amount of alkylaromatics. A large part of these alkylaromatics remain trapped within the inner micropores, causing a quasi-immediate blockage of their access by reactant molecules; this shows that only the protonic sites of the external hemicages seem to be active in ethylbenzene formation. The high selectivity of these sites to monoalkylates and their high

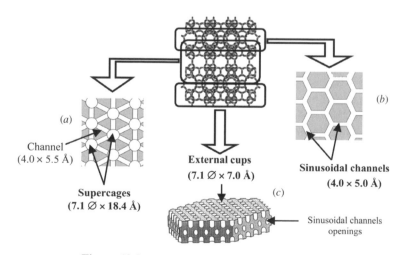

Figure 12.5 Pore structure of MCM22 (MWW).

stability are most likely due to the ease of desorption of product molecules from the large external hemicages.

12.2.3 Fine Chemicals Synthesis

Significant advances remain to be accomplished to make the fine chemical processes environmentally friendly: indeed, the so-called environmental factor[27] that is, the ratio between the amounts of waste and product, is still between 5 and 50, that is, 50 to 500 times greater than in refining and petrochemicals. Of course, this is because the fine chemicals synthesis requires a significant number of chemical and separation steps, but also because most of the chemical steps are carried out in batch reactors in homogeneous phase, either stoichiometrically or by using acid catalysts such as H_2SO_4 or $AlCl_3$. These acids, which are not reusable, have to be neutralized, which generates a large amount of valueless salts.

There is an urgent need to substitute cleaner technologies, such as those based on heterogeneous catalysis, for these polluting technologies. Indeed, solid catalysts such as zeolites offer many advantages, in particular, easy recovery of reaction products, safe handling, easy setup of continuous processes, and regenerability. Here we show that acid zeolites that, due to their shape-selective properties and the easy tailoring of their active sites and porosity, are the major catalysts in refining and petrochemicals should play a significant role in fine chemical synthesis in the near future.

However, some characteristics of the reactant and product (and solvent) molecules used in the reactions of fine chemical synthesis have to be considered: their large size, their low thermal stability, and their polarity. The small size of the micropores limits the use of zeolites to the synthesis of relatively small molecules; solutions to overstep this limitation have been found with the development of larger (mesoporous) molecular sieves and of nanocrystallite zeolites, the reaction then occurring on the large external surface. The low thermal stability of the molecules leads to operation at a low temperature, often in the liquid phase. Last, as will be shown in the first example (Section 12.2.3.1), the differences in polarity between reactant(s), product(s), and solvent molecules have to be considered for optimizing both the zeolite catalyst and the operating conditions.

12.2.3.1 *Acetylation of Arenes with Acetic Anhydride.* Arylketones are generally prepared by acylation of aromatics or by the related Fries rearrangement. These ketones are important intermediates in the synthesis of fragrances of the musk type and of pharmaceuticals such as paracetamol, ibuprofen, and S-naproxen.[28] Acylation processes are often carried out in the liquid phase by using batch reactors with corrosive metal chlorides, such as $AlCl_3$ as catalysts and acid chlorides as acylating agents. A characteristic of this reaction is the formation of a stable 1:1 molar adduct with the catalyst, which generates serious environmental problems: use of more than a stoichiometric amount of the "catalyst," necessary hydrolysis of the adduct for recovering the desired ketone with

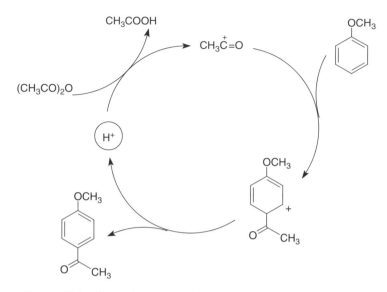

Figure 12.6 Mechanism of anisole acetylation over protonic zeolites.

destruction of the catalyst, and production of a large amount of HCl (more than 4 mol per mol of ketone produced) with corrosion problems and final production of valueless salts. All that makes the substitution of AlCl₃ by acid zeolites highly desirable, and also that of acetyl chloride by acetic anhydride or acetic acid.

The acetylation over protonic zeolites of aromatic substrates with acetic anhydride was widely investigated. Essentially HFAU, HBEA, and HMFI were used as catalysts, most of the reactions being carried out in batch reactors, often in the presence of solvent. Owing to the deactivation effect of the acetyl group, acetylation is limited to monoacetylated products. As could be expected in electrophilic substitution, the reactivity of the aromatic substrates is strongly influenced by the substituents, for example, anisole > *m*-xylene > toluene > fluorobenzene.[29] Moreover, with the poorly activated substrates (*m*-xylene, toluene, and fluorobenzene) there is a quasi-immediate inhibition of the reaction. It is not the case with activated substrates such as anisole and more generally aromatic ethers. It is why we have chosen the acetylation of anisole and 2-methoxynaphtalene as an example.

It is generally admitted that over zeolites, acetylation of aromatic substrates with acetic anhydride (AA) is catalyzed by protonic acid sites. The direct participation of Lewis sites was excluded by using two BEA samples with similar protonic acidities, but with very different Lewis acidities: indeed, these samples were shown to have quasi-similar activities.[30] The currently accepted mechanism is shown in Figure 12.6 for the anisole acetylation example. The limiting step of the process is the attack of anisole molecules by acylium ions.

Whatever the zeolite, 4-methoxyacetophenone is largely predominant (>98%), which indicates that this selective formation is not due to shape-selectivity effects, but is a characteristic of the reaction. In contrast, the selectivity of

2-methylnaphtalene (2-MN) acetylation depends very much on the zeolite employed: over FAU zeolites, acetylation of 2-MN occurs very selectively at the very active C1 position with formation of 1-acetyl-2 methoxynaphtalene (1-AMN), whereas with other large-pore zeolites with smaller-pore apertures (BEA, MTW, ITQ7), the desired 2-acetyl-6-methoxynaphtalene product, 2-AMN (precursor of the anti-inflammatory S-naproxen), also appears as a primary product. Moreover, at long reaction times, 1-AMN isomerizes into 2-AMN. This isomerization is much faster in the presence of 2-MN than in its absence, causing the following intermolecular (transacetylation) mechanism to be proposed:[31]

(12.7)

$$2\text{-MN} + 1\text{-AMN} \longrightarrow 2\text{-AMN} + 2\text{-MN}$$

The location of 2-MN acetylation over HBEA zeolites was largely debated. Some authors claimed that the bulkier isomer (1-AMN) could be formed only on the external surface, and the linear one (2-AMN) both within the micropores and on the external surface, while other than that both isomers were essentially formed within the micropores. The second proposal seems to be more likely; indeed, adsorption experiments showed that 1-AMN could enter the micropores of HBEA and even those of MFI, a medium-pore zeolite.[32]

Kinetic studies of the acetylation of several arylethers were carried out over HBEA zeolites. The main conclusion is that the rate and stability of the reactions are determined by the competition between reactant(s) and product(s) molecules for adsorption within the zeolite micropores.[28] This competition shows that the autoinhibition of arene acetylation, that is, the inhibition by the acetylated products, and also by the very polar acetic acid product is generally observed. This effect is much more pronounced with hydrophobic substrates such as methyl and fluoro aromatics than with hydrophilic substrates because of the larger difference in polarities between substrate and product molecules.

Various solutions to limit the negative effect of the polarity of the acetylated products were proposed:

- Choice of operating conditions favoring the adsorption of the reactant molecules and the desorption of the product molecules; fixed-bed reactor instead of batch reactor, high molar substrate/acylating agent ratio, low conversion, high temperature, use of solvents with adequate polarities.

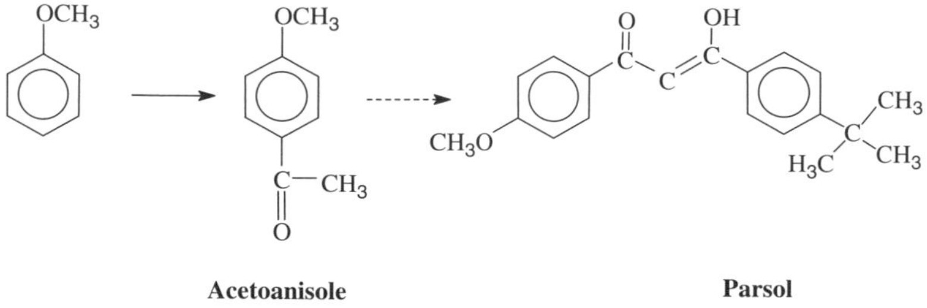

Acetoanisole   **Parsol**

Yield/ArH	AlCl₃	HBEA
	95 %	95 %
Chemical + Physical Steps	8	2
Aqueous effluents (t/t anisole)	4.5	0.035
Effluent composition (wt %)		
H_2O	68.7	99
Al^{3+}	5	0
Cl^-	24	0
Organics	2.3	1

Figure 12.7 Acetylation of anisole. Characteristics of the old (AlCl₃) and new (HBEA) technologies.

- Adequate choice and adjustment of the zeolite characteristics, small path of diffusion (small crystallites, mesopores), good balance between hydrophobicity and acidity.

Industrial processes were developed by Rhodia[33] for the selective acetylation of anisole into 4-methylacetophenone and veratrole into 3,4-dimethoxyacetophenone, which are, respectively, precursors of Parsol, used for sun protection, and of Verbutin, a synergist for insecticides. In both cases, the substitution of the new technology (fixed-bed reactor, HBEA or HFAU zeolites as catalysts, acetic anhydride as acylating agent) for the old technology (batch reactor, AlCl₃ as catalyst, and acetylchloride as acylating agent) constitutes a major environmental and economical breakthrough. The main improvements brought about by the new process of anisole acetylation are presented in Figure 12.7: this process is much simple, the consumption of water much smaller, and lower amounts of organic compounds and no mineral compounds in the aqueous effluents.[28]

12.2.3.2 One-Pot Multistep Synthesis of Ketones on Bifunctional Zeolite Catalysts.
One-pot multistep reactions constitute an elegant and efficient way to decrease the number of chemical and separation steps, hence, to develop greener synthesis processes. Bifunctional metal–acidic or metal–basic zeolite catalysts, which can be prepared easily with the desired properties (e.g., distribution of the

different types of active sites, characteristics of the diffusion path) are good candidates for these reactions.

A widely studied example is the synthesis of various ketones or aldehydes, such as methylisobutylketone from acetone, cyclohexylcyclohexanone from cyclohexanone, 1,3-diphenylbutan-1-one from acetophenone, 2-ethylhexanal from *n*-butyraldehyde. Thus methyl isobutyl ketone (MIBK), which is used as a solvent for ink and lacquers, was previously prepared through the three-step catalytic process: base catalyzed production of diacetone alcohol (DA) by acetone aldolization, acid dehydration of DA into mesityl oxide (MO), then hydrogenation of MO on a Pd catalyst (Figure 12.8). Since acetone aldolization also occurs by acid catalysis, it is possible to synthesize MIBK in one apparent step by combining the acidic and hydrogenation functions. Most of the studies were carried out in the gas phase by using fixed-bed reactors.

The best yield and selectivity to MIBK were obtained with PdMFI:[34] 28 and 98% at 453 K or PdCsMFI:[35] 35 and 82%. Pd supported over large pore zeolites such as HFAU are much less selective. The main by-products are propane and 2-methylpentane resulting from three-step bifunctional catalytic transformations of acetone and methylisobutylketone, respectively (C=O hydrogenation, dehydration, and C=C hydrogenation).[36] This is why Pd, which is known to be more selective for the desired hydrogenation of the C=C bond than for the hydrogenation of the C=O bond, is generally chosen. Diisobutyl ketone (DIBK) may also be formed from acetone and MIBK through a bifunctional scheme similar to the one involved in MIBK formation.[36] The significance of the secondary reactions of MIBK (methylpentane and DIBK formation) can be significantly reduced by an adequate choice of the zeolite pore structure (preferentially tridimensional, to facilitate the desorption of MIBK) and of the crystallite size.

Pd supported overlarge-pore tridimensional acidic zeolites such as HFAU are the more active and selective catalysts for the synthesis of bulkier ketones. Thus, in a 0.2% Pd-HFAU catalyst, yield and selectivity from cyclohexanone of 23 and 75% can be obtained in cyclohexylcyclohexanone synthesis.[37] Furthermore, the synthesis of aldehydes can only be made selective by joining the hydrogenating metallic sites (Pd) to basic sites (instead of acidic sites). Thus, 2-ethylhexanal, which is a component of perfumes and fragrances, can be synthesized with high yield and selectivity (64 and 91%, respectively) on a PdKX zeolite.[38] Much lower yields and selectivities are obtained over nonzeolitic materials, such as Pd/MgO.

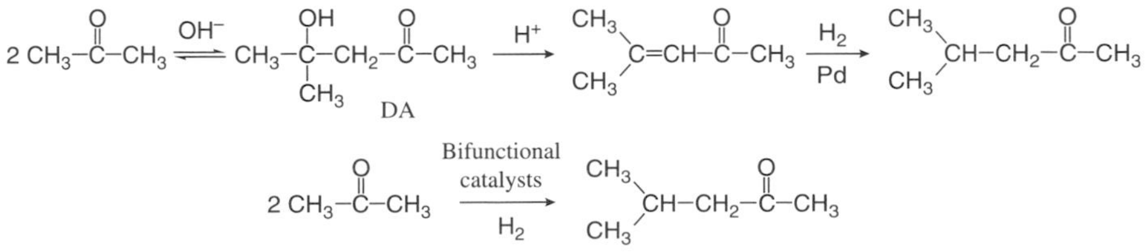

Figure 12.8 Synthesis of methyl isobutyl ketone (MIBK) from acetone.

12.3 CONCLUSIONS

As was shown here in some examples, the field of catalysis over zeolites, although mature, is still very much alive. The chemists who work with the synthesis zeolites continue to be very creative, the focus now being placed on the synthesis of materials that can catalyze reactions other than the acidic ones and/or reactions of bulkier molecules, that is, synthesis of zeolites with larger micropores or with a very large external surface, such as nanosize and delaminated[39] zeolites. New concepts related to the mode of action of zeolite catalysts continue to emerge, as shown here with the shape selectivity of the external surface. These concepts are particularly useful to scientifically design selective and stable catalysts.

Since zeolite catalysts are successfully introduced in the refining and petrochemical industries, it is not surprising that most of the recent advances concern incremental improvements of existing processes with the development of new generations of catalysts (e.g., dewaxing, ethylbenzene and cumene synthesis). The number of newer applications is much more limited, for example, direct synthesis of phenol from benzene and aromatization of short-chain alkanes, etc. However, both the improvement and development of processes contribute significantly to environmental advances.

In the field of fine chemical synthesis there is an urgent need to substitute the cleaner technologies for the old polluting ones. It is hoped that the large economic and environmental benefits brought by the recently developed catalysis processes—acetylation of anisole and of veratrole, Beckmann rearrangement, and so forth—will initiate great strides in this field.

REFERENCES

1. Crönstedt, A. F. *Akad. Handl. Stockholm*, 1756, **18**, 120.
2. Maesen, T.; Marcus, B. *Stud. Surf. Sci. Catal.*, 2001, **135**, 1.
3. Guisnet, M.; Gilson, J. P. (Eds.) *Zeolites for Cleaner Technologies*, Imperial College Press, Singapore, 2002.
4. Weisz, P. B.; Frilette, V. J. *J. Phys. Chem.*, 1960, **64**, 382.
5. Baerlocher, Ch.; Meier W. M.; Olson, D. H. (Eds.) *Atlas of Zeolite Framework Types*, 5th Rev. Ed. Elsevier, Amsterdam, 2001.
6. Barthomeuf, D. *Mater. Chem. Phys.*, 1987, **17**, 49.
7. Mirodatos, C.; Barthomeuf, D. *J. Chem. Soc. Chem. Commun.*, 1981, **39**.
8. Weisz, P. B. *Adv. Catal.*, 1962, **13**, 137.
9. Guisnet, M.; Alvarez, F.; Giannetto, G.; Perot, G. *Catal. Today*, 1987, **1**, 415.
10. Guisnet, M. *Pol. J. Chem.*, 2003, **77**, 637.
11. Chen, N. Y.; Garwood, W. E.; Dwyer, F. G. (Eds.) *Shape Selective Catalysis in Industrial Applications*. Chemical Industries 36, Marcel Dekker, New York, 1989.
12. Song, C.; Garces, J. M.; Sugi, Y. (Eds.) *Shape Selective Catalysis*, Am. Chem. Soc. Symp. Ser., vol. 738, 2000.

13. Degnan, T. F. Jr. *J. Catal.*, 2003, **216**, 32.

14. Derouane, E. G. *J. Mol. Catal. A: Chem.*, 1998, **134**, 29.

15. Daage, M. in *Zeolites for Cleaner Technologies*, Guisnet, M.; Gilson, J. P. (Eds.) Imperial College Press, Singapore, 2002, p. 167.

16. Alvarez, F.; Ribeiro, F. R.; Perot, G.; et al. *J. Catal.*, 1996, **162**, 179.

17. Martens, J. A.; Souverijns, W.; Verreslt, W.; et al. *Angew. Chem. Int. Ed.*, 1995, **34**, 2528.

18. Martens, J. A.; Parton, R.; Uytterhoeven, L.; Jacobs, P. A. *Appl. Catal.*, 1991, **79**, 95.

19. Guisnet, M. *J. Mol. Catal. A: Chem.*, 2002, **183**, 367.

20. Beck, J. S.; Haag, W. O *Handbook of Heterogeneous Catalysis*, Ertl, G.; Knözinger, H.; Weitkamp, J. (Eds.) VCH, Weinheim, 1997, p. 2123.

21. Beck, J. S.; Dandekar, A. B.; Degnan, T. F. in *Zeolites for Cleaner Technologies*, Guisnet, M.; Gilson, J. P. (Eds.) Imperial College Press, Singapore, 2002, p. 223.

22. Leonowicz, M. E.; Lawton, S. L.; Partridge, R. D.; et al. *Science*, 1994, **264**, 1910.

23. Lawton, S. L.; Leonowicz, M. E.; Partridge, R. D.; et al. *Micropor. Mesopor. Mater.*, 1998, **23**, 109.

24. Laforge, S.; Martin, D.; Guisnet, M. *Appl. Catal A: Gen.*, 2004, **268**, 33.

25. Du, H.; Olson, D. H. *J. Phys. Chem. B*, 2002, **106**, 395.

26. Rigoreau, J.; Laforge, S.; Gnep, N. S.; Guisnet, M. *J. Catal.*, 2005, **236**, 45.

27. Sheldon, R. A. *Chemtech*, 1994, **38**.

28. Marion, P.; Jacquot, R.; Ratton, S.; Guisnet, M. *Zeolites for Cleaner Technologies*, Guisnet, M.; Gilson, J. P. (Eds.) Imperial College Press, Singapore, 2002, p. 281.

29. Guidotti, M.; Canaff, C.; Coustard, J. M.; et al. *J. Catal.*, 2005, **230**, 375.

30. Berreghis, A.; Ayrault, P.; Fromentin, E.; Guisnet, M. *Catal. Lett.*, 2000, **68**, 121.

31. Fromentin, E.; Coustard, J. M.; Guisnet, M. *J. Catal.*, 2000, **190**, 433.

32. Moreau, V.; Fromentin, E.; Magnoux, P.; Guisnet, M. *Stud. Surf. Sci. Catal.*, 2001, **135**, 4113.

33. Spagnol, M.; Gilbert, L.; Benazzi, E.; Marcilly, C. European Patent WO9635655 (1996) (Patent to Rhodia).

34. Huang, T. J.; Haag, W. O. U.S. Patent 4,339,606 (1982) (Patent to Mobil Oil Corp.).

35. Chen, P.Y.; Chu, S. J.; Chang, N. S.; et al. *Stud. Surf. Sci. Catal.*, 1984, **46**, 231.

36. Melo, L.; Giannetto, G.; Alvarez, F.; et al. *Catal. Lett.*, 1997, **44**, 201.

37. Alvarez, F.; Silva, A. I.; Ramoa Ribeiro, F.; et al. *Stud. Surf. Sci. Catal.*, 1997, **108**, 609.

38. Ko, A. N.; Hu, C. H.; Chen, J. Y. *Appl. Catal. A: Gen.*, 1999, **184**, 211.

39. Corma, A.; Fornes, V. *Stud. Surf. Sci. Catal.*, 2001, **135**, 73.

13

ACID AND SUPERACID SOLID MATERIALS AS NONCONTAMINANT ALTERNATIVE CATALYSTS IN REFINING

José M. López Nieto

Intituto Tecnología Química, UPV-CSIC, Valencia, Spain

INTRODUCTION

New and pending industrial legislation throughout the world requires more stringent environmental protection. Thus, it will soon become illegal to release into the atmosphere some products (benzene, volatile hydrocarbons, carbon oxides, corrosive and reactive oxides of sulfur, and nitrogen) that contribute to the greenhouse effect. Due to new environmental legislation, many refiners will be required to make significant processing changes. In this way, new catalytic technologies will help to protect the ozone layer, to combat the greenhouse effect, and to solve environmental problems of energy, and so forth.

However, the catalyst is in other cases the noxious agent. A clear example is the use of liquid acids, that is, H_2SO_4, H_3PO_4, $HClO_4$, or HF, as catalysts in commercial processes: phenol production from cumene hydroperoxide (diluted H_2SO_4), production of caprolactame (H_2SO_4 oleum), isobutene/butanes alkylation (HF, H_2SO_4), polymers derived from aniline, and hydration of olefins. In 1989, 12 million tons of phosphoric acids and 44 million tons of sulphuric acid were used in United States. Albeit not all of these acids were used for catalysis, so the

Methods and Reagents for Green Chemistry: An Introduction, Edited by Pietro Tundo, Alvise Perosa, and Fulvio Zecchini

amount of liquid acids as catalysts should be greatly reduced in the succeeding years.

Classic superacid solutions are mixtures of strong acids like FSO_3H or HF with Lewis acids: SbF_5, TaB_5, or BF_3.[1] Besides their high acid strength, such liquids also present some difficulties (corrosion of the reactor and other parts of the plant, difficulty in catalyst regeneration, difficulty in separating the catalysts from the reaction mixture, high cost of environmentally safe disposal) that considerably limit their large-scale use.

The first step to reduce the negative impact of the liquid acids is to use the acid, superacid (as triflic acid, CF_3SO_3H), or Lewis acid (especially $AlCl_3$ or BF_3) impregnated on silica.[2] This is the case for paraffin isomerization (old technology), alkylation of benzene with olefins, production of polygasoline, some of the new isobutene/butanes alkylation technologies, and olefin hydration. This way not only could decrease the hazards, but also increase the catalyst life. However, the formation of the corresponding acid indicates that this is not a productive route.

The tendency in the past decades has been to replace them with solid acids (Figure 13.1).[3-5] These solid acids could present important advantages, decreasing reactor and plant corrosion problems (with simpler and safer maintenance), and favoring catalyst regeneration and environmentally safe disposal. This is the case of the use of zeolites, amorphous silico-aluminas, or more recently, the so-called superacid solids, that is, sulfated metal oxides, heteropolyoxometalates, or nafion (Figure 13.1).[3-12] It is clear that the well-known carbocation chemistry that occurs in liquid-acid processes also occurs on the solid-acid catalysts (similar mechanisms have been proposed in both catalyst types) and the same process variables that control liquid-acid reactions also affect the solid catalyst processes.

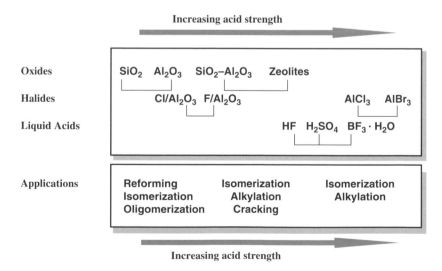

Figure 13.1 Industrial processes carried out on oxides, halides, and liquid acid catalysts (After Ref. 5).

However, the use of polyfunctional solid catalysts (involving both red-ox and acid properties) could modify, in some cases, the characteristics of some of the actual industrial processes.

Several metal oxides could be used as acid catalysts, although zeolites and zeotypes are mainly preferred as an alternative to liquid acids (Figure 13.1).[3,5] This is a consequence of the possibility of tuning the acidity of microporous materials as well as the shape selectivity observed with zeolites that have favored their use in new catalytic processes. However, a solid with similar or higher acid strength than 100% sulfuric acid (the so-called superacid materials) could be preferred in some processes. From these solid catalysts, nafion, heteropolyoxometalates, or sulfated metal oxides have been extensively studied in the last ten years (Figure 13.2). Their so-called superacid character has favored their use in a large number of acid reactions: alkane isomerization, alkylation of isobutene, or aromatic hydrocarbons with olefins, acylation, nitrations, and so forth.[3-6]

Nafion resin, which is a prefluorinated resinsulfonic acid-based catalyst (represented by $-CF_2-CF_2-SO_3H$ groups), has an acid strength similar to 100% sulfuric acid and shows interesting catalytic properties in acid reactions (Figure 13.2). However, they present an important disadvantage, the low surface area ($0.02 \ m^2 \ g^{-1}$), which makes it difficult for molecules to access their active sites.[7] The accessibility of acid sites in nafion was increased by entrapping nano-sized particles of nafion resin within a higly porous silica network using a sol–gel techniques.[8] However, silica with high surface areas are not required in this case, since the sulfonic groups of nafion could interact to a greater extent with the silanol groups of the silica, resulting in a decrease in their catalytic activity.[9] So, supported nafion catalysts present acid sites with high acid strength and relatively good stability, although with some problems in their ability to regenerate.

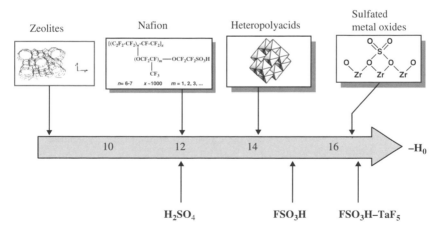

Figure 13.2 Acid strength of acid and superacid solids determined by Hammet method. For a comparable purpose, the acid strengths of some liquid acids are also included. (After Ref. 12.)

Heteropolyacids and related compounds have attracted increasing interest in catalysis owing too their ability to catalyze both acidic and red-ox processes.[3,10–12] They present strong acidities (the pH values of aqueous solutions of heteropolyacids indicate that they are strong acids) both in solid and in liquid solution (Figure 13.2). In addition, they can be prepared in an wide range of surface areas (partially salified heteropolyoxometalates permit the modification of the surface areas of these materials) or be supported in metal oxides.

At the end of the 1970s, new solids materials (especially sulphated metal oxides) with a hypothetical superacid character were reported (Figure 13.2).[13] Among these, zirconium oxide treated with sulfate anions is one of the most interesting catalysts.[14–16] However, it is now clear that the so-called superacid character of these materials is really related to the presence of both acid and red-ox properties. New catalytic reactions also have been proposed for these catalytic systems. However, it appears quite clear that the presence of sulfate will always be a limitation for their practical use. However, instead of sulphate, tungstate or hetepolyacids could also be interesting alternatives depending on the acidity required.[14–16]

In the case of C4-hydrocarbons, the use of acid or superacid solids will depend on both the acid strength required in each reaction and the reaction conditions required to optimize the thermodynamic equilibrium (Figure 13.3). For example, catalysts with very high acid strength could be substituted for a solid with a lower acidity by increasing reaction temperature. This has been proposed in both the isomerization of lineal alkanes and in the alkylation of isobutene with olefins, although the thermodynamic equilibrium should also be considered.

The modern gasolines are produced by blending products from crude oil distillation, that is, fluid catalytic cracking, hydrocraking, reforming, coking, polymerization, isomerization, and alkylation.[17] Two clear examples of the possible use of solid-acid catalysts in refining processes are the isomerization of lineal alkanes and the alkylation of isobutene with butanes. In both these cases, and due to the octane

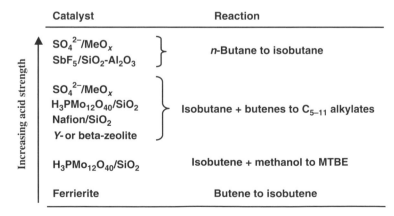

Figure 13.3 Possible catalytic uses of acid and superacid solids in the selective transformation of C$_4$-hydrocarbons by acid reactions.

requirements, the research octane numbers (RON) is an important factor to be considered. For this reason, there are very good reasons for replacing the liquid-acid catalysts with solids acids that have fewer adverse effects on the environment.

13.1 ALKANE ISOMERIZATION

Traditionally, the isomerization of light alkanes (n-C5 and n-C6, light straight run gasoline) have been carried out using bifunctional systems, that is, Pt-AlCl$_3$/Al$_2$O$_3$ (UOP, BP) or Pt-Mordenite (Shell, Mobil, Norton, Su-Oil, IFP) catalyst.[17,18] A chlorinated alumina-based catalyst act at a low operation temperature that is favorable in equilibrium.[17] This allow as to obtain isomerates of about 83 RON. However, it has the disadvantages of being highly sensitivity to water in the feed (requiring the use of expensive dryers) and having low catalyst stability (requiring the supplement of some organic chloride chemicals to preserve its catalytic activity).

The zeolite catalysts require high operating temperatures.[18] Although this reaction condition favors the production of an isomerate with a RON of about 78, these catalysts possess a tolerance to feedstock poisons such as sulfur or water.

These reactions can also be performed over a strong acid catalyst at reaction temperatures that are lower than over zeolites. Because of this, isomerization of n-butane over ZrO$_2$-supported sulfate catalysts was initially proposed by Hino and Arata.[13] They proposed these catalysts as being effective in butane isomerization at room temperature, a reaction that does not take place, even in 100% sulphuric acid. For this reason, these catalysts were considered as solid superacids, since they are active and selective in the isomerization of n-butane to isobutane at

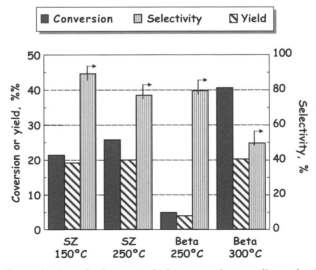

Figure 13.4 Isomerization of n-butane to isobutane on beta zeolite and sulfated zirconia catalysts at different reaction temperatures.

lower temperatures than zeolites (Figure 13.4).[19] In the last decade, many research workers have studied the improvement of catalytic behavior, catalyst decay, and the resistance of poisons.[13–15]

Thus, if the incorporation of some metal oxides indicated a notable improvement in the catalytic activity (permitting it to operate at lower reaction temperatures),[20,21] the incorporation of metals, especially Pt and working in the presence of H_2, has prolonged the life of the catalysts.[22,23] However, new catalyst formulations have recently increased the resistance of these catalysts to such poisons as water or sulfur during the isomerization of n-C5 and n-C6 paraffins.[24–26] Nevertheless, the use of other anions, by supporting WO_3 or MoO_3^{2-} [27] or heteropolyacids,[28] which have higher thermal stability, can also be interesting alternative routes to develop new catalytic systems.

On the one hand, high-throughput techniques can be used to achieve more specific catalysts,[25] while the use of conditions favoring a reduction in coke formation during the reaction (i.e., the use supercritical conditions)[29] could also be of crucial importance in the rapid incorporation of these catalysts into industrial processes.

In addition to this, solid acid catalysts can also be used in the hydroisomerization cracking of heavy paraffins,[30] or as co-catalysts in Fischer–Tropsch processes.[31] In the first case, it could also be possible to transform inexpensive refinery cuts with a low octane number (heavy paraffins, n-C_{8-20}) to fuel-grade gasoline (C_4–C_7) using bifunctional metal/acid catalysts. In the last case, by combining zeolites with platinum-promoted tungstate modified zirconia, hybrid catalysts provide a promising way to obtain clean synthetic liquid fuels from coal or natural gas.

TABLE 13.1 Alkylation of Isobutane with 2-Butenes on HF or H_2SO_4

Catalyst	HF	H_2SO_4	
Reaction Temperature, K	293	273	
Yield, in wt %			
Product:			
Isopentane	0.3	2.2	
Dimethylbutanes	—	—	
Methylpentanes	0.9	2.6	
Branched-C_7	2.2	2.7	
			2,2,4-
Trimethylpentanes	85.6	80. 0	2,3,4-
			2,3,3-
Dimethylhexanes	6.9	8.1	
Methylheptanes	—	—	
C_9^+-hydrocarbons	4.1	4.4	

Source: After Ref. 32.

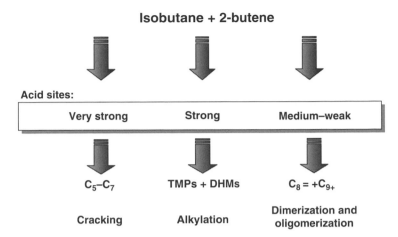

Figure 13.5 The influence of acid strength on the nature of the reaction products during the isobutane/2-butene alkylation.

TABLE 13.2 Isobutane Alkylation Catalyzed by Beta Zeolite and Sulfated Zirconia at Different Reaction Temperatures

Catalyst	SO_4^{2-}/ZrO_2		H-Beta	
Reaction temperature	293 K	273 K	293 K	273 K
2-Butene conversion, %	96.3	92.4	62.0	65.0
C_8-Selectivity	45.3	27.8	69.4	53.6
TMP/C_8	67.2	72.6	16.4	91.2
DMH/C_8	27.7	21.3	23.0	3.8
Alkenes/C_8	5.1	3.4	60.6	5.0
2,2,4-TMP/TMP	54.7	51.8	8.7	49.5

13.2 ISOBUTANE ALKYLATION

A clear example of the possible use of acid and/or superacid solids as catalysts is the alkylation of isobutane with butenes. Isobutane alkylation with low-molecular-weight olefins is one of the most important refining process for the production of high-octane number (RON and MON), low red vapor pressure (RVP) gasoline.[32] Currently, the reaction is carried out using H_2SO_4 or HF (Table 13.1), although several catalytic systems have been studied in the last few years.[32–34]

Adequate acid characteristics of solids are needed since the selectivity to trimethylpentanes (TMP), and especially the 2,2,4-isomer, strongly depends on the nature and strength of the acid sites of the catalyst. Thus, excessively strong acidity will drive the reaction toward cracking reactions, while an acidity that is too low will mostly produce oligomerization of the olefins (Figure 13.5).[9,34]

**TABLE 13.3   Isobutane/2-Butene Alkylation over Acidic Salts of
12-Tungstophosphoric Acids**

Catalyst	$H_3PW_{12}O_{40}$	$(NH_4)_2HP\,W_{12}O_{40}$	$Cs_{2.5}H_{0.5}P\,W_{12}O_{40}$	$Cs_{2.5}H_{0.5}P\,W_{12}O_{40}$
2-Butene conversion, %	56.9	45.3	73.1	85.7
Yield of TMP, %	35.4	34.4	38.5	41.1
TMP/DMH	4.5	3.7	2.8	3.4

Source: After Ref. 36.

Moreover, the efficiency of these catalysts could be modified by tailoring the nature of the metal oxide support and/or reaction conditions (especially the reaction temperature). In this way, interesting conclusions can be obtained when comparing the isobutane/2-butene alkylation catalyzed on two of the most studied catalysts, that is, beta zeolite and sulfated zirconia, when operating at different reaction temperatures. (Table 13.2).[19]

The catalyst with lower acid strength (beta zeolite) presents the higher selectivity to (TMP) at high reaction temperatures, while the sulfated zirconia presents an opposite trend: the lower the reaction temperature, the higher the selectivity to **TMP** is. In the case of sulfated zirconia catalysts, cracking rather than alkylation is favored at high reaction temperatures, while oligomerization rather than alkylation is favored on the beta zeolite at low reaction temperature.

Contrarily, the nature of the metal oxide in sulphate-based catalysts or the chemical composition of heteropolyoxometaltes also influences their catalytic

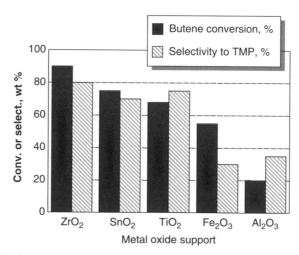

Figure 13.6   Variation of both the conversion of 2-butene and the selectivity to trimethylpentanes (TMP), with the nature of the metal oxide support obtained during the isobutene/2-butene alkylation at 32°C over sulfated supported on different metal oxides. (After Ref. 35.)

TABLE 13.4   Isobutane/2-Butene Alkylation over $H_3PW_{12}O_{40}/$ SO_4^{2-}/ZrO_2 Catalyst

Reaction Cycle	2-Butene Conversion, %	Selectivity to Saturated C_8, %	TMP Selectivity Isoctanes Fraction,[a]%
Fresh catalyst	93	95	76
First	92	55	69
Second	95	40	62
Third	91	91	70
Eleventh	83	78	84

Source: After Ref. 40.
[a] TMP $\times$ $100/C_8$.

behavior. Thus, during the isobutane/butenes alkylation over sulfated metal oxides at 309 K, both the conversion of 2-butene and the selectivity to **TMP** decrease with the lower acid character of the metal oxide support: $SO_4^{2-}/ZrO_2 >$ $SO_4^{2-}/TiO_2 > SO_4^{2-}/SnO_2 > SO_4^{2-}/Fe_2O_3 > SO_4^{2-}/Al_2O_3$ (Figure 13.6).[35] However, it has been observed that the last catalytic trend can be modified by changing the reaction temperature.[35]

In the case of heteropolyoxometalates, the compositions can also be tailored in order to achieve good catalytic properties. Table 13.3 shows the variation of the catalytic performance of acidic salts of 12-tungstophosphoric acids during the isobutene/2-butene alkylation.[36]

It can be seen that the catalytic activity strongly depends on the number and type of the incorporated countercation, which determines the number and strength of acid sites. In addition to this, the existence of mesoporosity (which also depends on the countercation) is also a key factor in the catalytic behavior of these catalysts.[37] In this way, SiO_{2-} or MCM-41-supported heteropolyacids also have been studied in order to increase catalytic activity, apparently without modifying the acid strength.[38]

Figure 13.7 shows an effective confrontation between the catalytic results observed on nafion/SiO_2, sulfated zirconia, and 12-tungstophosphoric acid supported on MCM-41 (HPW/MCM) at a reaction temperature of 50°C.[9] Under these conditions, the HPW/MCM catalyst presents the better yield of TMP. However, similar productivities to that could be obtained on nafion/SiO_2,[9] or sulfated zirconia,[19] at higher or lower temperatures, respectively, than those optimized for the HPW/MCM-41 catalyst.[38]

Moreover, the catalyst deactivation must also be considered in order to use these solid materials in industrial processes. Figure 13.8 shows the variation of catalytic activity (2-butene conversion) with the time on stream obtained under the same reaction conditions on different solid-acid catalysts. It can be seen how all the solid-acids catalysts studied generally suffer a relatively rapid catalyst deactivation, although both beta zeolite and nafion-silica presented the lower catalyst decays. Since the regeneration of beta zeolite is more easy than of nafion, beta zeolite was considered to be an interesting alternative.[39]

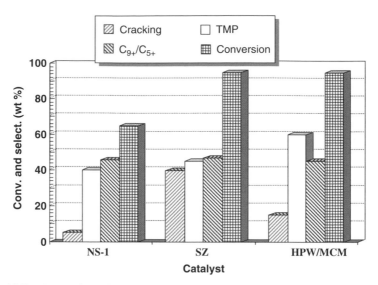

Figure 13.7 Conversion of 2-butene and the selectivities to cracking products, TMP, and C_{9+} hydrocarbons during the isobutane alkylation at 50°C on nafion/SiO_2 (NS-1), sulfated zirconia (SZ), and MCM-41-supported 12-tungstophosphoric acid (HPW/MCM). Experimental conditions: T = 32°C; TOS = 1 min; olefin WHSV = 1 h^{-1}: isobutane/2-butene molar ratio of 15.

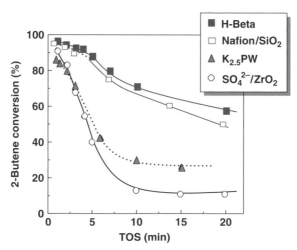

Figure 13.8 Catalyst decay during the isobutane alkylation on nafion/SiO_2, sulfated zirconia, beta-zeolite, and MCM-41-supported 12-tungstophosphoric acid.

The use of a polyfunctional catalyst could enhance the life of the catalyst. A clear example is the use of $H_3PW_{12}O_{40}-SO_4^{2-}/ZrO_2$ mixtures for isobutane/butenes alkylation (Table 13.4).[40] However, modifications of the type of reactor could also favor extended catalyst longevity.[41,42] During the last few years, other alternatives have been proposed that favor a better catalyst regeneration and/or lower catalyst deactivation: the use of supercritical isobutene regeneration[43] or dense-CO_2 enhanced the reaction media.[44]

Recently, Exelus Inc. has developed an innovative isobutene alkylation technology (ExxSact) as an economically viable alternative to the hydrofluoric acid or sulphuric acid processes.[45] In that case, both unique solid-acid catalyst (in which the strength and distribution of the acid sites reduce olefin oligomerization and paraffin cracking) and a novel fixed-bed reactor (using an innovative fluid dynamics and an unusual reactor residence time distribution) promises significantly improved yields and selectivity minimizing waste by-products and disposal problems associated with spent catalyst and regeneration of large quantities of liquid acids.

13.3 CONCLUSIONS

In conclusion, more efficient and clean solid (acid and superacid) catalysts will be used in the coming years to reduce not only the emission of environmentally harmful products but also the use of noxious catalysts. The optimal catalytic systems will be determined from the nature of acid strength of its active sites, the nature of the reaction, and the reaction conditions.

A first step to reduce the negative impact of liquid acids is to use them impregnated on a metal oxide support, although this is not the only alternative. Metal oxides, especially zeolites, could partially substitute liquid acids. The shape selectivity of zeolites can also help to develop new catalytic processes. However, acid solids or multifunctional acid-solid catalysts with an acid strength higher than or similar to zeolites, respectively, are required for reactions that need large amounts of acid. In this way, the study of sulfated metal oxides and/or heteropoly acids (in which both acid and red-ox mechanisms operate) make if possible to propose a suitable alternative to noxious acids, especially in high-acid-demanding reactions. However, it appears quite clear that the presence of sulfate will always be a limitation for their practical use. So, tungstate or heteropolyoxometalates rather than sulfate will make interesting alternatives. Moreover, nafion-resin–silica composites, with an effective surface area larger than pure nafion, also can be considered to be an effective solid-acid catalyst, especially in fine chemistry application.

In any case, the use of these catalytic systems will depend on their stability. In fact, the still unresolved problem with the application of these catalysts to industrial processes is whether to prolong the life of the catalyst or to continuously regenerate the catalyst in an efficient way.

In addition, a new physicochemical method to characterize the acid strength of acid and superacid solids also will be required in order to evaluate and compare acid strengths with those obtained with liquid acids.

ACKNOWLEDGMENTS

Financial support from CICYT PPQ2203-03946 project is acknowledged.

REFERENCES

1. Olah, G. A.; Prakhash, G. K. S.; Sommer, J. *Superacis*, Chap. 5, Wiley, New York, 1985.
2. (a) Tanabe, K.; Hölderich, W. F. *Appl. Catal.*, 1999, **181**, 399; (b) Yang, J.; Zheng, A.; Zhang, M. et al. *J. Phys. Chem. B*, 2005, **109**, 13124.
3. Corma, A. *Chem. Rev.*, 1995, **95**, 559.
4. Acid-base catalysis, *Special Issue of Catal. Today*, 1997, **38**, 255.
5. Corma, A. *Curr. Opin. Solid State Mater. Sci.*, 1997, **2**, 63.
6. Arata, K.; Matsubashi, H.; Hino, M.; Nakamura, H. *Catal. Today*, 2003, **81**, 81.
7. McClure, J. D.; Branderberger, S. G. U.S. Patent 4,038,213 (1977).
8. Harmer, M. A., Farnett, W. E., Sun, Q. *J. Am. Chem. Soc.*, 1996, **118**, 7708.
9. Botella, P.; Corma, A.; López Nieto, J. M. *J. Catal.*, 1999, **61**, 371.
10. Mizuno, N.; Misono, M. *Chem. Rev.*, 1998, **98**, 199.
11. Misono, M. *Chem. Commun.*, 2001, 1141.
12. Okuhara T.; Mizuno, N.; Misono, M. *Adv. Catal.*, 1996, **41**, 113.
13. (a) Hino, M.; Arata, K. *Chem. Lett.*, 1979, 477; (b) *J. Chem. Soc. Chem. Commun.*, 1980, 851.
14. Arata, K. *Adv. Catal.*, 1990, **37**, 165.
15. Song, X.; Sayari, A. *Catal. Rev.-Sci. Eng.*, 1996, **38**, 329.
16. Rossini, S. *Catal. Today*, 2003, **77**, 467.
17. Marcilly, C. *J. Catal.*, 2003, **216**, 47.
18. Cleveland, M. J.; Gosling, C. D. in *Proc. NPRA Annual Meeting*, 1999, p. 29.
19. Corma, A.; Juan Rajadell, M. I.; López Nieto, J. M. et al. *Appl. Catal. A: General*, 1994, **111**, 175.
20. Hollstein, E. J.; Wei, J. T.; Hsu, C.-J. U.S. Patent 4,198,041 (1990).
21. (a) Cheung, T.; D'Itri, J. L.; Gates, B. C. *J. Catal.*, 1995, **151**, 464; (b) Wan, K. T.; Khouw, Ch. B.; Davis, M. E. *J. Catal.*, 1996, **158**, 311; (c) Srinivasan, R., Keogh, R. A., Davis, B. H. *Appl. Catal. A: General*, 1995, **130**, 135.
22. Hino, M.; Arata, K. *Catal. Lett.*, 1995, **30**, 25.
23. Keogh, R. A.; Srinivasan, R.; Davis, B. H. *J. Catal.*, 1995, **151**, 292.
24. Corma, A.; Chica, A.; López Nieto, J. M. et al. WO Patent 02/074437 (2002).
25. Serra, J. M.; Chica, A.; Corma, A. *Appl. Catal. A: General*, 2003, 239, 35.
26. (a) Kimura, T. *Catal. Today*, 2003, **81**, 57; (b) Watanabe, K.; Kawakami, T.; Baba, K.; et al. *Catal Surv. Asia*, 2005, **9**, 17.
27. Yori, J. C.; Vera, C. R.; Parera, J. M. *Appl. Catal. A: General*, 1997, **163**, 165.
28. (a) Na, K.; Okuhara, T.; Misono, M. *J. Chem. Soc. Chem. Commún.*, 1993 1422; (b) Miyaji, A.; Hechicen, T.; Nagata, K. et al. *Appl. Catal. A: General*, 2003, **201**, 145.

29. Funamoto, T.; Nakagawa, T.; Segawa, K. *Appl. Catal. A: General*, 2005, **286**, 79.

30. Yori, J. C., Grau, J. M.; Benitez, V. M.; Sepulveda, J. *Appl. Catal A: General*, 2005, **286**, 71.

31. (a) Song, X.; Safari, A. *Chemtech.*, 1995 27; (b) Zhou, Z.; Zhang, Y.; Tierney, J. W.; Wender, I. *Fuel Proc. Technol.*, 2003, **83**, 67.

32. Corma, A.; Martínez, A. *Catal, Rev.-Sci. Eng.*, 1993, **35**, 483.

33. Corma, A.; López Nieto, J. M. in *Green Chemistry: Challenging Perspective*, Tundo, P., Anastas, P. (Eds.), Chap. 9, Oxford University Press, Oxford (2002).

34. (a) Albright, L. F. *Oil & Gas J.*, 1990, **88**, 79; (b) Weitkamp, J.; Traa, Y. *Catal. Today*, 1999, **49**, 193; (c) Hommeltoft, S. I. *Appl. Catal. A: General*, 2001, **221**, 421; (d) Feller, A.; Lercher, J. A. *Adv. Catal.*, 2004, **48**, 229.

35. Corma, A.; Martinez, A.; Martinez, C. *Appl. Catal. A: General*, 1996, **144**, 249.

36. Corma, A.; Martinez, A.; Martinez, C. *J. Catal.*, 1996, **164**, 422.

37. Essayen, K.; Kieger, S.; Coudurier, G.; Vedrine, J. C. *Stud. Surf. Sci. Catal.*, 1996, **101**, 591.

38. Blasco, T.; Corma, A.; Martinez, A.; Martinez-Escolano, P. *J. Catal.*, 1998, **177**, 306.

39. (a) Corma, A.; Martinez, A.; Arroyo, P. A. et al. *Appl. Catal. A: General*, 1996, **142**, 139; (b) Platon, A.; Thomson, W. J. *J. Catal.*, 2005, **282**, 93.

40. De Angelis, A.; Ingallina, P.; Berti, D.; Montanari, L. *Catal. Lett.*, 1999, **61**, 45.

41. Rao, P.; Vatcha, S. R. *Prep. ACS Symp., Div. Petrol. Chem.*, 1996, **41**, 685.

42. Barger, P. T.; Frey, S. J.; Gosling, C. D.; Sheckler, J. C. *ACS Symp., Div. Petrol. Chem.*, 1999, **44**, 134.

43. (a) Thompson, D. N.; Ginosar, D. M.; Burch, K. C.; Zalewski, D. J. *Ind. Eng. Chem. Res.*, 2005, **44**, 4534; (b) Thompson, D. N.; Ginosar, D. M.; Burch, K. C. *Appl. Catal. A*, 2005, **279**, 109; (c) Salinas, A. L. M.; Kong, D.; Taarit, Y. B.; Essayem, N. *Ind. Eng. Chem. Res.*, 2004, **43**, 6355.

44. (a) Sarsani, V. R.; Wang, Y.; Subramaniam, B. *Ind. Eng. Chem. Res.*, 2005, **44**, 6491; (b) Lyon, C.; Sarsani, V. R. *Ind. Eng. Chem. Res.*, 2004, **43**, 4809.

45. Mukherjee, M.; Dundaresan, S. *World Refining*, 2005, **15**, 22.

14

THE OXIDATION OF ISOBUTANE TO METHACRYLIC ACID: AN ALTERNATIVE TECHNOLOGY FOR MMA PRODUCTION

Nicola Ballarini, Fabrizio Cavani, Hélène Degrand, Eric Etienne, Anne Pigamo, and Ferruccio Trifirò

Dipartimento di Chimica Industriale e dei Materiali, Università di Bologna, Bologna, Italy

J. L. Dubois

ARKEMA, Pierre Bénite, France

14.1 THE TECHNOLOGIES FOR METHYLMETHACRYLATE PRODUCTION

Polymethylmethacrylate production currently amounts to more than 2.8 million tons per year on a worldwide scale, and the average yearly demand has been steadily increasing in recent years by more than 0.2 million tons. In contrast to the majority of bulk chemicals, for which the number of competitive processes is limited to one or two technologies, in the case of methylmethacrylate (MMA), the monomer for polymethylmethacrylate production, there are several different technologies that are currently successfully applied, and others have been claimed or are under investigation. The main characteristics of these technologies are reported in Table 14.1.[1]

The traditional acetone–cyanohydrin (ACH) process is the most widely used in Europe and North America, while other processes are more used in Asia. In the

Methods and Reagents for Green Chemistry: An Introduction, Edited by Pietro Tundo, Alvise Perosa, and Fulvio Zecchini

TABLE 14.1 A Summary of Processes for the Synthesis of Methylmethacrylate

Process	Raw Materials (without esterification)	Features	Producer
Commercial Processes			
ACH (acetone cyanohydrin)	Acetone, HCN, H_2SO_4	Toxic reactant Coproduct disposal	Several companies
New ACH	Acetone, methylformate	Three catalytic steps No coproducts, no supply of HCN	Mitsubishi Gas
Isobutene oxidation	Isobutene or t-butyl alcohol, air	Two catalytic oxidations	Several in Japan
Isobutene oxid with esterification	t-Butyl alcohol, air	Second oxidative step integrated with esterification	Asahi Kasei
Isobutene ammoxidation	Isobutene, air, ammonia, H_2SO_4	Catalytic ammoxidation to methacrylonitrile, and hydrolysis Coproduct disposal	Asahi Kasei
C_2 route	Ethylene, CO, H_2, H_2CO	Carbonylation of ethylene to propionaldehyde, and catalytic condensation with formaldehyde	BASF
Processes Under Development or Investigation			
Alpha process	C_2H_4, CO, CH_3OH, H_2CO	Catalytic carbonylation of ethylene to methylpropionate, and catalytic condensation with formaldehyde	Lucite (to become commercial in 2008)
Propyne process	C_3H_4, CO, CH_3OH	Catalytic carbonylation of propyne One-step reaction, high yield The limit is the supply of propyne	Shell
Isobutane process	i-C_4H_{10}, air	One-step catalytic oxidation	Several companies

ACH process, acetone and hydrogen cyanide react to yield acetone cyanohydrin; the latter is then reacted with excess concentrated sulphuric acid to form methacrylamide sulfate. In a later stage, methacrylamide is treated with excess aqueous methanol; the amide is hydrolyzed and esterified, with formation of a mixture of methylmethacrylate and methacrylic acid. The ACH process offers economical advantages, especially in Europe, where there are large plants in use, most of which have been in operation for decades, but the process also suffers from drawbacks that have been driving the need for the development of alternative technologies. Specifically, the process makes use of HCN, a very toxic reactant. Moreover, difficulties in its acquisition can be met; in fact, HCN is a by-product of propylene ammoxidation, and the integration of acrylonitrile and MMA products requires the two processes to be in balance. Alternatively, HCN can be produced on purpose, but this is feasible only for large production capacities. The second major drawback of the process is the disposal of ammonium bisulphate, the coproduct of the process. Additional costs are necessary for its recovery or pyrolysis.

The ACH process has recently been improved, as stated by Mitsubishi Gas.[2] Acetone–cyanohydrin is first hydrolized to 2-hydroxyisobutylamide with an MnO_2 catalyst; the amide is then reacted with methylformiate to produce the methyl ester of 2-hydroxyisobutyric acid, with coproduction of formamide (this reaction is catalyzed by Na methoxide). The ester is finally dehydrated with an Na-Y zeolite to methylmethacrylate. Formamide is converted to cyanhydric acid, which is used to produce acetone–cyanohydrin by reaction with acetone. The process is very elegant, since it avoids the coproduction of ammonium bisulphate, and there is no net income of HCN. Problems may derive from the many synthetic steps involved, and from the high energy consumption.

Other technologies, already commercially applied or under development, are summarized in Table 14.1. Among the latter, the method that uses isobutane as the raw material, directly transformed to methacrylic acid by a single oxidative step, is potentially advantageous in terms of energy and economics. The direct synthesis of methacrylic acid by the oxidation of isobutane looks particularly interesting because of (1) the low cost of the raw material, (2) the simplicity of the one-step process, (3) the very low environmental impact, and (4) the absence of inorganic coproducts. Several patents claiming the use of Keggin-type polyoxometalates (POMs) as heterogeneous catalysts for this reaction started appearing in the 1980s and 1990s.[3–13] Rohm & Haas was the first (1981), to claim the use of P/Mo/(Sb) mixed oxides for isobutane oxidation.[3] Even though no reference is given in the patent to POMs, the catalyst compositions claimed are clearly Keggin-type compounds. After this patent, several others that describe the use of modified Keggin-type POMs as catalysts have been issued to Asahi Kasei, Sumitomo Chem, Mitsubishi Rayon, and others.

Papers published in recent years[14–40] have tried to establish relationships between catalytic performance and the chemical–physical features of the POMs. Specifically, most of the attention has been focused on the possibility of improving the conversion of isobutane and the selectivity to methacrylic acid by developing POMs that contain specific transition metal cations. However, it seems that the further development of this process has met major obstacles in the preparation of a POM, which on the one hand, is active and selective, and on the other hand is structurally stable enough to withstand the reaction conditions necessary for the activation of the alkane, and the very high heat of reaction that develops.

14.2 THE SELECTIVE OXIDATION OF ISOBUTANE CATALYZED BY POMs: MAIN PROCESS FEATURES

A peculiarity of the processes described in the patents[2-13] is that all of them use isobutane-rich conditions, with isobutane-to-dioxygen molar ratios between 2 (for processes that include a relatively low concentration of inert components) and 0.8, and so closer to the stoichiometric value 0.5 (for those processes where a large amount of inert components is present). This is shown in Figure 14.1, which reports in a triangular diagram the feed composition claimed by the various companies, with reference to the flammability area at room temperature. Low isobutane conversions are achieved in all cases, and recirculation of unconverted isobutane becomes a compulsory choice. For this reason, Sumitomo claimed the oxidation of CO to CO_2 (contained in the effluents from the oxidation reactor) in

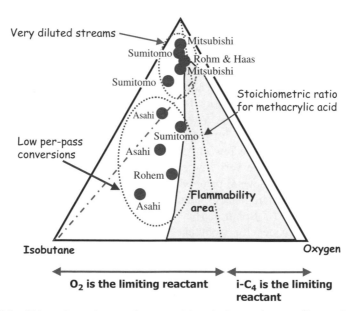

Figure 14.1 Triangular scheme of composition isobutane/oxygen/inert, showing the flammability area for mixtures at room temperature, and the feed composition claimed by several industrial companies.

a separate reactor with a supported Pd catalyst, after the condensation of methacrolein and methacrylic acid.[6,7]

In all cases steam is present as the main ballast. The role of steam is to decrease the concentration of isobutane and oxygen in the recycle loop and thus keep the reactant mixture outside the flammability region. Water can be easily separated from the other components of the effluent stream, and also plays a positive role in the catalytic performance of POMs. It is also possible that the presence of water favors the surface reconstruction of the Keggin structure, which decomposes during the reaction at high temperature, and also promotes desorption of methacrylic acid, saving it from unselective consecutive reactions.

Under the reaction conditions described in the patents, methacrolein is always present in nonnegligible amounts, and therefore a commercial process necessitates an economical method for recycling methacrolein. The patents assigned to Asahi Chemical Industry[12] claim the use of an organic solvent, a mixture of decane, undecane, and dodecane, which can efficiently absorb isobutane and methacrolein from the off-gas, with 99.5% recovery efficiency. Isobutane and methacrolein are then stripped with air and recycled.

Figure 14.2 shows the simplified flow sheet of the process, as reported in patents issued to Sumitomo.[1] CO_2 is maintained in the recycle loop to act as a ballast component; the desired concentration of CO_2 is obtained by combustion of CO, while excess CO_2 is separated. Methacrolein is separated and recycled to the oxidation reactor. An overall recycle yield of 52% to methacrylic acid is reported, with a recycle conversion of 96% and a per-pass isobutane conversion of 10%. The heat of reaction produced, mainly deriving from the combustion reaction, is recovered as steam.

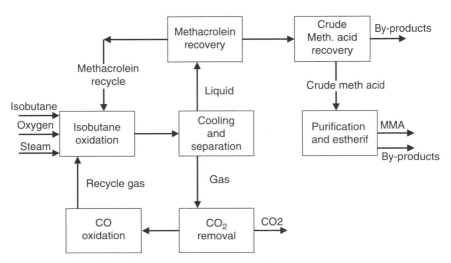

Figure 14.2 Simplified flow sheet for isobutane oxidation to methacrylic acid, as stated by Sumitomo.

Key points that limit the industrialization of the process were recently illustrated by researchers from Sumitomo.[38] Since the selectivity to methacrylic acid plus methacrolein typically decreases with temperature as the conversion increases, this implies that the rate of production of useful products increases only until the higher conversion compensates for the fall of selectivity. As a consequence of this, the maximum productivity value is reached at a specified temperature. For instance, when a selectivity of 45% is reached at 22% isobutane conversion, with a residence time of 5.4 s, a temperature of 370°C, and a feed containing 25% isobutane, 25% oxygen, and 15% steam, a productivity equal to 0.72 mmol/h/g_{cat} is obtained, which is one order of magnitude lower than the one needed to make the process industrially viable.[38] However, the productivity is limited by the oxygen conversion, the maximum concentration of which is dictated by the flammability limits (see Figure 14.1), and by temperature, since the POM decomposes above 380°C.

Possible solutions to overcome this problem are:[38] (1) decrease the residence time; the decrease of conversion is more than compensated by an increase of selectivity (due to the lower extent of methacrylic acid combustion), and in overall the productivity increases; (2) increase the total pressure, while simultaneously increasing both the oxygen and the isobutane partial pressure, as well as the total gas flow (so as to keep a constant contact time in the reactor). A higher pressure also implies smaller reactor volume, and hence lower investment costs. Under these circumstances, productivity as high as 6.4 mmol/h/g_{cat} was reached, which is acceptable for industrial production.[38] The additional heat required for the recirculation of unconverted isobutane and for increased pressure would be equalized by the higher heat generated by the reaction.

As concerns the isobutane-to-oxygen molar feed ratio, a value equal to 1 fulfills the requirements for the best catalytic performance: (1) it keeps the mixture outside the flammability area, (2) it develops isobutane-rich conditions, which imply a reducing environment, beneficial from the selectivity point of view, and (3) it approaches the stoichiometric ratio for methacrylic acid formation, in order to reach a higher per-pass isobutane conversion. The use of oxygen-enriched air creates an addition cost, but is necessary to have isobutane partial pressures that are as high as possible, while keeping a feed ratio equal to 1. The use of air would require isobutane and an oxygen partial pressures no higher than 15%.

An alternative process configuration includes the integration of isobutene oxidation with the process for the olefin production. The latter process may be a traditional dehydrogenation technology, or, alternatively, an oxidehydrogenation. In both cases, the integration between the two steps is different, as schematized in Figure 14.3. In the first case,[41] the exit stream of the dehydrogenation reactor is fed, after addition of air, to the first oxidation reactor, where isobutene is oxidized to methacrolein with a multimetal molybdate catalyst; the same catalyst is not active in hydrogen oxidation. The expensive separation of isobutane from isobutene is eliminated; however, the presence of hydrogen in the oxidation reactor may give rise to enhanced flammability problems. The final oxidation reactor is

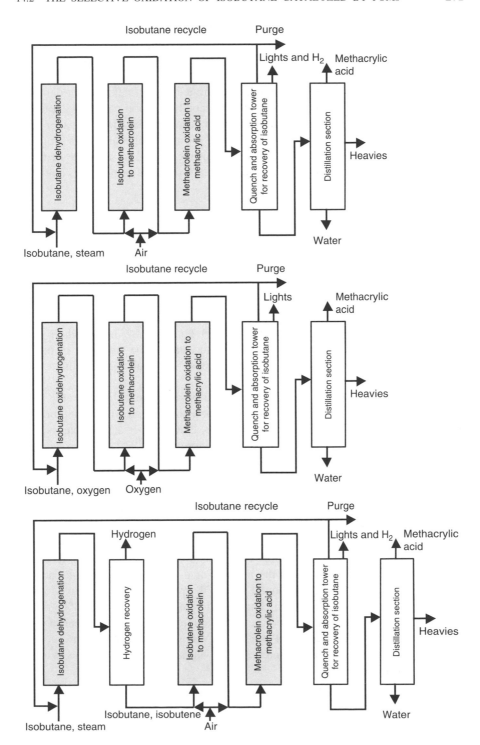

Figure 14.3 Different strategies for integration of isobutane (oxi)dehydrogenation to iso-butene and isobutene oxidation to methacrolein and to methacrylic acid.

aimed at the oxidation of methacrolein to methacrylic acid. The outlet stream is treated to recover methacrylic acid by water adsorption, as well as unconverted isobutane and isobutene from other uncondensable gases; the hydrocarbons are finally recycled.

In the second scheme, the alkane is transformed to the olefin by oxidehydrogenation, and the outlet stream is sent to the second oxidation reactor without any intermediate separation.[42] Isobutane and isobutene are recycled, together with oxygen, nitrogen, and carbon oxides. Finally, the third scheme differs from the first one in that hydrogen is separated from propane/propylene after the dehydrogenation step, and oxygen is preferably used instead of air in the oxidation reactor.[43]

14.3 THE SELECTIVE OXIDATION OF ISOBUTANE CATALYZED BY POMs: MAIN CATALYST FEATURES

There are a few features relative to POMs that are necessary for obtaining the best performance. In all cases, Vanadium is present in the structure of the P/Mo Keggin anion, while the cations include different components, that is, protons, divalent transition metal ions (preferably either Fe^{3+} or Cu^{2+}), and alkali metal ions (preferably Cs^+). The role of Cu ions is to catalyze the reduction of molybdenum, thus increasing the activity of the catalyst;[26,27] it also affects the surface acidity.

According to the Asahi patents,[4] in order to be active and selective, the POM has to be characterized by a cubic structure, and some degree of reduction (also achieved by in situ treatment with isobutene at 450C). Other authors claimed the importance of having the Keggin anion in reduced form.[20-22,32,33] It is possible that a more reduced catalyst leads to better selectivity of the product of partial oxidation, and is thus less active in full combustion. This might also explain why isobutane-rich conditions are claimed in most cases; since conditions with a high concentration of isobutane are more reducing than isobutane-leaner ones. A partially reduced catalyst can be achieved by preparing compounds that have organic cations, which during thermal treatment are oxidized at the expense of Mo^{6+}.[20-22] Another possibility is to prepare compounds that have cations that can exchange electrons with Mo^{6+} in the anion. For instance, it has been found that the presence of Sb^{3+} in the compound makes the reduction of Mo^{6+} to Mo^{5+} in the Keggin anion possible even under oxidizing conditions at 350–400°C (i.e., in the presence of air or reaction mixtures under hydrocarbon-lean conditions).[32,33,35,44] The reduced POM that develops is stabilized toward reoxidation, thus making it possible to maintain the partially reduced state even under oxidizing conditions.

Some of the results obtained with the various compositions for POM-based catalysts, found in the patent and scientific literature, are summarized in Table 14.2. An outstanding result is reported in Ref.40, for which a conversion higher than 20% is reported, with excellent selectivity to methacrylic acid, even in the

TABLE 14.2 Summary of Results Reported in the Scientific and Patent Literature for the Oxidation of Isobutane to Methacrolein and Methacrylic Acid Catalyzed by Keggin-Type POMs

Catalyst	T,°C	τ, s	$i\text{-}C_4/O_2/H_2O/N_2$ Molar Ratios	$i\text{-}C_4H_{10}$ Conv.%	Selectivity % MAC + MAA	Reference
$H_xPMo_{12}SbO_y$	340	6.1	10/13/30/47	10	20 + 50	3
$H_xP_{1.1}Mo_{12}V_{1.1}Cu_{0.1}Cs_{1.1}O_y$	320	3.6	30/15/20/35	10.3	16.3 + 55.7	4
$H_xP_{1.5}Mo_{12}V_1Cu_{0.2}Nd_{0.5}Cs_1O_y$	320	3.6	30/15/20/35	12.8	15.9 + 53.8	11
$H_xP_{1.5}Mo_{12}VO_{0.04}Cu_{0.2}Ba_{0.2}K_{0.5}Cs_{0.5}O_y$	320	2.4	10/16.8/10/63.2	16.3	10.0 + 50.1	9
$H_4PMo_{11}VO_{40}/Ta_2O_5$	350	2	4/8/0/88	28.5	41 + 13.3	12
$H_xP_{1.5}Mo_{12}V_{0.5}As_{0.4}Cs_{1.8}Cu_{0.3}O_y$	320	3.6	26/13/12/49	11.2	11.5 + 53.6	7
$H_xP_{1.5}Mo_{12}V_{0.5}As_{0.4}Cs_{1.4}Cu_{0.3}O_y$	330	5.4	6.5/15.5/15/63	25	42.6 + 2.5	13
$H_{3.6}Cu_{2.0}PMo_{11}VO_{40}/SiO_2^a$	348	^{b}	15.1/29.6/19.7/35.5	13.0	55.6 + 11.5	5
$H_{1.34}Cs_{2.5}Ni_{0.08}PMo_{11}VO_{40}$	340	2	17/33/0/50	31	29 + 8	14
$(Pyr)_3PMo_{12}O_{40}$	300	^{c}	2.2/13.7/33.5/50.6	22.2	51.1 + tr	21
$H_{2.4}Cs_{1.6}P_{1.7}Mo_{11}V_{1.1}O_{40}$	349	3.6	26/12/12/50	10.6	37.6 + 7.9	45
$H_xFe_{0.12}PVAs_{0.3}Mo_{11}O_y$	370	8.6	29/29/0/42	23.9	70 + 4	40
$H_xMo_{1.2}V_{0.5}P_{1.5}As_{0.4}Cu_{0.3}Cs_{1.4}O_y$	366	5.4	25/25/15/35 (P_{tot} 1.5 atm)	21.8	44.9	38

Abbreviations: MAC = methacrolein; MAA = methacrylic acid.

^{a} Forty-three percent of active phase.
^{b} W/F 2.1 g h mL^{-1}.
^{c} W/F 0.1 g min mL^{-1}.

273

absence of steam in the feed. The catalyst described therein is based on P/Mo/V POM, and contains As and Fe as dopants.

14.4 THE MECHANISM OF SELECTIVE ISOBUTANE OXIDATION CATALYZED BY KEGGIN P/Mo POM, AND THE NATURE OF THE ACTIVE SITES

The reaction network for isobutane selective oxidation catalyzed by POMs consists of parallel reactions for the formation of methacrolein, methacrylic acid, carbon monoxide, and carbon dioxide. Consecutive reactions occur on methacrolein, which is transformed to acetic acid, methacrylic acid, and carbon oxides.[30,31] Methacrylic acid undergoes consecutive reactions of combustion to carbon oxides and acetic acid, but only under conditions of high isobutane conversion.[38,39] Isobutene is believed to be an intermediate of isobutane transformation to methacrylic acid, but it can be isolated as a reaction product only for very low alkane conversion.[39]

From a kinetic point of view, the formation of methacrolein and methacrylic acid occurs through two parallel reactions. This has been explained by hypothesizing the mechanism illustrated in Figure 14.4, elaborated starting with reactivity studies on isobutene, methacrolein, and methacrylic acid, and from Fourier

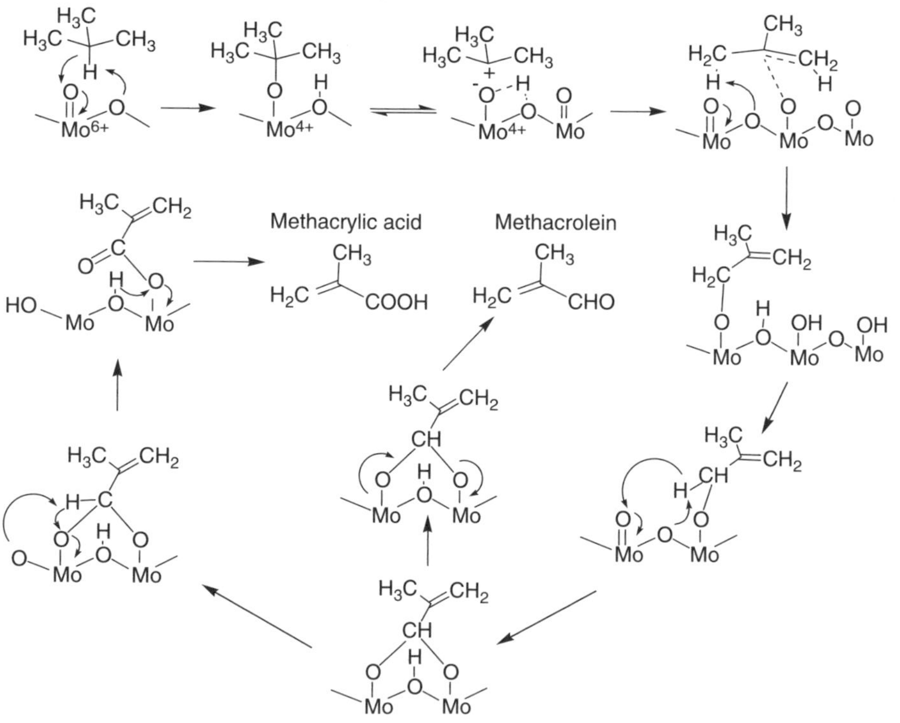

Figure 14.4 Mechanism of the oxidation of isobutane to methacrylic acid catalyzed by Keggin-type POMs.

transform infrared (FT-IR) measurements of adsorbed intermediate species, which develop from the interaction of isobutane and products with the catalyst.[30,31] The mechanism involves the initial separation of a H⁻ species at the tertiary C atom of the alkane; this is the rate-limiting step of the reaction. An adsorbed alkoxy species is thus formed, which is then converted to an allylic alkoxy species. A dioxyalkylidene species develops, where the primary carbon atom is connected to the catalyst surface via two C–O–Mo bridges. This intermediate is either transformed to methacrolein (through dissociation of a C–O bond), or to a carboxylate species (via oxidation on a Mo site), which is the precursor of methacrylic acid formation. Therefore, these two products have a common unsaturated intermediate, and from a kinetic point of view, this corresponds to the presence of two parallel reactions starting from either isobutene or isobutane, depending on whether the unsaturated intermediate desorbs into the gas phase to yield the olefin. The reduced catalyst is then reoxidized by oxygen, according to the classic redox mechanism.[23]

Several aspects of the reaction mechanism still need to be explained. For instance, nonnegligible amounts of C_2 compounds (acetic acid, acrolein) are obtained, the formation of which is not satisfactorily explained by reaction mechanisms proposed in the literature.

All Keggin-type POMs exhibit an initial unsteady catalytic behavior, which can last from a few hours up to 100 hours, depending on the composition of the POM and on the method employed for its preparation.[32,33] The progressive variation of catalytic performance occurring during this "equilibration" period is shown in Figure 14.5, where the conversion of isobutane and the selectiviy to the products are plotted as functions of the reaction time.[33,35] The catalyst was a

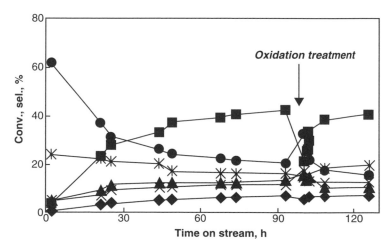

Figure 14.5 Catalytic performance of $(NH_4)_3PMo_{12}O_{40}$ prepared by precipitation at pH < 1 as a function of time-on-stream. T 380°C, τ 3.6 s; feed composition: 26% isobutane, 13% O_2, 12% H_2O, remainder He. *Symbols*: Isobutane conversion (◆), sel. to methacrylic acid (■), to methacrolein (▲), to acetic acid (×), to carbon monoxide (∗), and to carbon dioxide (●).

POM composed of $(NH_4)_3PMo_{12}O_{40}$; data were collected at a reaction temperature of 380°C, with an isobutane-rich feedstock (26 mol % isobutane, 13% oxygen, 12% steam, remainder helium), and a residence time of 3.6 s. At the very beginning of its lifetime, the fresh POM was completely unselective and inactive. After approximately 100 hours reaction time, it was 6.5% converted, with a selectivity to methacrylic acid of 42% and to methacrolein of 13%. The main by-product was carbon dioxide. Therefore, the equilibration time was necessary for the generation of the active and selective sites.

Along with variations in catalytic performance, the following phenomena occurred, which led to a substantial change in the chemical–physical features of the catalyst:

1. The partial structural decomposition of the POM, as evidenced by (i) the formation of small amounts of crystalline MoO_3, and (ii) the change in the cationic composition. In regard to the latter point, the ammonium content in the catalyst decreased, and its place in the cationic position of the POM was occupied by Mo ions. The formation of Mo dimeric species was shown in the downloaded catalysts, which were made of neighboring Mo ions located in the anion and in the counteranion cationic position.[46]

2. The progressive increase in the extent of POM reduction, as evidenced by the ultraviolet visible diffuse reflectance (UV-vis DR) spectra of catalysts downloaded after different reaction times during catalyst equilibration.

These phenomena occurred only when isobutane-rich conditions were used. Indeed, when the reaction was carried out under isobutane-lean conditions (e.g., with 1% isobutane in the feed), the partial structural decomposition and the reduction of the POM did not occur, and similarly the changes in catalytic performance also were not observed, but there was a minor change in selectivity to methacrylic acid.[32–35] This means that the reduction of the POM was due to the isobutane-rich conditions, and that the structural decomposition was due to the larger amount of reaction heat released at the catalyst surface under these conditions. Overheating of the catalyst particles took place, with temperatures that favored the incipient structural decomposition of the POM.

It was proposed that the increase in activity during the equilibration period was due to the generation of new active sites,[34,35] consisting of the Mo species located in the cationic position in the secondary framework of the POM. A similar hypothesis was formulated by other authors for the methacrolein oxidation to methacrylic acid.[47,48] More generally, it is currently believed that for exothermic reactions, and specifically for oxidations, the true working state of the POM, does not correspond to its crystalline form.[49] The presence of steam and the large amount of heat released provoke an incipient surface decomposition, which leads to the expulsion of the Mo species from the anion as a metastable defective compound is developed. The latter then evolves into an "oligomerized" surface form, which prevents the segregation of binary oxides.

The same is known to occur in the case of V-substituted POMs, in which the V^{5+} ions, originally located in the primary anion, are then transferred during the reaction into the cationic position of the POM framework, with generation of lacunary or decomposed Keggin units;[18,50,51] this phenomenon leads to a considerable increase in catalytic activity.

After a steady catalytic behavior was reached, the catalyst was treated in air at 350°C, in order to reoxidize it. Thereafter, the reaction was run again under isobutane-rich conditions (Figure 14.5), in order to understand the role of the POM reduction level on catalytic performance. The reoxidized catalyst exhibited a selectivity to methacrylic acid that was initially around 20%, and approximately 20–30 hours were necessary to recover the original performance of the equilibrated, reduced catalyst. On the contrary, the activity of the catalyst was almost the same as before the oxidizing treatment. This confirms that a partially reduced POM is intrinsically more selective to methacrylic acid than a fully oxidized one, and that one reason for the progressive increase in selectivity to methacrylic acid that occurs during the equilibration period was the increase in the POM reduction level, as a consequence of the operation under isobutane-rich conditions.

14.5 CONCLUSIONS

This chapter describes some aspects of the reactivity of POMs as catalysts for the selective oxidation of isobutane to methacrylic acid. If developed at the industrial level, this reaction could represent an alternative to the current production method via the ACH route. P/Mo Keggin-type POMs are active and selective catalysts for this reaction.

The following considerations are relevant:

1. The catalytic performance depends a great deal on the reaction conditions, and specifically on the isobutane-to-oxygen ratio in the feed. Usually isobutane-rich conditions are claimed to be more selective, and the reason for this is that under these conditions the operative POM is a partially reduced one, and a more reduced POM is intrinsically more selective than a fully oxidized one. High isobutane partial pressures help to improve the selectivity, avoiding further oxidation of methacrylic acid.
2. The presence of transition metal ions as cations for the Keggin polyanion is necessary in order to develop an active and selective catalyst.

Future perspectives in this field, and opportunities for the commercialization of the process of isobutane oxidation, depend on the development of more active and stable POMs-based catalysts.

REFERENCES

1. Nagai, K.; Ui, T. *Sumitomo Kagaku*, 2004, **II**, 1.

2. Abe, T. in *Science and Technology in Catalysis 1998*, Kodansha, Tokyo, 1999, p. 461.

3. Krieger, H.; Kirch, L. S. U.S. Patent 4,260,822 (1981), assigned to Rohm and Haas Company.

4. Yamamatsu, S.; Yamaguchi, T. European Patent 425,666 (1989), assigned to Asahi Chemical, Company; Japanese Patent 02,042,032 (1990).

5. Ernst, B.; Haeberle, T.; Siegert, H.J.; Gruber, W. DE Patent 42 40 085 A1 (1994), assigned to Röhm GmbH Chemische Fabrik; U.S. Patent 5,380,932 (1995).

6. Matsuura, I.; Aoki, Y. Japanese Patent 05,331,085 (1996), assigned to Nippon Catalytic Chemical Industry, CA 121:107989 (1996).

7. Nagai, K.; Nagaoka, Y.; Sato, H.; Ohsu, M. European Patent 418,657 (1990), assigned to Sumitomo Chemical Company.

8. Nagai, K.; Nagaoka, Y.; Ishii, N. European Patent 495,504 A2 (1992), assigned to Sumitomo Chemical Company.

9. Kuroda, T.; Okita, M. Japanese. Patent 04-128,247 (1991), assigned to Mitsubishi Rayon Company.

10. Ushikubo, T. Japanese Patent 06,172,250 (1992), assigned to Mitsubishi Chemical Industry.

11. Yamamatsu, S.; Yamaguchi, T. U.S. Patent 5,191,116 (1993), assigned to Asahi Kasei Kogyo Kabushiki Kaisha.

12. Kawakami, K.; Yamamastu, S.; Yamaguchi, T. Japanese Patent 03,176,438 (1991), assigned to Asahi Chemical Industry Company.

13. Imai, H.; Nakatsuka, M.; Aoshima, A. Japanese Patent 62,132,832 (1987), assigned to Asahi Chemical Industry Company.

14. Mizuno, N.; Tateishi, M.; Iwamoto, M. *Appl. Catal., A: General*, 1994, **118**, L1.

15. Mizuno, N.; Tateishi, M.; Iwamoto, M. *J. Catal.*, 1996, **163**, 87.

16. Mizuno, N.; Suh, D.-J.; Han, W.; Kudo, T. *J. Mol. Catal. A*, 1996, **114**, 309.

17. Mizuno, N.; Han, W.; Kudo, T. *J. Catal.*, 1998, **178**, 391.

18. Inumaru, K.; Ono, A.; Kubo, H.; Misono, M. *J. Chem. Soc., Faraday Trans.*, 1998, **94**, 1765.

19. Min, J. S.; Mizuno, N. *Catal. Today*, 2001, **66**, 47.

20. Ueda, W.; Suzuki, Y.; Lee, W.; Imaoka, S. *Stud. Surf. Sci. Catal.*, 1996, **101**, 1065.

21. Li, W.; Ueda, W. *Catal. Lett.*, 1997, **46**, 261.

22. Li, W.; Oshihara, K.; Ueda, W. *Appl. Catal., A: General*, 1999, **182**, 357.

23. Paul, S.; Le Courtois, V.; Vanhove, D. *Ind. Eng. Chem., Res.*, 1997, **36**, 3391.

24. Sultan, M.; Paul, S.; Vanhove, D. *Stud. Surf. Sci. Catal.*, 1999, **122**, 283.

25. Sultan, M.; Paul, S.; Fournier, M.; Vanhove, D. *Appl. Catal. A*, 2004, **259**, 141.

26. Langpape, M.; Millet, J. M. M.; Ozkan, U. S.; Boudeulle, M. *J. Catal.*, 1999, **181**, 80.

27. Langpape, M.; Millet, J. M. M.; Ozkan, U. S.; Delichère, P. *J. Catal.*, 1999, **182**, 148.

28. Langpape, M.; Millet, J. M. M. *Appl. Catal. A*, 2000, **200**, 89.

29. Cavani, F.; Etienne, E.; Favaro, M.; et al. *Catal. Lett.*, 1995, **32**, 215.

30. Busca, G.; Cavani, F.; Etienne, E.; et al. *J. Molec. Catal.*, 1996, **114**, 343.

31. Cavani, F.; Etienne, E.; Hecquet, G.; et al. in *Catalysis of Organic Reactions*, Malz, R. E. (Ed.) Marcel Dekker, New York, 1996, p. 107.

32. Cavani, F.; Mezzogori, R.; Pigamo, A.; Trifirò, F. C.R. Acad. Sci. Paris, Séerie IIc, *Chimie*, 2000, **3**, 523.

33. Cavani, F.; Mezzogori, R.; Pigamo, A.; Trifirò, F. *Chem. Eng. J.*, 2001, **82**, 33.

34. Cavani, F.; Mezzogori, R.; Pigamo, A.; Trifirò, F. *Stud. Surf. Sci. Catal.*, 2001, **140**, 141.

35. Cavani, F.; Mezzogori, R.; Pigamo, A.; Trifirò, F. *Topics Catal.*, 2003, **23**, 119.

36. Etienne, E.; Cavani, F.; Mezzogori, R. et al. *Appl. Catal. A*, 2003, **256**, 275.

37. Knapp, C.; Ui, T.; Nagai, K.; Mizuno, N. *Catal. Today*, 2001, **71**, 111.

38. Schindler, G.-P.; Knapp, C.; Ui, T.; Nagai, K. *Top. Catal.*, 2003, **22**, 117.

39. Schindler, G.-P.; Ui, T.; Nagai, K.; *Appl. Catal. A*, 2001, **206**, 183.

40. Deng, Q.; Jiang, S.; Cai, T.; et al. *J. Mol. Catal. A*, 2005, **229**, 165.

41. Khoobiar, S.; U.S. Patent 4,535,188 (1985), assigned to Halcon SD.

42. Brockwell, J. L.; Warren, B. K.; Young, M. A.; et al. WO Patent 97-36489 (1997), assigned to Union Carbide Chemicals and Plastics.

43. Hefner, W.; Machhammer, O.; Neumann, H. P.; et al. U.S. Patent 5,705,684 (1998), assigned to BASF.

44. Cavani, F.; Tanguy, A.; Trifirò, F.; Koutyrev, M. *J. Catal.*, 1998, **174**, 231.

45. Jalowiecki-Duhamel, L.; Monnier, A.; Barbaux, Y.; Hecquet, G. *Catal. Today*, 1996, **32**, 237.

46. Albonetti, S.; Cavani, F.; Trifirò, F.; et al. *J. Catal.*, 1994, **146**, 491.

47. Marosi, L.; Cox, G.; Tenten, A.; Hibst, H. *J. Catal.*, 2000, **194**, 140.

48. Mestl, G.; Ilkehans, T.; Spielbauer, D.; et al. *Appl. Catal A*, 2001, **210**, 13.

49. Jentoft, F. C.; Klokishner, S.; Kröhnert, J.; et al. *Appl. Catal. A*, 2003, **256**, 291.

50. Marchal-Roch, C.; Bayer, R.; Moisan, J. F.; et al. *Topics Catal.*, 1996, **3**, 407.

51. Ilkehans, Th.; Herzog, B.; Braun, Th.; Schlögl, R. *J. Catal.*, 1995, **153**, 275.

15

BIOCATALYSIS FOR INDUSTRIAL GREEN CHEMISTRY

ZHI LI, MARTIN HELD, SVEN PANKE, ANDREW SCHMID, RENATA MATHYS, AND BERNARD WITHOLT
Eidgenössische Technische Hochschule Zürich, Switzerland

INTRODUCTION

Biocatalysis can make an important contribution to green chemistry due to several distinctive features, such as mild reaction conditions, nontoxicity, high chemo-, regio-, and stereoselectivity, and high turnover frequency. The potential of biocatalysis for the synthesis of chemicals is evident[1-3] and examples of several industrial processes that are operational at BASF (Ludwigshafen, Germany), DSM (Geleen, the Netherlands), and Lonza (Visp, Switzerland) have recently been described.[4] These industries use enzymes for the production of medium- to high-priced compounds that cannot be produced equally well using chemical approaches.[5]

The key challenge in the development of a bioprocess is the discovery and development of an appropriate biocatalyst. The discovery of enantioselective bio-catalysts for a given transformation can be achieved by many methods, such as the screening of collections of wild-type microorganisms[6] or clonal libraries,[7] in vitro evolution,[8] and site-directed mutagenesis.[9] While the latter requires knowledge of the structure and preferably also of the catalytic mechanism of the enzyme, the first two methods need fast and specific detection systems. In vivo screening of microorganism is often the method of choice for the discovery of cofactor dependent multi-component enzymes. The range of reactions that can be

Methods and Reagents for Green Chemistry: An Introduction, Edited by Pietro Tundo, Alvise Perosa, and Fulvio Zecchini

carried out with microorganisms and the range of microorganisms that have already been isolated or remain to be discovered is enormous. As a result, much energy goes into the selection of new enzymatic activities by using natural isolates and/or their mutants.

15.1 BIOCATALYSTS FUNCTION IN THE PRESENCE OF ORGANIC SOLVENTS

A critical consideration in the development of biocatalytic systems is the form in which the enzyme or enzyme system is going to be used. There are two general approaches. One is to use isolated enzymes. If these are inexpensive, they can be used as disposable biocatalysts, as is the case for glucose isomerase,[10] which is the key biocatalyst in the production of high-fructose corn syrups from starch, or the lipases and proteases that are present in detergents. Alternatively, if enzymes are expensive to produce, they can be immobilized and used repeatedly by recovering the enzyme particles after each use.

The second general approach is to use whole cells that contain the enzyme or enzymes used in the biocatalytic process.[11] The use of whole cells has the added advantage that coenzyme-dependent enzymes can be used because it is possible to regenerate the relevant coenzyme, through metabolism of the whole cells. This, of course, requires that the whole cells are not only physically intact but also metabolically active. Since coenzymes are often involved in building new molecules, industrial biocatalysis typically uses whole-cell systems.

Much of industrial chemistry takes place in organic solvents, or involves apolar compounds. Biocatalysis, in contrast, typically involves aqueous environments. Nevertheless, enzymes and microorganisms do in fact encounter apolar environments in Nature. Every cell is surrounded by at least one cell membrane, and more complex eukaryotic cells contain large amounts of intracellular membrane systems. These membranes consist of lipid bilayers into which many proteins are inserted: present estimates, based on genomic information, are that about one-third of all proteins are membrane proteins, many of which are so-called intrinsic proteins that are intimately threaded through the apolar bilayer. These proteins are essentially dissolved in, and function partly within, an apolar phase.

The notion that enzymes might well function in apolar solvents has been explored in detail and confirmed by Klibanov and his followers during the past two decades.[12,13] Similarly, many microorganisms have been found to grow well in the presence of bulk solvents.[14–16] This has permitted the development of biotransformation systems in which water-insoluble compounds are dissolved in apolar phases, to be converted to products by whole cells present in an aqueous phase, the products generally then dissolving again in the apolar phase.[16–20]

In short, microbial cells can be employed as very effective reactors for the conversion of substrates to products, operating in mixed aqueous–apolar systems, optimized for the best space–time yields attainable at lowest cost.

15.2 CHEMISTRY WITH MONOOXYGENASES

The regio- and stereoselective hydroxylation of specific *non*activated carbon atoms is a very useful reaction. For example, it can be used for the functionalization of alkanes and for the preparation of chiral alcohols that are useful pharmaceutical intermediates. This type of transformation is, however, a significant challenge in classic chemistry. On the other hand, Nature has found a general solution to this challenge via the development of a large number of monooxygenases, some of which, such as the soluble cytochrome P450 monooxygenases (P450 cam, P450 BM-3), soluble methane monooxygenase (sMMO), and membrane-bound alkane hydroxylase (AlkB) of *Pseudomonas putida* GPo1, have been well investigated. Their synthetic applications are limited, however, due to narrow substrate ranges or poor selectivity. The filamentous fungus *Beauveria bassiana* ATCC 7159 is often used for laboratorial synthesis,[21] because it contains one or more unknown oxygenase systems, but these generally show low activities and selectivities. Thus, for specific hydroxylations, it is generally necessary to find appropriate biocatalysts.

To search for an appropriate whole-cell biocatalyst, it is necessary to identify an organism that contains large amounts of the desired enzyme. Equally important, the organism should not contain related pathway enzymes that modify or destroy the product synthesized by the desired enzyme.

In addition, the substrate and product should be transported through the cell membrane, either passively or actively, and necessary cofactors should be regenerated. Finally, the specific organism used should function well in an optimized bioreactor system.[22,23] All of these requirements can be met by using strains that contain the desired enzyme in question.

15.3 SCREENING FOR THE REGIO- AND STEREOSPECIFIC HYDROXYLATION OF PYRROLIDINE

We were interested in the discovery and development of biocatalysts for the regio- and stereoselective hydroxylation of pyrrolidine to 3-hydroxypyrrolidine, since both (*S*)- and (*R*)-3-hydroxypyrrolidines are useful intermediates for the preparation of several pharmaceuticals (Figure 15.1), and they are difficult to prepare by chemical syntheses. To avoid random screening of a large number of microorganisms, we concentrated on microorganisms with selected biodegradation abilities. Microorganisms that degrade *n*-alkanes often contain alkane hydroxylases that catalyze the hydroxylation of alkanes at a terminal position as the first degradation step. These alkane hydroxylases might also catalyze the desired hydroxylation of pyrrolidines. Moreover, the alcohol dehydrogenases in alkane-degrading microorganisms that catalyze the oxidation of terminal alcohols to the corresponding aldehydes during biodegradation may not be able to oxidize the desired product 3-hydroxypyrrolidine, which is a cyclic secondary alcohol. Therefore, we selected 70 *n*-alkane-degrading strains as the source of catalyst for

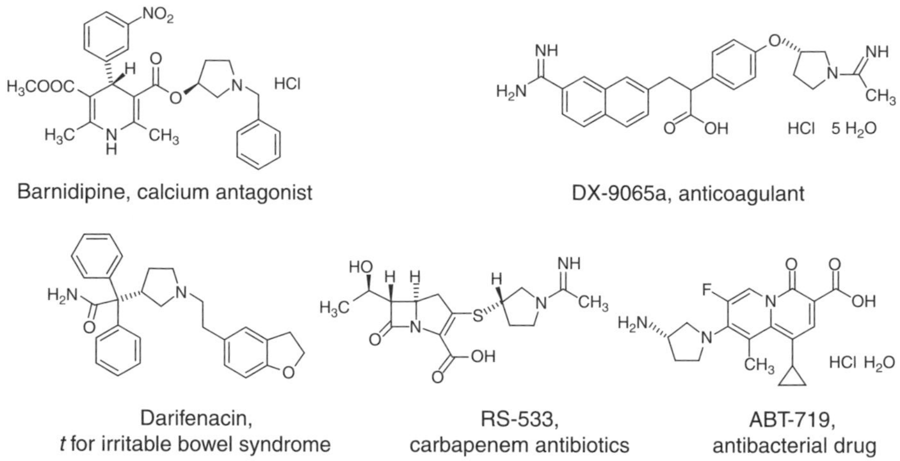

Figure 15.1 (S)- and (R)-3-hydroxypyrrolidines as pharmaceutical intermediates.

screening. By use of a miniaturized system, which allows for parallel inoculation, growth, and bioconversion on a microtiter plate,[24] coupled with a fast and sensitive Liquid Chromatography–Mass Spectroscopy (LC-MS) detection, we found that 25 of the 70 strains catalyze the 3-hydroxylation of N-benzylpyrrolidine.[25] Further investigation with 12 more active strains demonstrated complementary enantioselectivities (Table 15.1). Hydroxylation of N-benzylpyrrolidine with *Pseudomonas oleovorans* GPo1, the prototype alkane hydroxylating strain that contains a well-known membrane-bound alkane hydroxylase (alkB) afforded 62% of (R)-N-benzyl-3-hydroxypyrrolidine in 52% enantiomeric excess (ee), whereas

TABLE 15.1 Enantioselectivity and Activity of the Hydroxylation of N-Benzylpyrrolidine to N-Benzyl-3-hydroxypyrrolidine with Several Alkane-Degrading Strains

Entry	Microorganisms	Product ee (%)	Relative Activity[a]
1	HXN-1100	70 (R)	4
2	HXN-400	65 (R)	0.5
3	*P.putida* P1	62 (R)	1
4	*P. oleovorans* GPo1	52 (R)	1
5	BC20	40 (R)	1
6	HXN-1500	25 (R)	3
7	HXN-500	10 (R)	11
8	HXN-200	53 (S)	6
9	HXN-100	10 (S)	3
10	HXN-1900	<10 (S)	10
11	HXN-1000	<10 (S)	1
12	HXN-600	0	3

[a]Relative activity was based on the activity with *P. oleovorans* GPo1.

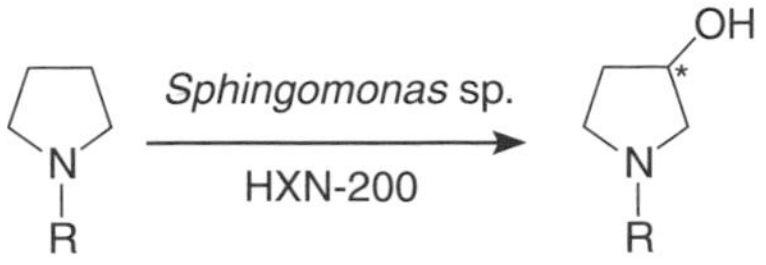

	Activity (U/g cdw)	ee (%)	ee after crystalization (%)
R = CH_2Ph	5.8	53 (S)	
R = COPh	2.2	52 (R)	95 (R)
R = CO_2Ph	14	39 (S)	96 (S)
R = CO_2CH_2Ph	16	75 (R)	98 (R)
R = CO_2t-Bu	24	23 (R)	

Figure 15.2 Improvement of the hydroxylation activity and enantioselectivity with *Sphingomonas* sp. HXN-200 by substrate modification.

hydroxylation with *Sphingomonas* sp. HXN-200 gave 62% of (*S*)-enantiomer in 53% ee with six times higher activity.

It was found that *Sphingomonas* sp. HXN-200 contains a soluble alkane monooxygenase and accepts a broader range of substrates. The enantioselectivity and activity were further improved by introducing a "docking/protecting" group into the pyrrolidine substrates. As shown in Figure 15.2, changing the *N*-substitution from a benzyl to a benzyloxycarbonyl group resulted in a three-fold increase of the activity and an improvement of product ee from 53% (*S*) to 75% (*R*).[26] We found that the ee can be further increased by simple crystallization of the bioproduct from a mixture of *n*-hexane and ethyl acetate. Thus, it is possible to produce (*R*)- and (*S*)-3-hydroxypyrrolidine in 98% and 96% ee, respectively.

We optimized the growth of strain HXN-200 on *n*-octane and produced the cells in a large amount. It was found that the cells can be stored at −80°C for two years without significant loss of activity. The frozen/thawed cells, that are easy to handle for the organic chemist, can be used for routine hydroxylation in an organic chemistry laboratory.

15.4   SYNTHESIS OF HYDROXYPYRROLIDINES, HYDROXYPYRROLIDINONES, AND HYDROXYPIPERIDINES

3-Hydroxypyrrolidines were easily prepared in 62–94% yield and in 1–2 g/L by hydroxylation with the frozen/thawed cells. Even higher productivity was obtained by hydroxylation with growing cells of HXN-200,[26] which is obviously the best choice for an industrial hydroxylation. It was found that *Sphingomonas* sp. HXN-200 can also catalyze the hydroxylation of 2-pyrrolidinone.[27] As shown in Figure 15.3, hydroxylation of *N*-benzylpyrrolidinone gave the corresponding (*S*)-4-hydroxypyrrolidine in >99.9% ee, demonstrating the best enantioselectivity ever reported for biohydroxylations. Changing the "docking/protecting" group to a *tert*-butoxycarbonyl group increased the activity to 11 U/g cell dry weight (cdw). Although the enantioselectivity was slightly lower, simple crystallization

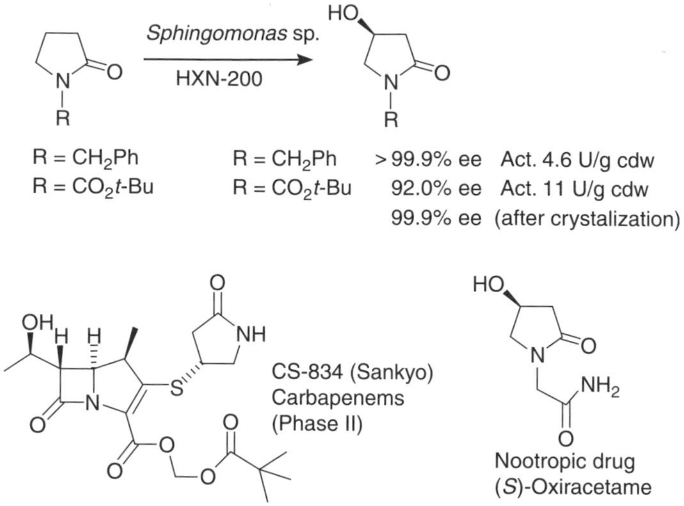
Sphingomonas sp.
HXN-200
HO
R = CH2Ph
R = CO2t-Bu
R = CH2Ph >99.9% ee Act. 4.6 U/g cdw
R = CO2t-Bu 92.0% ee Act. 11 U/g cdw
99.9% ee (after crystalization)
CS-834 (Sankyo)
Carbapenems
(Phase II)
Nootropic drug
(S)-Oxiracetame

Figure 15.3 Biohydroxylation of 2-pyrrolidinones with *Sphingomonas* sp. HXN-200.

from *n*-hexane/ethyl acetate improved the ee from 92% to 99.9%. Here again, (*S*)-4-hydroxypyrrolidines are useful intermediates for the preparation of several pharmaceuticals.

Sphingomonas sp. HXN-200 was also able to accept a six-member ring substrate. Hydroxylation of *N*-benzyl- and *N-tert*-butoxycarbonyl-2-piperidinone gave the corresponding (*R*)-4-hydroxy-piperidin-2-ones in 31% and 68% ee, respectively[28] (Figure 15.4). This provides a simple synthesis of such types of useful synthons.

Further investigation demonstrated that *Sphingomonas* sp. HXN-200 is also the best biocatalyst known thus far for the hydroxylation of *N*-substituted piperidines.[29] Even small cyclic compounds such as azetidines, which are difficult to be hydroxylated by other system, are also good substrates for strain HXN-200.[29] Excellent regioselectivity and high activity were obtained in all hydroxylations shown in Figure 15.5. This provides simple and efficient synthesis of 4-hydroxypiperidines and 3-hydroxyazetidines, which are useful pharmaceutical intermediates.

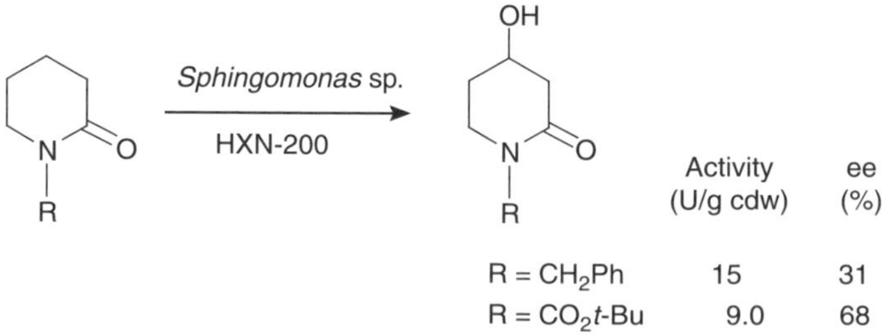
Sphingomonas sp.
HXN-200
OH
Activity (U/g cdw)
ee (%)
R = CH2Ph 15 31
R = CO2t-Bu 9.0 68

Figure 15.4 Biohydroxylation of 2-piperidinones with *Sphingomonas* sp. HXN-200.

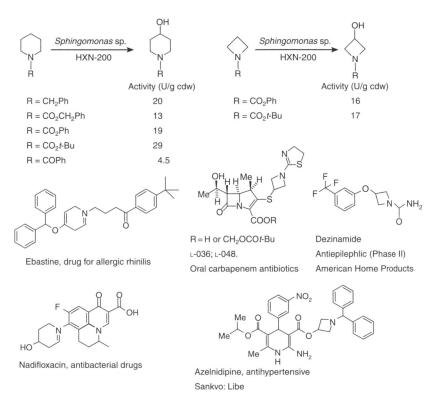

Figure 15.5 Biohydroxylation of piperidines and azetidines with *Sphingomonas* sp. HXN-200.

15.5 BIOCATALYST FORMULATION AND TUNING

To facilitate its application in organic synthesis, we developed a lyophilized cell powder of *Sphingomonas* sp. HXN-200 as a biohydroxylation catalyst.[29] Hydroxylation of *N*-benzyl-piperidine with such catalyst powder showed 85% of the activity of a similar hydroxylation with frozen/thawed cells, shown in Figure 15.6. The fact that rehydrated lyophilized cells are able to carry out such a reduced nicotinamide adenine dinucleotide (NADH)-dependent hydroxylation indicates that these cells are capable of retaining and regenerating NADH at rates equal to or exceeding the rate of hydroxylation. To our knowledge, this is the first example of the use of lyophilized cells for a cofactor-dependent hydroxylation.

For an industrial biotransformation, it is often necessary to further optimize an appropriate biocatalyst. This includes the elimination of the follow-up enzymes in the wild-type strain by mutations and the improvement of other characteristics by additional mutations and the selection of improved strains. Alternatively, the genetic information for a desired enzyme might be introduced in a host that has many of the preceding characteristics, and that has no enzymes that could modify or degrade the desired product.

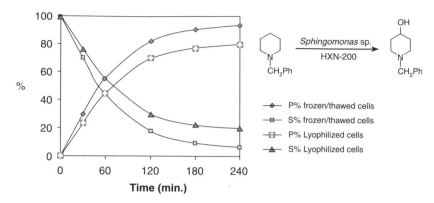

Figure 15.6 Hydroxylation of *N*-benzylpiperidine (5 mM) with lyophilized cell powder and frozen/thawed cells of *Sphingomonas* sp. HXN-200 at cell concentration of 4.0 g cdw/L.

15.6 AN IDEAL BIOCATALYST BASIS: *E. coli*

Perhaps surprisingly, we have found in our work that *E. coli* is an excellent whole-cell biocatalytic host.[23,30–32] It is relatively easy to introduce new desired enzymes into various *E. coli* strains. Generally, for many of the hydroxylation reactions that we have studied, degradative enzymes or downstream pathway enzymes that could modify or eliminate desired hydroxylation products, are not present in *E. coli*. In working with two liquid-phase systems, *E. coli* is more sensitive to apolar solvents than *Pseudomonas* strains. However, we have developed mixed apolar phase systems, based on highly apolar solvents such as hexadecane or substituted phthalates,[19] which are highly compatible with *E. coli*. Thus, it is possible to use very toxic substrates, and produce equally toxic products, which dissolve in the hexadecane or phthalate phase and have very little effect on the host organism present in the aqueous phase.[33]

The beauty of working with only a few host strains is that these can be optimized for growth and growth medium, expression system for a wide range of biocatalysts, behavior in the presence of solvents, regeneration of coenzymes, downstream processing, recovery of cells for further biocatalysis use, and waste treatment. Thus, function can be optimized generally, cost can be minimized generally, and different enzymes can be introduced from a wide range of sources. In addition, each enzyme to be used can be optimized further for top biocatalytic performance by mutagenesis, directed evolution,[34] or gene shuffling.[35]

15.7 SYNTHESIS OF CATECHOLS BY A RECOMBINANT WHOLE CELL BIOCATALYST

An interesting example of the application of recombinant whole-cell biocatalysis is the conversion of 2-hydroxybiphenyl (2-phenylphenol) to 2,3-dihydroxybiphenyl

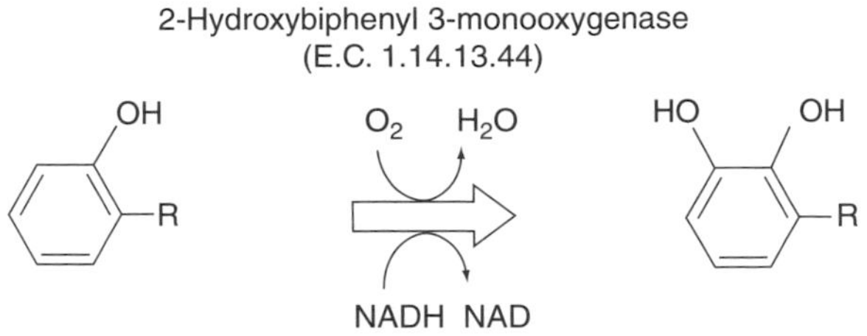

2-Hydroxybiphenyl 3-monooxygenase
(E.C. 1.14.13.44)

Figure 15.7 Production of 3-substituted catechols using a designer biocatalyst.

or 3-phenylcatechol (Figure 15.7). Catechols are important building blocks for the chemical and pharmaceutical industries. However, their chemical synthesis is cumbersome. Especially the synthesis of 3-substituted catechols requires numerous agents with low environmental compatibility (organometallic reagents, HBr) and energy intensive reaction conditions (low temperature). At the same time, catechol and some of its derivatives are central metabolites in the microbial catabolism of aromatic compounds.[36] Besides naturally occurring aromatics such as tyrosine and phenylalanine, nonbiological aromatic solvents and numerous polycyclic aromatic hydrocarbons are also readily degraded or modified by bacteria, with the genus *Pseudomonas* playing an important role in such turnover of aromatics in nature.

Pseudomonas azelaica HBP1 is a prominent example of interesting aromatic compound degradation.[37] The strain readily degrades 2-phenylphenol—a man-made compound that has been widely used as a food-protecting agent and as a germicide. The initial step of 2-phenylphenol transformation by the strain is formation of 3-phenylcatechol by a 2-hydroxybiphenyl 3-monooxygenase (Figure 15.7).[38] This enzyme has a broad substrate range and oxidizes numerous other 2-substituted phenols to corresponding 3-substituted catechols in a highly regioselective cofactor-dependent reaction. However, the strain cannot be used for catechol synthesis, because reaction products are instantly broken down. Hence, the 2-hydroxybiphenyl 3-monooxygenase gene was cloned and expressed in *E. coli* JM101.[39] The resulting biocatalyst *E. coli* JM101 [pHBP461] efficiently overproduces the monooxygenase but does not degrade the products formed, which makes this strain a promising candidate for synthesis of 3-substituted catechols.

A major challenge that had to be met arose from the extreme bactericidal properties of phenols and catechols. Most microorganisms are poisoned at phenol or catechol concentrations in the 0.1–1 g/L range. The biocatalyst *E. coli* JM101 [pHBP461] is no exception, and was inactivated by 200 mg/L of both 2-phenylphenol and 3-phenylcatechol.[40] Furthermore, 3-substituted catechols are of limited stability in aerated aqueous solutions and form multimeric humic-acid-like structures as unwanted side products. 3-Phenylcatechol, for instance, has a half-life time of only 14 h at pH 7.2.

Last but not least, catechols are highly water-soluble (the water solubility of catechol is approximately 1 g per 2.3 mL of water), which makes it difficult to directly extract them in situ from reaction media with organic, water immiscible solvents. Nevertheless, extraction of catechols from aqueous systems with hydrophobic polymers such as the polystyrene-based resin Amberlite XAD-4 is

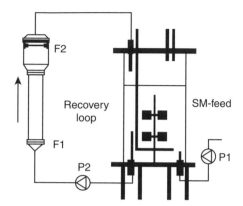

Figure 15.8 The liquid-stream-driven process (LSDP) principle.

straightforward. We have therefore employed XAD-4 to combine biocatalytic synthesis with simultaneous product extraction. The system (Figure 15.8) comprises a continuously stirred tank reactor, a starting material feed pump, a product recovery loop with a (semi-) fluidized bed of XAD-4, and a pump to circulate the entire reaction mixture through the loop.[40] Preliminary studies indicated that XAD-4 had no detrimental effects on *E. coli* JM101 (pHBP461), hence, separation of biomass and reaction liquid prior to catechol extraction was not required. The biocatalytic reaction was carried out at very low concentrations of the toxic substrate and product. This was achieved by feeding the substrate at a rate lower than the potential bioconversion rate in the reactor.

This assured that all substrate fed to the bioreactor was instantly converted to product: no substrate accumulated in the bioreactor. To prevent accumulation of product in the reactor, the reactor contents were circulated through the external fluidized-bed module, which contained Amberlite XAD-4 resin. All product adsorbed to the resin, while cells and medium components passed through the bed and back into the bioreactor, ready to convert more of the substrate newly fed into the bioreactor.

Figure 15.9 shows that this approach worked quite nicely: the substrate was added to a total concentration of 2.5 g/L, but neither substrate nor product accumulated in the bioreactor medium.[30] Without a product recovery loop the product concentration (3-phenylcatechol) did not exceed 0.4 g/L, because of biocatalyst deactivation (results not shown). With the loop, 2-phenylphenol and 3-phenylcatechol concentrations remained below 0.1 g/L. Therefore, cell viability and biocatalytic activity were maintained, as indicated by the constantly low dissolved oxygen tension in the aerated reactor. As a result product yields (based on the 3-phenylcatechol eluted from the product sink) increased by one order of magnitude.[40]

The system was used for the preparative scale synthesis of numerous catechols. Table 15.2 shows that a half dozen other phenols could be converted to the corresponding catechols. In each case, 1 to 2 g of substrate was converted with yields

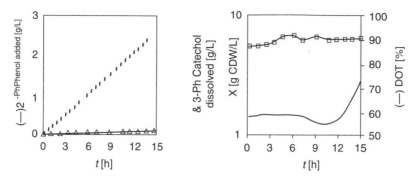

Figure 15.9 Synthesis and in situ recovery of 3-phenylcatechol.

of 80 to 90%, and after desorption and isolation of product, overall yields of 60 to 80% were estimated.[41] The entire process was scaled up to the 300 L scale, resulting in the formation of a total of 1.1 kg 3-phenylcatechol, at a purity of 77%. One crystallization was sufficient to bring 25 g of the material to a purity of >98%; this compound is now for sale by Fluka, in Switzerland.

Though a detailed life-cycle analysis has not yet been made, the reagents that go into these biocatalytic reactions already indicate a certain trend. Besides starting materials the route requires only salts, glucose, glycerol and air, XAD-4 for product extraction, acidified methanol for product elution, and a solvent for product polishing (hexane or methanol/acid mixtures). With the exception of the XAD-4 resin, all reagents are from renewable or almost inexhaustible stocks. With sufficient XAD-4 reuse—exceeding a hundred or so cycles—waste streams are minor and comparatively green since they consist mainly of biomass and water.

This bioconversion illustrates several points. First, *E. coli* is a highly adequate host for the bioconversion of toxic compounds. Second, it is possible to attain adequate fluxes at low concentrations of substrate. Third, it was relatively easy to scale up the system from laboratory (several liters) to pilot scale (several hundred liters).

TABLE 15.2 Production Data for the Bioconversion of Substituted Phenols to the Corresponding Catechols

Compound	Starting Material Added (g)	Yield (Total) (mole%)[a]	Yield (Adsorbed and Eluted) (mole%)[a]
3-Et-catechol[b]	2.0	90	82
3-Pr-catechol	1.2	81	69
3-i-Pr-catechol	2.6	91	76
3-sec-Bu-catechol	1.8	84	78
3-Cl-catechol	1.1	91	71
3-Br-catechol	2.6	84	73

[a]Mole % of total starting-material added.
[b]No standard available.

15.8 SYNTHESIS OF EPOXYSTYRENE WITH VERY HIGH ENANTIOSELECTIVITY

Another useful bioconversion is the epoxidation of styrene to epoxystyrene. Styrene oxide is a valuable building block, because the epoxide function allows versatile synthetic chemistry, and the benzene ring is part of the majority of today's drugs (Figure 15.10). It is used, for example, in the production of the anti-helmintic drug Levamisole.[42] However, to be an attractive building block for drug synthesis, the styrene oxide needs to be enantiopure. There are some chemical asymmetric synthesis routes described, but they usually deliver only moderate to good enantiomeric excesses. Kinetic resolution of styrene oxide employing the Jacobsen catalyst affords enantiopure styrene oxide, but this method has the inherent drawback of being limited to a maximum chemical yield of 50%.[43] In order to circumvent these various limitations, we attempted to design an asymmetric synthesis route from styrene to styrene oxide using biocatalysis. Asymmetric synthesis in principle allows a 100% ee/100% yield process, and the enantioselectivity of enzymes should aid in overcoming the problem just mentioned in achieving satisfactory ee's.

We have used two enzyme systems for this purpose. The first is the xylene monooxygenase system of *P. putida* mt2, which is capable of utilizing xylene and toluene derivatives for growth. In the first enzymatic step, xylene monooxygenase introduces an oxygen atom in a toluene or xylene methyl substituent group. This monooxygenase can also introduce an epoxide in the vinyl double bond of styrene and substituted styrenes.[44] The reaction was carried out using *E. coli* recombinants carrying only the xylene monooxygenase system, encoded by xylMA, resulting in efficient formation of epoxystyrene with an enantiomeric excess of 92%[45] (see Figure 15.10).

Still, a biocatalyst that performs this reaction with an even higher ee was desirable, which led us to investigate styrene degradation in various bacterial soil strains. One of these, *Pseudomonas* sp. strain VLB120 was selected for further

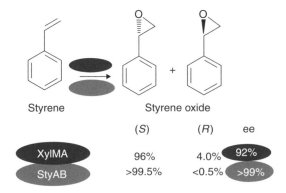

Figure 15.10 Biological oxidation of styrene to (*S*)-styrene oxide.

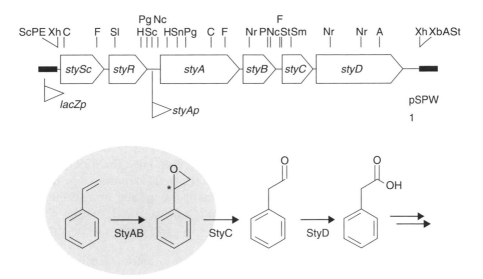

Figure 15.11 Styrene degradation in *Pseudomonas* sp. strain VLB120.

study, because it appeared to degrade styrene via styrene oxide and phenylacetaldehyde (Figure 15.11). After preparation of a genomic library in *E. coli*, clones that could convert indole to indigo were selected and analyzed for their ability to transform styrene to styrene oxide. One such clone contained a 5.7-kb DNA fragment that encoded the major part of a styrene degradation pathway, the first step of which consists of the oxidation of styrene to epoxystyrene with a cytoplasmic two-component styrene monooxygenase (Figure 15.11). The genes encoding the styrene monooxygenase were used to construct a recombinant biocatalyst in *E. coli* JM101, which achieved the conversion of styrene into (*S*)-styrene oxide with an ee of more than 99% (Figure 15.10). In other words, natural biodiversity was sufficient to increase the ee to more than satisfactory levels.[33,46]

In designing a production process for (*S*)-styrene oxide, several issues must be considered. First, styrene monooxygenase activity depends on the availability of NADH, making the use of an enzyme reactor with (partially) purified enzyme complex and expensive because the cofactor needs to be regenerated while product is produced. Second, the substrate as well as the product of the reaction are not very soluble in water and, even at such low concentrations, toxic to living cells, complicating the development of a whole-cell biocatalytic system. One way to still create a potentially economically attractive process is to use growing cells in a two-liquid phase culture. The partition coefficients of substrate and product dictate that both compounds remain preferentially in the organic phase and the aqueous concentrations remain below toxic levels, while the overall concentrations of substrate and product in the reactor can be increased far beyond what would be possible if only the aqueous phase was present. In the present example, the overall styrene concentration could be increased from 2 mM for a one-liquid phase culture to 135 mM for a two-liquid phase culture that consisted of 50 vol % of aqueous medium with the biocatalyst and 50 vol % of the organic phase with 2 vol % styrene. During the reaction, a small amount of the styrene partitions out of the organic phase, is oxidized, and is then reextracted into the organic phase. The organic phase serves as substrate pool and product sink at the same time. We investigated this mode of production on a 2-L scale with recombinant *E. coli* JM101 cells that carried the styrene monooxygenase genes on the expression vector pSPZ10 under control of the alk regulatory system. This vector carries the styrene monooxygenase genes under control of the alkBp promoter, which is induced by octane and is not repressed by glucose in *E. coli*. The pBR322-based vector has been optimized by introducing transcriptional terminators to translationally shield regions important for plasmid propagation and by replacing the tetracycline and ampicillin resistance genes with a kanamycin resistance gene.[33] Such a genetic system allows easy induction in a two-liquid-phase culture and the use of a cheap carbon source for the cultivation.

We optimized the cultivation protocol with respect to aqueous medium composition, organic phase, and phase ratio (the ratio of the volume of the organic phase to the total liquid volume). The best system consists of a defined mineral medium, with glucose as the carbon source and diethylhexylphthalate as the organic phase at a phase ratio of 0.5. The organic phase contained 2 vol % styrene and 1 vol % of octane, which we added as an inducer for gene expression.

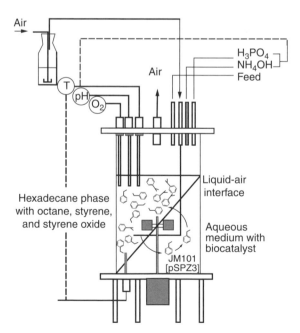

Figure 15.12 Two-liquid-phase-based biooxidation of styrene to styrene oxide with a recombinant whole-cell biocatalyst.

With this system we converted 135 mM styrene (relative to the total liquid volume) to styrene oxide in 10 h at a cell dry weight of around 10 g/L aqueous phase, with an average activity of 152 U/L total liquid volume. This corresponds to a space–time yield of 1.1 g (*S*)-styrene oxide per liter and hour. These are the highest specific activities reported thus far for a microbial epoxidation process.[33]

The bioconversion was carried out in a two-liquid phase system (Figure 15.12), which was developed at the 2-L level, and scaled up to the 30-L level to produce almost 400 g of product. Several apolar phases were used, of which bis(2-ethylhexyl)phthalate (BEHP) was preferred because it showed a better partitioning of epoxystyrene toward the apolar phase and away from the aqueous phase than did hexadecane. This was important because the product was quite toxic to the recombinant biocatalyst when it appeared in the aqueous phase. This bioconversion illustrates that apolar compounds like styrene and its epoxide, which are quite toxic to microorganisms, can be handled successfully in two-liquid-phase cultures. The toxicity of the substrate and product are not significant issues here.

Again, a recombinant *E. coli* strain performs quite well in this system. Subsequent phase separation and distillation permit the simple purification of the product. The final product had an ee >99%, which is significantly better than that seen for the XylMA based system (Figure 15.10).

15.9 LIKELY COSTS OF LARGE SCALE HYDROXYLATION OR EPOXIDATION PROCESSES WITH WHOLE CELL BIOCATALYSTS

To get some idea of the prices to be expected for compounds produced with these approaches, we have estimated the total cost of producing 10,000 tons per annum of 1-octanol from *n*-octane, based on data collected for this conversion by *P. oleovorans*, during growth in a two-liquid-phase system containing 15% (v/v) hexadecane as a carrier phase.[47] *n*-Octane is dissolved in the carrier phase to a concentration of 5–10% (v/v), converted by the *P. oleovorans* cells in the aqueous phase, and the product 1-octanol dissolves in the hexadecane phase once more. Downstream processing consists of a phase separation, followed by two distillation steps. In the first step, the C8 alkane/alkanol are separated from the hexadecane, which is recycled into the bioreactor. In the second step, the *n*-octane is distilled off the *n*-octanol: the octane is recycled to the bioreactor, and the octanol is collected as the desired product. This approach leads to a very clean product stream of >98% pure 1-octanol.[48]

The total cost of the process was estimated to be about 8 US$ per kg product (Figure 15.13). The most significant cost item (ca. 40% of the total) was due to the glucose and salts necessary to support cell growth; that is, these are part of the biocatalyst formation costs. Biocatalyst costs can be reduced by increasing specific activities per g cells, and by extending the useful lifetime of whole-cell biocatalysts. We estimate that reductions in biocatalyst cost and associated process costs might lower the preceding estimate to 5 US$ per kg product. Although the estimate described here was carried out for a specific

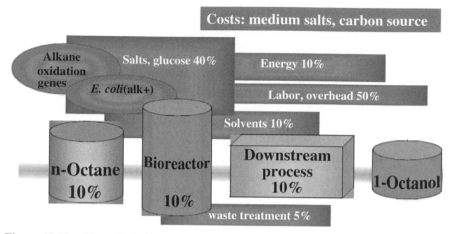

Figure 15.13   Biocatalysis for the oxidation of *n*-octane to 1-octanol at 10,000 tons/year.

alkane-to-alkanol conversion, it applies to similar hydroxylations or oxidations of other apolar compounds. A key parameter in all such estimates is the specific activity per g cells that can be achieved. Typical activities are in the order of 5 to 50 U/g cell dry mass, where U is the international enzyme activity unit, expressed as μmol of product formed or μmol substrate utilized per minute.

The preceding estimate suggests that products valued at more than 10 US$ per kg are potential targets for biocatalytic production, provided the market is sufficiently large. Clearly, compounds valued at 50–100 US$ or more per kg are interesting biocatalysis targets. However, high tonnage targets of 5–7 US$ per kg are still worth investigating.

15.10   CONCLUSION

It is interesting to speculate on the development of such a focused biocatalysis-based chemical industry. It is likely that at least one more decade will pass before a significant biocatalysis-driven company emerges. A very important attribute of such a company will be the ability of management not to be side-tracked by nonissues, examples of which are:

- Toxic compounds: educts, products
- Polar, apolar, mixed-phase biotransformation media
- Reactor systems: batch or continuous
- Product recovery: batch or integrated
- Waste handling
- Scale
- Geopolitics
- Biosafety regulation

What will matter is the extent to which a biocatalysis company manages to focus on the market, on the development of biocatalysts to address market needs, and on the ability to compare the potential of biocatalysts and chemical alternatives. Finally, it is possible to list some of the properties of that company that is likely to emerge as the premier biocatalysis company in the next decade:

- Understands the chemicals market
- Solid experience in synthetic chemistry: deep understanding of its potentials and limitations
- Biocatalyst and process development: done by organic chemists, chemical engineers, as well as molecular biologists

REFERENCES

1. Drauz, K.; Waldmann, H. *Enzyme Catalysis in Organic Synthesis: A Comprehensive Handbook*, VCH, Weinheim, New York, 1995.
2. Faber, K. *Biotransformations in Organic Chemistry: A Textbook*, 3rd, completely rev. ed., Springer-Verlag, Berlin, New York, 1997.
3. Wubbolts, M. G.; Witholt, B. *Selected Industrial Biotransformations*, Montie, T. C. (Ed.) Plenum Publishing Co. New York, 1998, pp. 271–329.
4. Schmid, A.; Dordick, J. S.; Hauer, B.; et al. *Nature*, 2001, **409**, 258–268.
5. Tramper, J. *Biotechnol. Bioeng.*, 1996, **52**, 290–295.
6. Duetz,W. A.; Minas, W.; Kuhner, M.; et al. *Bioworld*, 2001, **2**, 8–10.
7. DeSantis, G.; Zhu, Z.; Greenberg, W. A.; et al. *J. Am. Chem. Soc.*, 2002, **124**, 9024.
8. Reetz, M. T.; Zonta, A.; Schimossek, K.; Lebeton, K.; et al. *Angew. Chem. Int. Ed. Engl.*, 1997, **36**, 2830.
9. Peters, M. W.; Meinhold, P.; Glieder, A.; Arnold, F. H. *J. Am. Chem. Soc.*, 2003, **125**, 13442.
10. Jensen, V. J.; Rugh, S. *Meth. Enzymol.*, 1987, **136**, 356–370.
11. Wubbolts, M. G.; Favre-Bulle, O.; Witholt, B. *Biotechnol. Bioeng.*, 1996, **52**, 301–308.
12. Klibanov, A. M. *Acc. Chem. Res.*, 1990, **23**, 114–120.
13. Laane, C.; Boeren, S.; Vos, K.; Veeger, C. *Biotechnol. Bioeng.*, 1987, **30**, 81–87.
14. De Smet, M. J.; Witholt, B.; Wijnberg, H. *J. Org. Chem.*, 1981, **46**, 3128–3131.
15. De Smet, M. J.; Kingma, J.; Wijnberg, H.; Witholt, B. *Enzyme Microbiol Technol.*, 1983, **5**, 352–360.
16. Witholt, B.; Van Beilen, J.; Wubbolts, M. G.; et al. *Bioforum*, 1997, **1**, 32–36.
17. Witholt, B.; de Smet, M. J.; Kingma, J.; et al. *Trends Biotechnol*, 1990, **8**, 46–52.
18. Witholt, B.; Wubbolts, M. G. *Ind. Bioprocess.*, 1996, **18**, 6–7.
19. Schmid, A.; Kollmer, A.; Mathys, R. G.; Witholt, B. *Extremophiles*, 1998, **2**, 249–245.
20. Schmid, A.; Sonnleitner, B.; Witholt, B. *Biotechnol. Bioeng.*, 1998, **60**, 10–23.
21. Grogan, G. J.; Holland, H. L. *J. Mol. Cat. B: Enzymatic*, 2000, **9**, 1–32.

22. Rothen, S. A.; Sauer, M.; Sonnleitner, B.; Witholt, B. *Biotechnol. Bioeng.*, 1998, **58**, 356–65.

23. Favre-Bulle, O.; Schouten, T.; Kingma, J.; Witholt, B. *Biotechnology (NY)*, 1991, **9**, 367–371.

24. Duetz, W. A.; Rüedi, L.; Hermann, R.; O'Connor, K.; et al. *Appl. Environ. Microbiol.*, 2000, **66**, 2641–2646.

25. Li, Z.; Feiten, H.-J.; van Beilen, J. B.; Duetz, W.; Witholt, B. *Tetrahedron: Asymmetry*, 1999, **10**, 1323–1333.

26. Li, Z.; Feiten, H.-J.; Chang, D.; Duetz, W. A.; et al. *J. Org. Chem.*, 2001, **66**, 8424–8430.

27. Chang, D.; Witholt, B.; Li, Z. *Org. Lett.*, 2000, **2**, 3949–3952.

28. Chang, D.; Feiten, H.-J.; Witholt, B.; Li, Z. *Tetrahedron:Asymmetry*, 2002, **13**, 2141–2147.

29. Chang, D.; Feiten, H.-J.; Engesser, K.-H.; et al. *Org. Lett.*, 2002, **4**, 1859–1862.

30. Favre-Bulle, O.; Witholt, B. *Enzyme Microbial Technol.*, 1992, **14**, 931–937.

31. Favre-Bulle, O.; Weenink, E.; Vos, T.; et al. *Biotechnol. Bioeng.*, 1993, **41**, 263–272.

32. Schneider, S.; Wubbolts, M. G.; Sanglard, D.; Witholt, B. *Tetrahedron Asymmetry*, 1998, **9**, 2833–2844.

33. Panke, S.; Wubbolts, M. G.; Schmid, A.; Witholt, B. *Biotechnol. Bioeng.*, 2000, **69**, 91–100.

34. Arnold, F. H.; Volkov, A. A. *Curr. Opin. Chem. Biol.*, 1999, **3**, 54–59.

35. Stemmer, W. P. C. *Nature*, 1994, **370**, 389–391.

36. Harayama, S.; Timmis, K. N. *Catabolism of Aromatic Hydrocarbons by Pseudomonas*, Hopwood, D.; Chater, K. (Eds.) Academic Press, London, 1987, pp. 151–174.

37. Kohler, H.-P. E.; Kohler Staub, D.; Focht, D. D. *Appl. Environ. Microbiol.*, 1988, **54**, 2683–2688.

38. Suske, W. A.; Held, M.; Schmid, A.; et al. *J. Biol. Chem.*, 1997, **272**, 24257–24265.

39. Schmid, A.; Kohler, H.-P. E.; Engesser, K.-H. *J. Mol. Cat. B.: Enzymatic*, 1998, **5**, 311–316.

40. Held, M.; Schmid, A.; Kohler, H. P.; et al. *Biotechnol. Bioeng.*, 1999, **62**, 641–648.

41. Held, M.; Suske,W.; Schmid, A.; et al. *J. Mol. Cat. B.: Enzymatic*, 1998, **5**, 87–93.

42. Hasegawa, J.; Ohashi, T. *Chemistry Today*, 1996, **14**, 44–48.

43. Tokunaga, M.; Larrow, J. F.; Kakiuchi, F.; Jacobsen, E. N. *Science*, 1997, **277**, 936–938.

44. Wubbolts, M. G.; Reuvekamp, P.; Witholt, B. *Enzyme Microbial Technol.*, 1994, **16**, 608–615.

45. Wubbolts, M. G.; Hoven, J.; Melgert, B.; Witholt, B. *Enzyme Microbial Technol.*, 1994, **16**, 887–894.

46. Panke, S.; Witholt, B.; Schmid, A.; Wubbolts, M. G. *Appl. Environ. Microbiol.*, 1998, **64**, 2032–2043.

47. Mathys, R. G.; Schmid, A.; Witholt, B. *Biotechnol. Bioeng.*, 1999, **64**, 459–477.

48. Mathys, R. G.; Kut, O. M.; Witholt, B. *J. Chem. Technol. Biotechnol.*, 1998, **71**, 315–325.

INDEX

Methods and Reagents for Green Chemistry: An Introduction, Edited by Pietro Tundo, Alvise Perosa, and Fulvio Zecchini
Copyright © 2007 John Wiley & Sons, Inc.